Manned Spaceflight Log II

David J. Shayler and Michael D. Shayler

Manned Spaceflight Log II—2006–2012

 Springer

Published in association with
Praxis Publishing
Chichester, UK

David J. Shayler, F.B.I.S
Hurst Green
Halesowen
West Midlands
U.K.

Michael D. Shayler
Birmingham
U.K.

SPRINGER–PRAXIS BOOKS IN SPACE EXPLORATION

ISBN 978-1-4614-4576-0   ISBN 978-1-4614-4577-7 (eBook)
DOI 10.1007/978-1-4614-4577-7
Springer New York Heidelberg Dordrecht London

Library of Congress Control Number: 2006937359

Cover design: Jim Wilkie
Project management: OPS Ltd., Gt. Yarmouth, Norfolk, U.K.

Printed on acid-free paper

Springer is part of Springer Science+Business Media (www.springer.com)

Contents

Preface

Following the publication of the first edition of *Praxis Manned Spaceflight Log 1961–2006* we were pleased with the response and feedback from readers. It was always our intention to produce subsequent works covering the years beyond 2006 and it was rewarding to know that the earlier volume provided such a handy quick reference guide to each manned space mission; a guide that could lead to further research, or just answer a query. The work was never intended to be a definitive account of every mission, and it remains doubtful if such an encyclopedic series of volumes will ever be written, and in the scope of the original project there was simply not enough pages to expand each entry further than a brief summary.

For the current work we decided to cover the missions flown to the end of the fifth decade and those in the opening two years of the sixth decade of human orbital space flight operations. We were therefore able to update the original entry for Soyuz TMA-9 which had just been launched when the book went to press. By including the flights in 2011 we could introduce the first year in a new decade which also marked a number of milestones in space exploration.

In April 2011 the 50th anniversary of Vostok, the first space flight by Yuri Gagarin was celebrated around the world. That month also saw the 40th anniversary of the launch of Salyut, the world's first space station and the 30th anniversary of the first Space Shuttle mission by Columbia. In October the 20th anniversary of the launch of Soyuz TM-13, the final manned space flight under the Soviet Union was quietly achieved, whilst back in March the 10th anniversary of the first exchange of ISS expeditions took place with the ISS-1 crew handing over to ISS-2. So, in the history books at least, 2011 was a banner year for manned space flight anniversaries, and also added a new entry with the final flight of the Space Shuttle in July and the mission of STS-135.

With the Americans flying the final Shuttle missions in 2011 so the Russians completed the final evaluations of the upgraded Soyuz TMA-M spacecraft. With the retirement of the Shuttle the TMA-M became the primary crew transport and rescue vehicle for ISS operations for the next few years. In September the Chinese began their next phase of manned space operation with the launch of their first Salyut-class space station called Tiangong ("Heavenly Palace").

The following year saw the first operational launches of the latest version of the venerable Soyuz and the initial crew to visit the Chinese space station. The Shenzhou 9 crew featured a three-person team including Jing Haipeng, the first Chinese taikonaut to make a second flight, having flown on Shenzhou 7 in 2008 and Liu Yang the first Chinese woman to fly in space. The year also saw the first (unmanned) commercial mission dock with the ISS, the SpaceX Dragon spacecraft, which gave a successful demonstration of the capacity of the vehicle to support future manned space flights as one of several American aerospace company designs hoping to replace the Space Shuttle.

This book, therefore, focuses upon the latest developments covering September 2006 through December 2012, a period of six years in which the main space station assembly was completed, yet another lease of life was given to Soyuz after over 40 years of operational service, and a significant new step in the development of Chinese permanent presence in space. In this time period there were over 40 new space flights to insert into the log, an impressive total.

We have not, however, neglected the rich ancestry of these recent missions. In the earlier volume we explained the methods and systems available to reach space across the 1961 to 2006 period and, as we focused upon orbital space flight, we included a section that explained the quest for space in which efforts were made in ballistic suborbital flight by space capsules or astro-flights by winged vehicles, all within the confines of the Earth's atmosphere. This included the 13 X-15 rocket plane astro-flights between 1962 and 1968 which surpassed the 80 km (50 mile) altitude, the two 1961 Mercury–Redstone suborbital missions that both surpassed the 185 km (115 miles) altitude, and all three 2004 SpaceShipOne X-Prize flights in excess of 100 km altitude (62 miles).

For this current edition we have amended the opening two chapters to overview those earlier efforts and incorporated the latest development and achievements to bring the story up to date and cover 50 years of operations. For the missions flown between 1961 and 2012 we have presented a short overview which describes the main milestones and advances within the first five decades of human space flight activities and what lessons were learned or experiences gained within each decade.

From Chapter 3 we cover the new space flights completed between September 2006 and December 2010 completing the fifth decade of operations including an update to conclude the Soyuz TMA-9 entry begun in the earlier edition. Chapter 4 commences the story of the sixth decade for missions flown in 2011 and to the end of September 2012. We have also included a brief entry for Soyuz TMA-06M launched and docked with the station in October 2012. These entries follow the format adopted in the earlier volume to provide continuity. A summary is then added for Chapter 5 which includes a resume of Soyuz TMA-07M, the final mission planned for 2012, looks forward to the remaining years of the sixth decade of operations, and discusses what may be the future of human space activities beyond that.

A series of appendixes and a bibliography complete the work offering further reference to the main text. The updated tables cover the period 1961–2012 and we have included a full EVA log which originally appeared in the companion title *Walking in Space*.

We hope that this volume will be as useful and as popular as the earlier volume and offers the reader an opportunity to look up the basic facts for each mission and provide a link to further research on the missions in other titles of the Praxis/Springer Space Library as listed in the bibliography.

Dave Shayler
Mike Shayler
December 2012

Acknowledgments

As with the compilation of the earlier *Praxis Manned Spaceflight Log 1961–2006* this work would not have been possible without the continued support and encouragement of a worldwide network of fellow researchers and authors over a period of many years including

- Australia: Colin Burgess;
- Europe: Brian Harvey (Ireland), Bart Hendrickx (Belgium), and Bert Vis (The Netherlands);
- U.K.: Phil Clark, David Harland, Gordon Hooper, Dominic Phelan, Tony Quine, Andrew Salmon, George Spiteri, Andrew Wilson, and Robert Wood;
- U.S.A.: Michael Cassutt, John Charles, James Oberg, Curtis Peebles, and Asif Siddiqi.

We must not forget our fellow researchers, colleagues, and good friends Rex Hall, M.B.E. and Neville Kidger, who provided an unselfish wealth of information, support, suggestions, and cooperation over a 30 yr period. Some of their research is incorporated in this current volume and brings back fond memories of them.

U.K. space journalist Tim Furniss, our fellow author on the first edition of this work, deserves special thanks for his contributions to the reporting and documentation of space exploration over the years. His earlier works *Space Shuttle Log* and *Manned Spaceflight Log* were the inspiration for the *Praxis Manned Spaceflight Log*, which included unpublished updates to his earlier works through 1990. Tim's contributions to the earlier edition of this book were generous and key to its format and success. Unfortunately, whilst fully endorsing the new work, Tim was unable to devote as much attention as he would have wished but his encouragement and support are much appreciated.

The authors also wish to express their appreciation for the ongoing help and support of the various public affairs departments of NASA, ESA, the Canadian

Space Agency, the Japanese Space Agency JAXA, and the Russian Space Agency. Though a significant amount of information is placed on the various websites it still requires personal contact to refine details and to conduct personal research and study—truly "digging in the dust"—and for this we thank the above departments for their online assistance and efforts over the past 10–15 years or so.

The staff, publications, and archives of the British Interplanetary Society in London, especially those involved in its magazine *Spaceflight*, continue to be supportive of our research, for which we are grateful.

The online reference websites of *Collect Space* and *Space Facts* provide a valuable quick look source for the latest news and developments supporting the news releases from the various space agencies and are recommended for further research.

Once again we express our thanks to Colonel Al Worden (Command Module Pilot for Apollo 15) for his generous and updated Foreword.

We thank the generous assistance of Brian Harvey for the photo of the first Chinese female in space—Liu Yang—taken at the 2012 IAF Congress in Naples, Italy, October 2012. We also appreciate the use of Mark Wade's image of a display model of a Chinese Feitian EVA suit from his website *Encyclopedia Astronautica*.

All other images are via NASA, unless otherwise stated, via the AIS photo archive.

As always we appreciate the help, support, and understanding of our families during the time it took to compile and prepare this book. Special thanks to our wives Bel (who helped update all the tables) and Ruth for allowing us time to devote to the project when there were still pressing domestic projects and chores to complete. It is also worth mentioning the input of our mother Jean Shayler on this project. At the age of 83 Mom mastered the modern art of inputting data on a computer and transferred all the new log entries from handwritten notes.

Last but by no means least we appreciate the continued enthusiasm, support, encouragement, and professionalism of Clive Horwood of Praxis Books. Special mention should be made to Clive's team at Praxis: Sue, Romy, and Harry, who over the years fully supported both Clive and his authors worldwide.

Thanks also to the staff of Springer-Verlag, New York and Germany, for pre and post-production support; to Neil Shuttlewood and staff at OPS Ltd. for their editing and typesetting skills; to Jim Wilkie for his efforts in preparing our original cover design to the finished article; and the book printers and binders for the end result. Without such a team effort from Praxis and Springer this series would not be as popular as it has become.

Foreword

It was a sound heard around the world, a faint beep-beep-beep from an object in orbit around the world named Sputnik. The date was October 4, 1957 and the U.S.S.R. had just launched the first satellite into space. The United States launched Vanguard shortly after but it was a complete failure and a humiliation to a proud nation. Then, on January 31, 1958, the United States successfully launched the Explorer into orbit on a Jupiter C rocket.

The space race was on and would have the world's attention that has lasted to this day. Each side built and launched rockets and men into orbit in quick succession. Yuri Gagarin was the first human to almost orbit the Earth, reentering just short of one complete orbit. He was the hero of the time, and the U.S.S.R. was clearly the leader in space travel. The United States could not allow the Soviets to hold that lead because of the political climate. The result was that the United States embarked on a very ambitious program, not only to catch up with the Soviets but to show technical and operational superiority in this new arena. The country that controlled space would be the envy of the world and hold a giant edge over everyone else. The military considers the "high ground" very important and many battles have been fought over it. Space is the ultimate high ground, and the Soviets made other countries and especially the United States very nervous because of Sputnik. They had the high ground and the United States could not live with the idea that they did not control that vantage point.

So, the American Space Program was born. In 1957 the old National Advisory Committee for Aeronautics (NACA), formed in 1915, was dissolved and the National Aeronautics and Space Administration (NASA) created to replace it. The base for NASA was the ongoing research centers at Langley in Virginia, Lewis in Cleveland, Ohio, and Dryden and Ames in California. There would soon be several additional centers dedicated to the space side of the effort: in Houston, Texas; Huntsville, Alabama; and at Cape Canaveral, Florida; plus some smaller centers around the country for specialized work. The groundwork was laid for a launch center at Cape Canaveral, and a massive building program commenced. It was decided to keep all the space centers close to the southern edge of the country to facilitate the movement of huge structures by barge to the launch site. The "Cape", which later became Kennedy Space Center, would become the national launch center for American astronauts to fly into space.

The U.S. Space Program was not just about placing a man in space. It was also about the technology that would allow the U.S. to do things in space that no other country could accomplish. Solid state technology replaced vacuum tubes and resulted in lower weight and size for any given device. It was also more reliable and less energy consuming. This was the kind of thinking and machinery that would get astronauts from the U.S. to the Moon and back safely and consistently. It would, incidentally, also provide the technical edge for American industry to build and sell products around the world.

The human side of the space equation was also getting started early in 1958 and 1959 when NASA selected the first seven astronauts. They were a diverse group, mostly test pilots or engineers, and in great physical shape. They were also very media friendly, and NASA made much of the Boy Scout aura around them. The first group of 7 (except for Deke Slayton who had a heart problem) went into space as part of the Mercury Project, the first step in the program to place a man on the Moon. Other groups were picked to provide more astronauts for future flights that would be more complicated and designed to mimic the maneuvers and procedures that would be required to make a lunar landing.

The Gemini Program was established to prove that the elements of a lunar landing flight could be accomplished in Earth orbit. The final plan for the lunar landing was to fly two vehicles to the Moon: a Command Module (CSM) that would serve as a mother ship and a Lunar Module (LM) that would be used to actually descend to the Moon and land. The two vehicles would separate while in lunar orbit for the LM to make the landing, and then the two vehicles would join in lunar orbit after the landing crew finished on the surface. It was a risky but high-reward plan. The maneuvers that had to be developed to make all this happen were incorporated into the Gemini program. So, the Gemini flights would test and confirm that lunar orbit rendezvous and docking could be accomplished, that space walks (EVAs) could be accomplished safely and that man could survive okay for two weeks in space. All of these objectives were accomplished and the stage was set for Apollo.

The Apollo program, the most ambitious endeavor ever attempted by mankind, was born in 1961 when President John F. Kennedy challenged the American people to land a man on the Moon and bring him back safely within 10 years. It electrified the country and the world. NASA was ready and the funding was available quickly. America went into space. The original seven astronauts flew on the Mercury project and the next groups flew the Gemini and finally the Apollo to the Moon. Neil Armstrong was the epitome of the space traveler. He was quiet, unassuming, and the perfect person to be the role model for the space program, and that is to say nothing of his competence to do the job. The steps leading up to Apollo 11, the infrastructure of the various centers, the selection and training of astronauts from Mercury to Apollo, and the strategy designed to put a man on the Moon were all

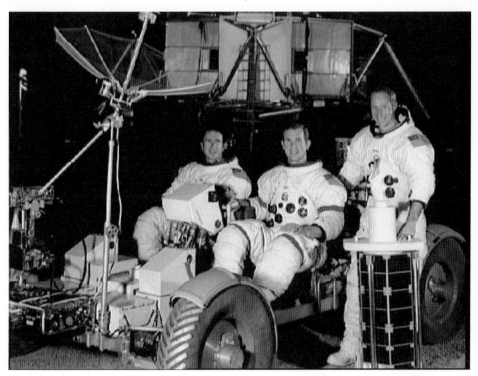

Apollo 15 crew photo. (Left to right) Jim Irwin, Dave Scott, and Al Worden.

accomplished with great success, providing the confidence for NASA managers to go forward to the Moon.

After the successful Apollo program, NASA conducted the Skylab space station and Space Shuttle programs. The Shuttle was designed to complete multiple missions into Earth orbit by flying a reusable vehicle much like an airplane. A wide range of scientific payloads were carried into orbit, satellites and planetary probes deployed, and servicing of the Hubble Space Telescope accomplished. There were also a number of scientific missions flown in pressurized laboratory modules, leading to long missions on future space stations. The Shuttle was a remarkable vehicle, much ahead of its time.

Over in the Soviet Union the emphasis focused on long-duration space flight in a series of Salyut space stations. By the 1970s there was a thaw in relations between the United States and the Soviets which resulted in a joint demonstration mission—the Apollo Soyuz Test Project—and 20 years later to further cooperation in the Shuttle–Mir space station program. This partnership has evolved today into working together on the International Space Station (ISS) built with the resources of the Space Shuttle and foreign partners. The ISS has been a truly international project, with astronauts getting there via a Russian spacecraft. Soon they will be delivered to the ISS by a civilian spacecraft. Resupplies arrive on freighters from Europe and Japan

Alfred M. Worden, Apollo 15 Command Module pilot.

and crew members come from many different nations. It is at the ISS that the new skills necessary to venture away from Earth are gained.

Low Earth orbit (LEO) will become a commercial venture in the near future. The growth of space tourism will open up opportunities for many to experience, the thrill and wonder of a flight into space, but it will be up to governments and perhaps new international partnerships to do the long-term funding for flights into deep space.

I was part of the great Apollo experiment, flying as the Command Module Pilot of Apollo 15. By the time we flew in July of 1971, we were very confident in the system, the machinery, and the people in mission control. In fact we were quite comfortable doing many of the things in flight that had been done by mission control before us. For instance, we did most of our own navigation during the flight. We trained mostly for those things that could go wrong, believing that if everything went okay it would be easy. We were getting the hang of space flight, and it was beyond comprehension for a guy that grew up on a farm in Michigan. This country accomplished six lunar landings without an accident. Remarkable!!

There have been unbelievable feats achieved by many nations to get into space. There are now many countries that have the capability to send humans into space, and at least one country that could probably go to the Moon in a few years. The United States does not have long-term space capability right now, so we will be in the bleachers until this country gets back on track. In the meantime, the development of technology will be important to whatever we do in space.

The past 50 years have been rapid-growth years for the manned and unmanned space programs of many countries. We have learned much about planet Earth, ourselves, and from the robotic spacecraft sent to explore our nearest planets. We will, as a community of nations, continue to move outward to the planets and eventually to other solar systems where there might be intelligent life. Someday we might need to go there to escape from a planet that is no longer habitable, but that is far into the future.

As the number of missions into space increase, so their exploits naturally disappear from the headlines. Only when we return to the Moon or venture to Mars will human space exploration once again be at the forefront of the world's media or, of course in times of tragedy, as there surely will be. It is important to record the development, activities, and results from our still relatively few journeys into space.

With each mission accomplished today, a new topic of history is created for tomorrow. In decades to come, when the dawn of the space age is reflected upon, our achievements and disappointments will be scrutinized in detail. From records kept at the time future generations will learn how our adventure in space began.

Colonel Alfred M. Worden USAF Retd.
NASA Group 5 (1966) Pilot Astronaut
Command Module Pilot Apollo 15, 1971

To all the members of each space flight crew who leave the pad—
safe landings

Leaving Gagarin's Start.

In recognition of all former space explorers and their efforts in taking
small steps and giant leaps away from our planet
Also
To the memory of fellow space authors, colleagues, and great friends
Rex Hall (1946–2010)
Neville Kidger (1953–2009)
Always fondly remembered

Figures (and tables)

Completing the fifth decade: 2006–2010

Commencing the sixth decade: 2011–2012

The immediate future: 2012–2020

Tables

Abbreviations and acronyms

AAP	Apollo Applications Program
ACCESS	Assembly Concept for Construction of Erectable Space Structure
AHMS	Advanced Heath Management System
ALSEP	Apollo Lunar Scientific Experiment Package
ANDE-2	Atmospheric Neutral Density Equipment 2
ARED	Advanced Resistive Exerciser Device
ASCAN	AStronaut CANdidate
ASI	*Agenzia Spaziale Italiana* (Italian Space Agency)
ASTP	Apollo–Soyuz Test Project
ATM	Apollo Telescope Mount
ATV	Automated Transfer Vehicle
CBM	Cameras Berthing Mechanism
CCDev	Commercial Crew Development Program
CDR	Critical Design Review
CDRA	Carbon Dioxide Removal Assembly
CETA	Crew and Equipment Translation Aid
CLB	Crew Lock Bay
CMG	Control Moment Gyroscope
CNES	*Centre National d'Etudes Spatiales* (French Space Agency)
COLBERT	Combined Operational Load Bearing External Resistance Treadmill
COSTAR	Corrective Optical Space Telescope Axial Replacement
COTS	Commercial Orbital Transportation Services
CSM	Command and Service Module
DIU	Data Interface Unit

DLR	*Deutsches Zentrum für Luft und Raumfahrt* (German Aerospace Center)
DOD	Department of Defense
DOS	*Dolgovremennaya Orbitalnaya Stanziya* (Permanent Orbital Station)
DRAGONSat	Dual RF Astrodynamic GPS Orbital Navigation Satellite
EAS	Early Ammonia Servicer
EASE	Experimental Assembly of Structures in Extravehicular Activity
EDFT	EVA Development Flight Test
ELC-4	ExPRESS Logistic Carrier-4
ELM-PS	Experiment Logistics Module–Pressurized Section
ELV	Expendable Launch Vehicle
EMU	Extravehicular Mobility Unit
ERA	European Robotic Arm
ESEF	European Space Exposure Facility
ESP	External Stowage Platform
ET	External Tank
EuTEF	European Technology Exposure Facility
EVA	Extra Vehicular Activity
EWIS	External Wireless Instrumentation System
EXPOSE-R	Exposed to vacuum (Russian segment)
ExPRESS	Expedite the PRocessing of Experiments for Space Station
FAI	*Fédération Aéronautique Internationale*
FD	Flight Day
FE	Flight Engineer
FGB	*Funktsionalno Gruzovoy Blok* (Functional Energy Block)
FGS	Far Guidance Sensor
FRF	Flight Readiness Firing
FSS	Flight Support System
GATOR	Grappling Adaptor To On-Orbit Railing
GSFC	Goddard Space Flight Center
GTA	Generation II Astronaut
GUCP	Ground Umbilical Carrier Plate
HTV	H-II Transfer Vehicle
ICBM	InterContinental Ballistic Missile
ICC	Integrated Cargo Carrier
ICC-VLD	Integrated Cargo Carrier–Vertical Light Deployable
ISLE	In-Suit Light Exercise
ISP	Intermediate Stowage Platform
ISS	International Space Station
IUA	Interface Umbilical Assembly

IVA	Intra Vehicular Activity
JEF	Japanese Exposed Facility
JEM	Japanese Experiment Module
JSC	Johnson Space Center
KAP	Korean Astronaut Program
KRT	*Kosmichesky Radioteleskop* (Space Radio Telescope)
LDEF	Long Duration Exposure Facility
LM	Lunar Module
LOR	Lunar Orbital Rendezvous
LRV	Lunar Roving Vehicle
MAXI	Monitor of All-sky X-ray Image
MBS	Mobile Base System
MBSU	Main Bus Switching Unit
MCC-H	Mission Control Center-Houston
MEEP	Moving Electrode–type Electrostatic Precipitator
MIRAS	Microwave Imaging Radiometer with Aperture Synthesis
MISSE	Materials International Space Station Experiment
MLP	Mobile Launch Platform
MMOD	MicroMeteoroid and Orbital Debris
MMT	Mission Management Team
MMU	Manned Maneuvering Unit
MOL	Manned Orbiting Laboratory
MPAC	Micro-PArticles Capturer
MPLM	Multi-Purpose Logistics Module
MPS	Main Propulsion System
MS	Mission Specialist
MT	Mobile Transporter
NASA	National Aeronautics and Space Administration
NASDA	NAtional Space Development Agency of Japan
NICMOS	Near Infrared Camera and Multi-Object Spectrometer
OBSS	Orbiter Boom Sensor System
OCE-EK	Optical Control Electronics–Enhancement Kit
OMS	Orbital Maneuvering System
OPF	Orbiter Processing Facility
ORU	Orbital Replacement Unit
OSCE	Organization for Security and Cooperation in Europe
OTD	ORU Transfer Device
OWS	Orbital Workshop, Skylab
P3	Port 3 (truss segment)
PAS	Payload Attach System
PDGF	Power Data Grapple Fixture
PMA	Pressurized Mating Adapter
PMM	Permanent Multipurpose Module
Prox-Ops	Proximity Operations

RCS	Reaction Control System
RKK	*Raketno-Kosmicheskaya Korporatsiya* (Rocket and Spacecraft Corporation)
RMS	Remote Manipulator System
ROKVISS	Robotic Component Verification on the ISS
RPC	Remote Power Controller
RRM	Robotic Refueling Mission
RSLS	Redundant Set Launch Sequencer
RSP	Resupply Stowage Platform
RSR	Resupply Stowage Rack
SADE	Solar Array Drive Electronics
SAFER	Simplified Aid For EVA Rescue
SARJ	Solar Alpha Rotary Joint
SEED	Space Exposed Experiment Developed for Students
SFP	Space Flight Participant
SIC&DHU	Science Instrument Command and Data Handling Unit
SIM	Scientific Instrument Module
SKK	Replaceable cassette container
SLF	Shuttle Landing Facility
SM	Servicing Mission
SOLAR	ESA science laboratory
SPDM	Special Purpose Dexterous Manipulator
SRB	Solid Rocket Booster
SSF	Space Station Freedom
SSME	Space Shuttle Main Engine
SSPTS	Station-to-Shuttle Power Transfer System
SSRMS	Space Station Remote Manipulator System
STA-54	Shuttle Tile Ablator-54
STEM	Science Technology, Engineering, and Math
STIS	Space Telescope Imaging Spectrograph
STORRM	Sensor Test of Orion Relative Navigation Risk Mitigation
STS	Space Transportation System
TBA	Trundle Bearing Assembly
TDRSS	Tracking and Data Relay Satellite System
TORU	Teleoperated mode of spacecraft control
TPS	Thermal Protection System
TTM	Tip-tilt Mirror Telescope
TVIS	Treadmill with Vibration Isolation Stabilization System
UPA	Urine Processor Assembly
URI	*Universalnui Rabotschi Instrument* (Universal Working Tool)
VAB	Vehicle Assembly Building

VDU	*Vynosnaya Dvigatyelnaya Ustanovka* (Outer Engine Unit)
VSSA	Video Stanchion Support Assembly
WRS	Water Recovery System

Other works by the authors

Other space exploration books by David J. Shayler

Challenger Fact File (1987), ISBN 0-86101-272-0
Apollo 11 Moonlanding (1989), ISBN 0-7110-1844-8
Exploring Space (1994), ISBN 0-600-58199-3
All About Space (1999), ISBN 0-7497-4005-X
Around the World in 84 Days: The Authorized Biography of Skylab Astronaut Jerry Carr (2008), ISBN 9781-894959-40-7

With Harry Siepmann

NASA Space Shuttle (1987), ISBN 0-7110-1681-X

Other books by David J. Shayler in this series

Disasters and Accidents in Manned Spaceflight (2000), ISBN 1-85233-225-5
Skylab: America's Space Station (2001), ISBN 1-85233-407-X
Gemini: Steps to the Moon (2001), ISBN 1-85233-405-3
Apollo: The Lost and Forgotten Missions (2002), ISBN 1-85233-575-0
Walking in Space (2004), ISBN 1-85233-710-9
Space Rescue (2007), ISBN 978-0-387-69905-9

With Rex Hall

The Rocket Men (2001), ISBN 1-85233-391-X
Soyuz: A Universal Spacecraft (2003), ISBN 1-85233-657-9

With Rex Hall and Bert Vis

Russia's Cosmonauts (2005), ISBN 0-38721-894-7

With Ian Moule

Women in Space: Following Valentina (2005), ISBN 1-85233-744-3

With Colin Burgess

NASA Scientist Astronauts (2006), ISBN 0-387-21897-1

Other books by David J. Shayler and Mike Shayler in this series

With Andy Salmon

Marswalk One: First Steps on a New Planet (2005), ISBN 1-85233-792-3

With Tim Furniss

Praxis Manned Spaceflight Log: 1961–2006 (2007), ISBN 0-387-34175-7

Prologue

MANNED SPACEFLIGHT LOG—A USER'S GUIDE

To provide a continuation to the earlier book *Praxis Manned Spaceflight Log 1961–2006* the basic layout for each new log entry remains essentially the same. Each mission is given its formal designation but not chronologically numbered. It is recognized that there are variations in the way a manned space flight is defined. It is much easier to document a national mission when a crew is launched that lands in the same spacecraft. With an increasing tendency to launch international crews on one spacecraft, we have found it far simpler to present the missions in launch sequence and to go on to describe the achievements made on that mission than to present a formal world or national mission sequence.

For the purposes of this work covering the period September 2006 to September 2012 we will summarize the reasoning behind the format and categories used on each log entry. Most of this was explained in the earlier work but it is worth refreshing these notes for the benefit of new readers and as a reminder for those who have the earlier volume. The purpose of this book is not to replace the earlier work but to supplement and expand the earlier publication as the program itself expands.

Briefly, the guide for each entry is explained below. For a more in-depth explanation please refer to the 2007 edition.

International designator

This is the official orbital international designator number as issued by the International Committee on Space Research (COSPAR). This system assigns a designation to all satellites and fragments from that satellite and is based upon the year of launch and the number of *successful* launches within that calendar year (January 1–December 31). For example, Apollo 11 received the international designator 1969-59A, which indicates it was the 59th successful launch in the year of 1969.

The letter code at the end of the designator refers to the main spacecraft; the letter "B" is usually assigned to the launch vehicle stage that placed the spacecraft in orbit, subsequent letters usually refer to ejections or fragments and can exceed the letter "Z", after which the sequence becomes "AA", "AB", "AC", and so on up to "AZ", then "BA", "BB". etc. Letters "I" and "O" are not used. The system was introduced in January 1963, orbital payloads prior to that date were assigned identifications utilising the 24 letters of the Greek alphabet. For this edition the international COSPAR designators for 2006 through 2012 are used.

Launched

"Launched" indicates the date the vehicle was launched at local time, but not converted to GMT or UT. Exact times are omitted for clarity, although for those missions where the date changes we have given both local time and GMT or UT.

Launch site

Where possible we have included precise pad and launch complex details as well as the launch site.

Landed

"Landed" indicates the date the vehicle landed at local time at the landing site, but not converted to GMT or UT. Exact times are omitted for clarity, although for those missions where the date changes we have given both local time and GMT or UT.

Landing site

The exact landing site is given as accurately as possible but the coordinates are omitted for clarity.

Launch vehicle

Launch vehicle details are as complete as possible including the serial numbers used in the actual rocket, and in the case of the STS the serial numbers of its components such as Solid Rocket Boosters, External Tank, and Main Engine.

Duration

Durations are given as accurately as possible from official sources. For the Shuttle this is normally from liftoff to wheel stop. For Soyuz and Shenzhou this refers to liftoff to touchdown.

Call sign

Call signs were originally used by the first Mercury astronauts to identify their spacecraft. Officially they were not used on Gemini or the first two Apollo missions, but from Apollo 9 (1969) they were revived to identify separate spacecraft over the communication links during docking missions. This ceased with Apollo 17. The call signs "Skylab" or "Apollo" were used for the 1973–1975 period. The call signs for Shuttle missions were related to the specific Orbiter being flown, such as "Atlantis". For Soyuz missions the commander has a personal call sign he uses every time he commands a Russian mission: for example, cosmonaut Gennady Padalka has the call sign of Altair and has used this call sign (Altair-1) on all his missions between 1998 and 2012. Anyone who flies with him adopts the call sign "Altair" and adds "-2" or "-3" as necessary. For the ISS Soyuz call signs are used when the space station passes over Russian ground stations, whereas a system of CDR (commander) and FE1 (Flight Engineer-1) through FE5 are generally adopted by the other crew members in U.S. documentation. It is still unclear if individual call signs are used during Shenzhou missions. If they are used, and there is apparently no clear evidence that this is the case, it seems reasonable that they would adopt a similar pattern to the Soyuz system.

Objective(s)

The main mission objective.

Flight crew

"Flight crew" indicates the *prime crew* or *flight crew* only and is presented in the order: commander, pilot, missions specialists or flight engineers. Full name, age at launch, affiliation, and crew position are given along with the details of previous flights as a quick cross-reference. When an ISS crew member was launched by or landed in the Shuttle, then they are listed as either (up) or (down) as appropriate.

Space agency abbreviations

NASA (American)	National Aeronautics and Space Administration
RSA	Russian Space Agency
ESA	European Space Agency
CSA	Canadian Space Agency
JAXA	Japanese Space Agency

Flight log

This records the key events on each mission, including where necessary any pre and post-mission information to supplement the main account. When a crew was launched on one vehicle but landed on a separate vehicle (such as the space tourist flights to the ISS) then the main details of their activities are presented on their *launch* mission and only briefly mentioned on their *landing* mission.

Milestones

"Milestones" present the main sequence listings, celebrations, anniversaries, and notable achievements within that given mission.

APPENDIXES

Appendixes A to E review the orbital records from April 12, 1961 through September 30, 2012. Any crew member in space on this date will find their accumulative time for their "current" mission *is not included* for clarity, but their "flight in progress" is noted. We have included a complete listing of all manned space flights, the time space explorers have spent accumulatively on their missions, a list of space walks and EVA experience, and a brief timeline of historic milestones in the human exploration of space.

GENERAL NOTES

Due to the complexity of this presentation of data and events, references have not been listed due to lack of physical space within the confines of the book. We have expanded the bibliography section and full references are available by contacting the authors if desired. Equally, an index is not provided for the same reasons of room and clarity, continuing the format as presented in the earlier volume. Wherever possible we have followed both the metric and imperial system of weights and measures. This edition has been compiled with American English spelling and date format in mind.

About the authors

Spaceflight historian David J. Shayler, F.B.I.S. (Fellow of the British Interplanetary Society or—as Dave likes to call it—Future Briton In Space!), was born in England in 1955. His lifelong interest in space exploration began by drawing rockets aged 5, but it was not until the launch of Apollo 8 to the Moon in December 1968 that the interest for human space exploration became a passion. He fondly recalls staying up all night with his grandfather to watch the Apollo 11 moonwalk. His first articles were published by the British Interplanetary Society in the late 1970s and in 1982 he created Astro Info Service (*www.astroinfoservice.co.uk*) to focus his research efforts. Dave's first book was published in 1987 and has been followed by over 20 other titles, featuring works on the American and Russian space programs, the topics of spacewalking, women in space, and the human exploration of Mars. Dave's authorized biography of Skylab 4 astronaut Jerry Carr was published in 2008.

In 1989 Dave applied as a cosmonaut candidate for the U.K. Project Juno program in cooperation with the Soviet Union (now Russia). The mission was to spend seven days in space aboard the space station Mir. Dave did not reach the final selection but progressed further than he expected. The mission was flown by Helen Sharman in May 1991. In support of his research Dave has visited NASA field centers in Houston and Florida in the United States and the Yuri Gagarin Cosmonaut Training Center in Russia. It was during these trips that Dave was able to conduct in-depth research, interview many space explorers and workers, tour training facilities, and handle real space hardware. He also gained a valuable insight into the activities of a space explorer and the realities of not only flying and living in space but also what goes into preparing for a mission and planning future programs.

Dave is a friend of many former and current astronauts and cosmonauts, some of whom have accompanied Dave on visits to schools across the country. For over 30 years Dave has delivered space-themed presentations and workshops to children and social groups across the U.K. This program is intended to help the younger generation develop an interest in science and technology and the world around them, in addition to informing the general public and interested individuals about the history and development of human space exploration.

Dave lives in the West Midlands region of the U.K. and enjoys spending time with his wife Bel, a rather large white German shepherd called Jenna, and indulging in his love of cooking. His other interests are in reading about military history, visiting historical sites and landmarks, and Formula 1 motor racing.

Michael Shayler. Most of Mike Shayler's knowledge and experience connected with this publication has come about through David's work with Astro Info Service. A certified proof reader and editor, Mike has performed initial proof reading and editing of David's work for Astro Info Service and commercial publishers since 1992, as well as a period working freelance for Springer-Praxis. This included work on the titles *Women in Space*, *Walking in Space*, *The Scientist-Astronauts*, and *Russia's Cosmonauts*. As a former member of the British Interplanetary Society, Mike was also editor and article writer for their quarterly junior publication, *Voyage*, for its initial five issues until it was taken in-house. Mike co-authored the Springer-Praxis book *Marswalk One* and the initial edition of this log and has also created, edited, and co-written many articles, website pieces and publications for Astro Info Service. Mike also served as the liaison between David and Skylab astronaut Jerry Carr for the latter's biography, editing and coordinating the material produced. He has also conducted several lectures at educational establishments on behalf of Astro Info Service, as well as research trips to the Johnson Space Center in Houston, Texas. Mike lives with his wife, Ruth in Birmingham, England and spends his spare time developing his artistic talents in painting and drawing.

December 2012

1

Reaching the heavens in the quest for space

There are moments in human history that define significant change and development. One of the more positive moments occurred on October 4, 1957. On that day, the world's first artificial satellite was placed into Earth orbit, signifying the dawn of what became known as the "space age". That defining moment, when Sputnik attained a successful orbital insertion, finally unlocked the age-old secret of flight beyond the confines of our planet and brought human dreams, stories, and plans to explore "space" closer to fruition and reality. The rapid pace of advancement has become a feature of space exploration since that date. After those centuries of hopes and desires, another significant development occurred but a short two and one half years later.

WHERE BLUE SKIES TURN BLACK

On the morning of April 12, 1961, a Soviet pilot named Yuri Gagarin sat strapped to an ejection seat in the cockpit of his craft awaiting the start of his next flight. As a serving air force officer there was nothing strange about that, except that the "craft" was a spacecraft, not an aircraft, and Gagarin was lying on his back, not sitting upright in the normal flying position. One other significant difference in his pending flight was that Gagarin was sitting on a rocket standing on a launchpad instead of in a military jet on the end of a runway. It was true that Gagarin was about to fly into the atmosphere once more, but not for long. Within minutes of leaving the launchpad he was passing through the upper reaches of the planet's atmosphere, where blue skies turn black and "air" becomes "space", pioneering a new mode of transport, that of manned space flight.

Since that day, small steps and giant strides have been made in the exploration of space, be it by automated satellites, space probes, or crewed vehicles. Each and every mission adds to the database of experience, whether successful or not. There had been arguments for and against human space exploration even before Gagarin left the pad. Over the decades, these have expanded to include debates on the militarization of space; on the importance of scientific objectives; the direction of future space efforts and how they could be financed; the drain on science and unmanned programs to fund manned activities; and whether the budget should be spent on space at all when there are so many problems here on Earth. These arguments will certainly continue for many decades to come, countering the natural global evolution of the program, but on the 50th anniversary of Gagarin's mission (April 2011), the time was right to take stock and review what had been achieved in human space flight, how we had arrived there, and where to go in the future.

There have been countless volumes over the previous six decades or so that have provided an exhaustive narrative of various manned space flight programs, the hardware, operations, and results. The basic records of these missions were recorded in the earlier edition of this log (*Praxis Manned Spaceflight Log 1961–2006*, Tim Furniss and David J. Shayler with Michael D. Shayler, Springer-Praxis 2007) providing both a handy reference in its own right, and a companion volume to the various titles that examined each program and mission in more detail.

Where blue skies turn black, Earth orbit.

Where space begins

Living on a planet means "space" is all around us, a fact most people often overlook or are even still unaware of. Standing on "terra firma", it is easy to forget that this "firm ground" is actually traveling through space on its own journey around the star we know as "sol", or the Sun. To journey "into space" generally means traveling throughout the atmosphere to a point where you briefly see the blue sky turn black and into a condition where normal aerodynamic control surfaces such as wings, rudder, and ailerons are useless; only rocket engines can provide the impulse to move under a vacuum condition. It is a condition controlled by the forces of gravity pulling on objects in perpetual free fall.

The barrier between blue sky and space has been defined at different altitudes above the Earth. For some it is 50 miles (80.45 km); others state 62 miles (or 100 km), while still others claim you are not in space until you are in orbit and you cannot call yourself a true space explorer unless you have completed at least one orbit of Earth. With even more flights to the edge of the defined atmosphere planned, such as the Virgin Galactic SpaceShipTwo program, so the debate of what is or is not a true flight into space will continue.

Cosmonautics Day

The achievement of Gagarin's historic flight became an annual celebration in the Soviet Union, known as "Cosmonautics Day". Though this continued in the modern Russia following the collapse of the communist system, it was with perhaps less enthusiasm. However, with the growth of the internet and social media, international events under the banner "Yuri's Night" are now being celebrated across the world, from small gatherings to larger official functions. For the 50th anniversary of Gagarin's flight, this took on even more importance, not only for citizens in general, but for those directly associated with carrying his memory into space on each flight.

It has been a tradition of cosmonaut mission commanders, whether under the Soviet or Russian banner, to use a personal radio call sign for communication identification. As part of the celebrations of the previous 50 years of sending humans into space came the decision to use the call sign "Gagarin" for Soyuz TMA-21 (ISS Expedition 28) in April 2011. In recognition of both the American and Russian pioneers of space flight, the names of both the first Soviet Union cosmonaut (Yuri Gagarin) and the first American astronaut (Alan B. Shepard) in space were included in the design of the mission emblem. Radio call signs and mission emblems have become recognized as the flight crews' personal input into each highly technical and scientific mission into space, humanizing the nuts and bolts of the hardware in which the crew members fly and continuing a long tradition of emblazoning aircraft with nose art or naming oceangoing vessels with a personal identity.

Behind the headlines, the press kits, news releases. and postflight debriefings are personal stories of triumph and achievement and, at times, of tragedy and disappointment. Of course the science of flying into space, like most technology, is prone to failure and unforgiving disaster, but when things work as designed the results can be both spectacular and awe inspiring. When reading the accounts of each mission into space, the fine line between success and failure each crew faces through every second of the mission should be remembered. Whether these missions last a few minutes or hours, days, or months, the ever present threat of disaster and failure remains. Space is an unforgiving environment. No matter how many times humans venture there, neither the depth of training nor experience can totally eliminate the potential for system failure and risk to the health and safety of the crew. This is a specter constantly riding with every crew on each mission, but the same experience and frequency of exploring space does reduce such risks in that challenging environment. The true legacy of all previous flights into space is that they add to the ever expanding database of knowledge which allows future missions to push farther and deeper in space than ever before and to do so with added safety and with more confidence.

As we stand at the beginning of the second half of the initial hundred years of human space exploration, it is appropriate to review what has been achieved and what lessons have been learned. All this experience, both good and bad, can then be applied to current operations and future plans. How those plans turn out in reality will be the responsibility of future generations, all of whom will follow in the trail of Yuri Gagarin. Even though his total space flight experience was just 108 minutes, he pioneered the journey from Earth to space. Yuri Gagarin will forever remain in the annals of human exploration as *the first* to leave the Earth, blazing a trail in the conquest for space for others to follow.

REACHING THE HEAVENS

Though it is not evident to everyone, we are all explorers of space, moving with the Earth on a continuous journey within our galaxy in the universe. For a fortunate few in recent years, that journey in space has become very personal as they have left Earth for a short time, paving the way for longer and even farther journeys to the planets and beyond. It can be a privilege to witness defining eras of history and that is certainly true of following the first human steps towards the stars. Though we will never witness those journeys in our lifetime, we have witnessed the start of that adventure, and that is rewarding in itself. Half a century into that journey, there have been countless reviews of our progress in space exploration; what has been achieved to date, how this experience can be useful in understanding where we wish to focus future directions, and why those decisions are made. Part of this decision making is to look back at the rich heritage of today's truly international manned space program.

The quest for space

There is not the scope to provide a detailed analysis in this present volume but for the benefit of understanding the more recent space flights, a brief overview of our steps to space is presented here (see the first three chapters in *Praxis Manned Spaceflight Log 1961–2006*, Springer Praxis 2007 for additional details). For every flight into space there has to be the journey from the surface of our planet, through our atmosphere, and into the vacuum beyond, at all times increasing in velocity to overcome the pull of gravity which keeps us on the planet in the first place. As humans, space is not our natural environment and it takes an enormous effort to get us there, sustain us while we explore, and then protect us to get safely home again. Of course, the farther we explore from our planet, the more support we will need and the more difficult the return to Earth becomes. Learning to sustain ourselves in space has always been a challenge, from the first flights of a few minutes or hours to the current months spent on board the space station.

Towards the heavens

For centuries, humans dreamed of exploring the heavens above the blue sky, creating vivid stories of exploring the void amongst the stars. But the first hurdle was to devise a system for leaving the ground and surviving the adventure. Our atmosphere was the first barrier, together with the substantial hurdle of gravity. Understanding the "science" was beyond the early dreamers and planners. For centuries, observations of the night sky were based upon myth, religion, legend, belief, fear, and the imagination. One of the first "space sciences", without actually being known as such, was the study of the stars, our Sun and Moon and the movement of the "heavenly bodies" we know as the planets. This was all grouped under what became known as "astrology" before serious scientific study of the cosmos became "astronomy". Even today, we continue to use the same names for constellations and planets first derived in both "astrology" and "astronomy" by observers from ages past.

The early writers scribed stories of ascending to the heavens by means of woven baskets carried by flocks of geese, or wearing wings of feathers. Eventually, we began utilizing the new art of "balloons" to ascend though the atmosphere, hoping to reach as high as the Moon to visit the inhabitants thought to live there. As strange and weird as these writings are to the modern world, they can be interpreted another way, in that these studies helped to discover new "sciences". We began to study the atmosphere more closely, develop telescopes to look at the moon, planets, and stars, evaluate materials, and determine why it was so difficult to follow the ideas of dreamers and writers to reach the heavens. Gradually, the sciences refined these early dreams into the theories, principles, and understanding we needed to develop the right method of reaching the stars—that of rocketry.

That method was first discovered by the ancient Chinese in the gunpowder projectiles used in conflict and celebrations. By the end of the 19th century, the theorists and dreamers had started realizing that the rocket would indeed be the best vehicle to explore the known vacuum of space and to get humans off the ground, to attain the required height and speed to circle the Earth and not fall

Ascent to orbit.

immediately back to the ground. The problem then was to create rockets capable of carrying not only instruments and biological samples, but the first humans beyond the atmosphere.

As the 20th century dawned, so did the understanding of the relatively new science of rocketry under the banner of "astronautics". The huge developments that came over the relatively short period between 1900 and 1960 were amazing. True, rockets had been around for some time, but as weapons of war. Turning that effort towards the stars was, to a degree, another method of waging war, a strategic race to win the high ground. Scientific exploration was only a secondary consideration. But the "space race" brought the rocket that led us to orbit and we have not looked back since.

The new sciences

The "science" of space flight is often perceived by the general public to be a "new" skill, but is in fact one that is centuries old and comprises a melting pot of past experiences and developments. Today's studies in, for example, materials and fluid physics, biochemistry, and medicine have evolved from the basic questions posed since ancient times: "How does this work and why?" The desire of human nature to "find out" and "experience" things, to address the unknown, has driven us from caves and "dark ages" to where we are today. True, not every development or advance can be called positive, but generally we have advanced in the understanding of our planet and our place in it and, in recent decades of course, how to leave the planet and explore beyond its boundaries.

In developing the sciences, there is often feedback and applications that can develop other fields. Such can be said for space exploration, though this is not always highlighted or promoted. This is a shame, as it offers an insight to those who do not understand the larger picture, who question the huge investments made in exploring away from Earth when there are still so many problems around us. Equally, advances in science, medicine, technology, sport, and even military operations can feed back to assist developments in space exploration. It is a two-way flow of knowledge.

As the science of space flight evolves, the division between automated and human space exploration will become less obvious as financial and other considerations push the requirements for such complex programs farther beyond national pride towards international cooperation. The farther we venture from Earth, so the need to sustain the crew by means of self-contained systems will increase, as will the reliance on automated systems. Robotic spacecraft will support and complement human endeavors and, in turn, human intervention will help sustain and maintain robotic operations.

BALLISTIC, SUBORBITAL, OR ORBITAL

The first development in rocketry which enabled space flight capability to be considered were missiles categorized as "ballistic" in that, according to the dictionary, it is "a missile which is initially powered and guided but which falls under gravity on its target." In other words it does not have enough velocity to overcome the pull of gravity to attain orbital flight in space.

The first such vehicles were the German V-2 missiles used during WWII. One of these on October 3, 1942 attained an altitude of 60 miles (97 km). During the 1950s both the United States and Soviet Union used adapted V-2 missiles upon which to base their development for longer range intercontinental ballistic missiles (ICBMs) deployed during the Cold War. From those developments emerged the first launch vehicles to place payloads into space and eventually orbit. The Soviet-developed R-7 ICBM became the workhorse of the Russian space program, and has remained the mainstay of that country's launch operations for over 55 years.

It is currently (2012) used in the launch of Soyuz manned and Progress resupply craft to the ISS.

Suborbital trajectory

In the early days of manned space flight both the Soviets and the Americans planned for a series of suborbital flights before committing their crew members to the more challenging orbital space flight trajectories. A suborbital flight path is similar to a ballistic trajectory to the upper reaches of the atmosphere and then falls back due to insufficient velocity to attain orbital flight. This type of space-flight was flown during the first American manned (Mercury) space shots in 1961. Launched from a carrier aircraft, the 13 X-15 rocket research aircraft flights which were similar but termed "astro-flights" rather than suborbital as they reached lower peak altitudes. It was this type of trajectory that was achieved by the three SpaceShipOne flights in 2004 to claim the "X-Prize". The failed Soyuz launch in 1975 was also high enough to follow a suborbital trajectory.

Suborbital transportation

Research conducted over many years indicates that using a semi-ballistic, suborbital trajectory may be a future possibility for traveling between the United States and Europe in just one hour, or to Australia in less than two hours. At present this is being investigated for unmanned, courier, business, or for military quick-response applications rather than passenger travel, due to the high costs envisaged. In theory future transportation systems could develop suborbital "space-lines" carrying 50 passengers from Europe to Australia in 90 minutes or 100 passengers to California in an hour. A distant dream today, perhaps, but the thought of such hypersonic suborbital spaceplanes means that you could have breakfast in London, lunch on the beach in California or Australia, time for shopping and back home to catch an evening show! Of course the price of such a trip and what it would do to the jet lag effect would put all but the very rich, or higher society, off such a regular adventure, but eventually . . .

Orbital space flight

In simple terms to attain true flight into space and achieve at least one orbit of Earth requires sufficient velocity and altitude to "fall" towards the planet in a closed orbit, which, as the surface curves away, is sufficient to loop around as long as the velocity and altitude is sustained. If not gravity will take over and the increasing density of the layers of the atmosphere drag the vehicle down towards the Earth once again. To escape low Earth orbit requires additional velocity "boosts" to increase speed to higher orbits or out towards the Moon or planets, using the gravitational forces of those celestial objects and the momentum of the vehicle to allow the spacecraft to journey towards them.

Yuri A. Gagarin, the first human space explorer.

METHODS OF SPACE FLIGHT

Having realized that the most efficient method to punch through the atmosphere, gain the velocity to counter the pull of Earth's gravity, and "fly in space" was the rocket, the next challenge was to devise the best way to harness that power for both unmanned and manned operations. The main difference between those operations was, of course, the safety of the human crew being carried. As missile rocketry was still in its infancy, the chances of a vehicle blowing up on the launch-pad or in the early stages of flight were very high. The development of rescue systems and striving to ensure the safety of the launch vehicle was paramount. This effort became known as man-rating.

By the 1950s, only two nations had the capability and infrastructure to support a concerted effort to explore space; the U.S.S.R. and the United States. It was here that the military consideration of "securing the high ground" developed the Cold War "arms race" into a "space race" and, eventually, a "moon race".

Capsules or wings?

During the 1950s, it was recognized that the most efficient method of conducting flights in space was by means of rocket power. What was less clear was the design of the vehicle that would be lifted into space carrying a crew. There were two main schools of thought. One considered using converted intercontinental ballistic missiles with sufficient power to lift a small pressurized compartment into orbit. The other advocated rocket-powered aircraft, which could either be air-launched from a carrier aircraft flying in the higher reaches of the atmosphere or rocket-launched from a launchpad. Once the missions had been completed, either system would need to be able to return to Earth and here again the two systems would differ.

The pressurized compartments, usually referred to as "capsules", could descend using retro-rockets and an ablative covering, protecting the structure of the vehicle (and its precious human cargo) from incineration during entry. They could then deploy a series of parachutes to affect a softer landing either on land or water. These types of spacecraft would be single use and relatively small, to ensure that the capabilities of the parachutes were not compromised. In the case of water recovery from the ocean, they would require a fleet of naval craft to support the retrieval of crew and hardware.

The rocket-powered aircraft would have the capability of landing on a runway, which offered a more cost-effective method of recovery. It would also be possible to reuse the same vehicle after a period of turnaround. Both methods would be developed to pioneer the first decade of human space exploration.

Rocket planes

The American series of rocket research aircraft was pioneered by the historic X-1 (which broke the sound barrier for the first time in October 1947) and culminated in the 199 flights of the X-15 by 1968. These aircraft investigated high-altitude and high-speed flying up to Mach 6 from the 1940s through the 1960s, pioneering the engineering and technology which assisted in the development of both the Space Shuttle and the hardware intended for the Apollo program. Such experiments and studies were not confined to the United States, but were developed across Europe from the 1930s. They continue in many proposals for future space operations across the world.

Although none of these rocket plane flights would orbit the Earth, 13 of the X-15 missions, flown by 8 different pilots, surpassed the USAF-recognized altitude threshold of 50 miles (80.45 km) between "air" and "space". These were designated "space flights" by the American Air Force, and five USAF officers (Mike Adams, Joe Engle, Pete Knight, Bob Rushworth, and Bob White) who flew these missions were awarded "astro-wings". The remaining three (Bill Dana, John McKay, and Joe Walker) were civilian pilots who did not qualify for the USAF title. To complicate the matter further, the Fédération Aéronautique Internationale (FAI) who maintained international records in aviation achievements decided that only flights over 100 km (62 miles) would be classed as a space flight. Joe Walker made two flights above that limit and still did not receive his "astro-wings". This debate over when an astro-flight is or is not a space flight continues to the present day.

The X-plane rocket aircraft series was supplemented by another series of strange "research craft" called "lifting bodies" which, during the 1960s and early 1970s, investigated the flying characteristics of blunt-bodied, wingless vehicles. They were used to develop the technologies for protecting a vehicle during reentry, flying at subsonic speeds, and completing a controlled horizontal landing on a runway. This work would also provide very useful data for the development of what became known as the Space Shuttle.

Some 50 years later, the development of the SpaceShipOne and SpaceShipTwo vehicles, which pioneered private access to the very fringes of space, continued the legacy of these programs. Such studies have also continued in supersonic and hypersonic commercial passenger aircraft programs, which in the future may shorten the flight time for commercial passengers across the globe and develop future aircraft-type vehicles to "shuttle passengers, payloads, and experiments to and from Earth orbit." Despite the abandonment of the Soviet Buran vehicle and the retirement of the American Space Shuttle, the story of winged vehicles in space is not at an end, it is just in hiatus for now. The list of paid-up and waiting "space tourists" for short commercial trips to the edge of space grows and the inaugural flights approach. Perhaps the wider interest and investment in such access to space will finally be realized as commercial opportunities arise.

Suborbital space flight

The first rockets to explore space flew a basic up-and-down trajectory and did not have the velocity (of 18,000 mph or 29,000 kph) to attain Earth orbit. In their plans to place a man in space first, both the Americans and the Soviets investigated a program of suborbital space shots, launching on smaller rockets to about 100 miles (approximately 160 km) before separating for landing several hundred miles downrange. Clearly these rockets could not attain the thrust required for orbital flight and the vehicles they carried (commonly called capsules) would not have the capacity for a controlled land landing similar to the X-15, so they were termed suborbital missions. While the Russians abandoned the suborbital program in favor of proceeding directly to orbital missions, NASA chose to send its first two astronauts on suborbital hops down the Atlantic Missile Range in 1961, ending up in the Atlantic Ocean barely 15 minutes after launch. These two missions have always been accredited as America's first two space flights, but in some more recent online listings, while the pilots Alan Shepard and Gus Grissom are accredited with their "suborbital space missions", they are not listed under manned space flights. Instead, the three-orbit mission of John Glenn in February 1962 is credited as America's first (orbital) flight in space, so the debate continues—and will probably do so for some time.

When is a space flight not a space flight?

Over a decade later, a third suborbital trajectory was inadvertently flown during an aborted Soyuz launch in April 1975. In this case, a spent stage of the R-7 launch vehicle failed to separate cleanly and caused the remaining launch vehicle to veer off course, triggering an abort just a few minutes into the mission before completing an emergency parachute recovery just over 21 minutes after launch. As a result of the failure, the Soviets did not assign an official Soyuz designation to the flight (it should have become Soyuz 18). Instead, they termed the event "the April 5 anomaly". In the West, this "mission" is often referred to as Soyuz 18-1, to distinguish it from the successful replacement Soyuz 18 mission flown with a different crew a few weeks later. As the aborted flight attained a peak altitude of 119 miles (192 km) it was officially credited as a space flight in progress, becoming the highest altitude suborbital trajectory to date.

In September 1983, a second Soyuz launch was aborted seconds before release from the pad when the carrier rocket suddenly exploded. The emergency escape rocket fired and propelled the crew to a safe, if rapid recovery five minutes later. As the maximum altitude attained was just a few thousand feet and the "launch" had not taken place, this did not become an accredited space flight or official "mission". It was an unwelcome and surprising experience. In July 1985, Shuttle mission 51F suffered a main engine failure which threatened its ascent to orbit. Fortunately, sufficient velocity and altitude had been attained at the time of the failure and the loss of the engine could be compensated by those remaining. Abort-To-Orbit (ATO) mode was followed, resulting in a lower-than-planned Earth orbit which was modified over the next few days.

The question of when a space flight is not credited as a flight into space, but is instead deemed a mission in progress, was demonstrated tragically in January 1986 with the STS-51L mission and the tragic loss of the Space Shuttle Challenger and its crew of seven. The vehicle exploded just 73 seconds after leaving the pad at an altitude of 14,020 m (45,997 ft.), far below the recognized altitude to be termed a true space flight. However, in respect to the lost crew, NASA credited them posthumously with a mission duration of 1 minute 13 seconds to the point of the disaster and officially termed the ill-fated flight a "space mission in progress".

This same classification was attributed to the STS-107 mission in February 2003, when that vehicle and its crew of seven were lost during a high-altitude breakup just 16 minutes from the planned landing in Florida. Again as a mark of respect to the crew, they were credited a mission duration to the point of loss of signal. Unlike STS-51L, they had completed a 16-day flight and were coming home from orbit when disaster struck. In most records of space flight missions, the flights of Soyuz 18-1 and STS-51L tend to be listed as attempted orbital missions which fell short of that goal during flight. Had all gone well, they would have both attained sufficient velocity to have been accredited as true orbital space flights.

PHASES OF A SPACE FLIGHT

In all space flights, there are three phases that are common to each mission. Though the profile, objectives, and even the hardware changes, the sequence remains the same for human missions—launch, inflight, and landing.

Launch sites

There have only been three launch sites used to send humans into orbit since 1961, one each in the Soviet Union, the United States, and China. In addition, Edwards

Launch Complex 39, Kennedy Space Center, Florida, U.S.A.

Air Force Base in California was the home of the X-15 program from which a series of "astro-flights" were conducted in the 1960s. Nearby Mojave Airport was the departure point for the 2004 SpaceShipOne flights. Despite plans, there were no launches of military Manned Orbiting Laboratory missions in the 1960s, nor classified Shuttle missions during the 1980s, from the Vandenberg Air Force Base Space Launch Complex 6 (SLC-6, also known as "Slick-6") in California.

The first launch of a manned orbital space flight, on April 12, 1961, was from Pad 1 at Site 5 at the huge Baikonur Cosmodrome, in what was Soviet Central Asia (now known as the Republic of Kazakhstan). All subsequent Soviet/Russian manned launches have taken place from the same cosmodrome, though a few have used Pad 31 at Site 6.

Similarly, all American manned space flights have begun from the extensive launch complex at Cape Canaveral in Florida. The early missions launched from Pad 5 (used for the suborbital Mercury-Redstone launches), 14 (Mercury Atlas), 19 (Gemini Titan), or 34 (Apollo Saturn 1B), with the Pad A or B sites at Launch Complex 39 serving as the launch site for all Apollo Saturn V and Skylab/ASTP Saturn 1B missions. The LC-39 pads were subsequently converted to launch all 135 Shuttle missions. As changes take place once more in Florida to remove Shuttle-related launch systems and install facilities for the next generation of U.S. launch vehicles, these historic pads will be used to place new American vehicles into orbit. Across the world in China, the third launch site for manned spacecraft is located in Jiuquan and is used to support the launch of Shenzhou missions.

Flight profiles

There are no "typical" space flights, as each mission is different by objective and flight profile. These profiles are determined by the geographical location of the launch site, the direction (azimuth) of launch, and the particular inclination and the height of the orbit above the planet. This can result, for example, in orbits over the polar caps, or out to a distance of 35,000 km (22,000 miles) which matches the rotation of the planet. This is where many of the communications and weather satellites are located, over the relevant part of the Earth, to maximize their capacity.

Of course, trajectories for leaving Earth orbit add to the complexity of the mission. For the Apollo lunar flights, even reaching for the Moon involved studies of three main trajectories. If the launch vehicle assigned had been large enough, then a direct flight to the Moon and back could have been flown—a profile termed "direct ascent". As this was beyond the capabilities and funding to meet the 1970 deadline, the Americans chose an alternative route. The second option was to launch elements of the lunar spacecraft on separate, smaller rockets, and then bring the spacecraft together in orbit before continuing out towards the Moon. Called "Earth orbital rendezvous", this method was far more challenging because it would have meant developing the capability for several on time launches, rendezvous and docking, and proximity operations that would be required to make the method a success. Any delays could seriously have threatened President Kennedy's deadline and fallen behind the expected Soviet competition to the Moon.

The third option was to launch a two-part spacecraft, one of them a separate lander, on one large launch vehicle into Earth orbit. This landing vehicle would then be extracted unmanned from the top of the launch vehicle where it was stored for launch. Once joined up, the combination would be sent to the Moon with a crew of three astronauts and, once in lunar orbit, two of the crew would separate the lander to complete the surface exploration program with the third astronaut remaining in orbit. Creating a lighter vehicle to land on the Moon meant a lighter vehicle to lift off from the surface and a smaller engine required to be able to do so. After rendezvous, the main spacecraft would return to Earth with the crew of three. This system, though still risky, raised questions over whether to proceed and the capability of making various safety decisions throughout the mission. This method was called "lunar orbital rendezvous" and was the method chosen for the American Apollo missions, to great success. To gain the necessary experience in orbital rendezvous, longer duration missions, and spacewalking techniques NASA created the two-man Gemini program which completed 10 highly successful manned missions in Earth orbit during 1965 and 1966.

Riding the stack.

Circumlunar missions (flights around the far side of the Moon without entering orbit and then heading on back to Earth) were flown by unmanned Soviet Zond vehicles to test their lunar capabilities. Although no Soviet manned flights ever flew in competition to Apollo, it is known that the Soviets were developing their own manned lunar program. Initially, cosmonauts would have followed an Earth orbital rendezvous profile, but in 1964 this was amended to a probable lunar orbital rendezvous profile. Unfortunately, while the Americans chose LOR as the best way to develop Apollo, the argument in Russia was actually over which designers would be the lead bureau rather than which method to follow. Coupled with hardware failures and the success of the Americans, this led the Soviets to cancel their manned lunar program long before any cosmonauts left the launchpad to test the lunar hardware in space.

Since Apollo 17 in 1972, every manned orbital mission to date has been in Earth orbit, either on independent flight profiles or as a docking mission to a space station. Significant experience in orbital rendezvous was achieved by the Americans during the Gemini program (1965–1966) and with the Apollo era missions (1968–1975) but despite several rendezvous with satellites by the Shuttle, actual docking experience for Shuttle crews did not begin until the Shuttle–Mir program of 1995–1998, extending to the ISS assembly missions between 1998 and 2011.

For the Russians, the first manned docking occurred between Soyuz 4 and 5 in 1969, but it was as part of regular Soyuz–Salyut operations between 1971 and 1986, followed by extensive Mir operations during 1986 through 2000, that they gained significant experience of automated and manual docking. This was supplemented by experience with docking unmanned Progress resupply craft to Salyut, Mir, and then ISS from 1978. After over 40 years of docking Soyuz (and Progress) variants to space stations, the reliable Soyuz continues to be the mainstay of Earth orbital operations. In 2011, the Shenzhou 8 demonstrated Chinese unmanned space station docking capability and was followed the next year by the first docking of a crew, who completed both a manual and automated docking to Tiangong 1 aboard Shenzhou 9. Rendezvous and docking, plus station-keeping and proximity operations will remain a focal point of Earth orbital operations for the rest of this decade, 50 years after they were first demonstrated during Project Gemini.

Landing methods

Having completed their mission crews prepare for the return to Earth. The method of crew recovery depends upon the design of the vehicle and the location of the landing area. This is usually a barren expanse of land or one of the planet's vast oceans, both of which give a wide margin for error. The occupants of the spacecraft always hope for a safe and as soft a landing as possible.

For the Soviets, and subsequently the Russian planners, the preferred landing site has always been on soil, usually the immense expanse of Kazakhstan. One of the main reasons for this in the early days was the then secret nature of the whole

Water impact for Orion mock-ups.

Land impact tests for Orion revealing the planned dual landing capability.

Soviet space project; keeping the returning crews and vessels away from the eyes of the outside world. It also avoided diverting naval resources into recovering crews from distant oceans. The early Vostok missions were unable to support a returning cosmonaut landing inside the spacecraft and thus provided an ejection seat for separate parachute descent. However, since 1964 all returning Soviet/ Russian spacecraft with a human crew on board have featured retro-rockets to soften the impact on the arid steppe land.

All Soyuz crews train for water recovery and, though none have been planned, there was one "splashdown", on a frozen lake in 1976. The recovery proved to be a challenge for both the crew on board and the rescue teams. An earlier mission also just missed landing in a lake by just 50 meters in 1971. With the advent of international cooperation, a number of international backup landing sites have been established for Soyuz spacecraft emergency landing situations, foremost of which are sites in Canada.

The ground landing is commonly called a "dust down". Also termed a "soft landing" method, a landing in a Soyuz is never "soft", with some dramatic impacts reported over the years, and with spacecraft bouncing upon landing or dragged by parachute. Crews have frequently described the impact as similar to a car crash. The earlier Vostok missions were also planned with ground landings, although cosmonaut training included water recovery techniques as a precaution. But with Vostok, the landing speed of the capsule was higher than any crew would have been able to survive, so the solo cosmonaut used an ejection seat system to vacate the capsule and descend by personal parachute.

One problem that this caused was in officially verifying the early missions. The Soviets had to state officially that each cosmonaut had launched and landed in their spacecraft in order to qualify for the new, internationally recognized aero-nautical record. One of the criteria for this was that an occupant had to be inside the vehicle from the moment of leaving the pad to it touching down back on Earth. The Soviets quietly had to ignore this, particularly for Gagarin's first mission, and maintained this pretense until 1978 when reports emerged that the first cosmonaut had indeed used the ejection system and parachute at the end of his flight. By then, of course, it hardly mattered.

When the Vostok was "upgraded" to fly a larger crew the ejector seat was removed. But this brought back the problem of the higher landing velocity and no suitable escape system. It was for these Voskhod craft that the retro-rocket package was first introduced, located in the recovery parachute system to slow the descent enough for the crew to survive the landing. Fortunately, only two Voskhod were flown, so the risk was minimal before the advent of Soyuz. The Chinese Shenzhou is similar to Soyuz. It is also designed for ground landings and follows a similar profile to the Russian craft.

All American manned space flights under Mercury, Gemini, and Apollo ended with parachute recoveries in the ocean and were retrieved with the assistance of the U.S. Navy. This costly exercise was one of the reasons the Shuttle vehicle was designed with ground landings in mind, reducing NASA's bill from the U.S. Navy. The possibility of an orbiter ditching on water was still feasible in emergency situations and all crews did train for water egress up to the vehicle's retirement. Conversely, land recovery techniques were also studied for both Gemini and Apollo although it was never adopted beyond tests and demonstrations. The vehicles currently in development to replace the Shuttle are being designed with both land and water-landing capabilities in mind.

From 1981 through 2011, 133 of the 135 Shuttle missions launched ending with a landing inside the continental United States, on runways in Florida (78 landings), California (54 landings), or New Mexico (1 landing). The Shuttle also had the capacity to land at a number of contingency landing sites around the world, although these were never called upon. Neither were the various transatlantic landing sites that were on standby during each launch in case the mission was terminated early. The Shuttle also had the capacity (in theory) to return to its launch site if necessary in an emergency situation, but again this was never required, much to the relief of each Shuttle crew!

EMERGENCY ESCAPE

Every crew aims to complete their mission as planned, safely and efficiently. While hoping for the best, they are all certainly aware of the dangers and uncertainties of space flight and train hard for emergency, contingency, and alternative missions, should things go wrong. For any space flight, "things" could go wrong during training, on the launchpad, during the ascent to orbit, in the flight itself, during reentry, on landing, or during recovery. Such contingencies will also form part of the "commercial and tourist" space flights that are likely to begin in the next few years. Those who wish to pay for this experience will have to train for and be aware of such situations as part of their acceptance for the flight.

Preparing for every eventuality

Though each space flight is unique in its content, the profile for its preparation and execution is essentially the same: A mission is identified and assigned its objectives; a flight plan is created and the hardware prepared; the flight crew is selected and trained; then eventually the vehicle is launched, flies its mission, and then returns to Earth. All this is then followed by evaluations of the crew, research, and mission performance and examination of returned hardware in preparation for the next mission. This is a basic overview but it stands true for all successful missions flown to date, regardless of the country of origin, or mission objectives.

While this may be the plan, it does not always turn out that way. Human space flight is a high-performance, high-risk endeavor, which will always carry an element of danger for the mission, hardware, and crew. It has been demonstrated several times that accidents can occur during any of the stages of a mission, from training to recovery. For each of these potential risk areas, safety features, systems, and procedures were built in to help protect or possibly rescue a crew.

Some of these were introduced or modified for use on future flights only after an emergency had occurred during a previous mission.

Each crew is trained in emergency or contingency procedures and is provided with medical kits, escape equipment, and alternative flight plans to help deal with off-nominal stations. Mission planners develop alternate mission profiles to gain at least something from the mission should the primary objective have to be abandoned or curtailed, but this has to be done with crew safety in mind at all times. Though mission success is at the forefront of each crew in their execution of their flight, crew safety is the overriding factor and the primary responsibility of the mission commander. As much as each crew member would want to perform to the maximum and contribute as much as they could as a team member, they all have homes and families to return to. The expectation, excitement, and personal rewards of flying in space run strong in each crew member, but never as strongly as the desire to come home safely. Accepting the risk is part of being a space explorer, but these are not foolhardy individuals willing to risk their own lives or threaten the safety of others.

Escape tower, ejection seat, or luck?

Though accidents and tragedy have occurred in preparations for space flight, the most likely accident scenarios occur during the mission itself, the first of which is the ascent from the launchpad into orbit. Sitting inside a spacecraft, strapped to thousands of gallons (or liters) of highly explosive fuel, the crew needs some assurance that if things go wrong there is at least a chance to get out, however unlikely the disaster or slim the chance of surviving it.

Crew escape systems varied with the design of each spacecraft. There were pad escape systems, incorporated into the launchpads to allow the crew either to vacate the pad inside the spacecraft or quickly exit the vehicle and clear the launch area. Slide wires and escape chutes were incorporated into the towers built for Apollo, Shuttle, and Buran, while escape from any potential explosion could be achieved by escape tower on Mercury, Apollo, Soyuz, and Shenzhou, or by ejection seats on Gemini. Each countdown process includes periods of evaluation built into the launch preparations as well as options to abort the launch before the critical time. Such safeguards continue to feature in the operational launch procedures of both Soyuz and Shenzhou.

During the Shuttle program, several launch attempts were abandoned due to the weather or over equipment concerns. Most of these were canceled long before the vehicle was committed to engine ignition. On five occasions, there was a "Redundant Set Launch Sequencer" (RSLS) abort called, which occurred between the ignition of the three main engines at 6.6 seconds before liftoff and the lighting of the solid rocket boosters at $T - 0$ seconds. If computers (not humans) sensed a problem in the main engines, the launch would be aborted, preventing the SRBs from igniting. The SRBs could not be turned off once ignited, thus committing the Shuttle to launch and at least 123 seconds of flight since no abort was possible prior to SRB separation, even if a main engine failed. Fortunately, the RSLS

Artist impression of the launch abort profile for the Orion spacecraft.

system worked as designed on all five occasions (STS-41D in 1984, STS-51F in 1985, STS-55 in 1993, STS-51 also in 1993, and STS-68 in 1994).

Once a vehicle has left the pad, the options for escape from a pending explosion are more limited. There must at least be time to identify and react to the problem in the first place. Things happen rapidly in space flight and the journey from pad to orbit only takes eight to ten minutes. Events can occur in seconds, or even microseconds, so the technology has to be able to react quickly. Sometimes, there simply is no time to react and tragedy occurs.

The first manned spacecraft featured two methods of crew escape during launch. Vostok carried an ejection seat for its solo pilot, while Mercury incorporated a launch escape tower. The tower idea was continued for Apollo and was later incorporated on Soyuz and, more recently, Shenzhou. One reason for this is simply that there was insufficient room or mass capacity to provide an ejection system for every crew member once the single-seat spacecraft were phased out. Ejection seats were retained for Gemini as it was thought more suitable than an escape tower for that program. Like Vostok, these seats could be used for crew escape during recovery operations as well as for problems occurring during ascent.

For those programs using the escape tower, it was available for emergencies on the pad, but during ascent it was jettisoned at ballistic recovery altitude or once orbital speed was attained, which rendered the tower system unusable. For recovery, these spacecraft relied on parachutes (and, in the case of Soyuz and Shenzhou, the retro-rockets) to reduce the landing impact velocity.

These options were not available for Voskhod and Shuttle. For the amended Vostok, flying as Voskhod, crew escape was virtually impossible and the landing risky. Though promoted as an "improved", or "upgraded" spacecraft, Voskhod was in fact nothing more than a stripped-down Vostok. It was designed to carry a crew of up to three instead of one, mainly for the purpose of achieving spectacular space firsts ahead of the Americans. It was a risky, and lucky, two-mission program.

With additional crew members, the ejection seat had to be removed, leaving the crew no method of pad or launch escape. This also affected the safe recovery of the two crews, as they could not eject prior to landing. The retro-rocket package that was added was essential for the crew to survive the landing impact. Both missions were completed without any serious incidents, but it was fortunate that nothing went wrong. This could have been a prime reason for the cancellation of the program after only two flights, moving on to the more capable Soyuz with its built-in soft-landing system.

For the American Shuttle, things were more complicated. With crews numbering up to seven, crew escape had limited options. For both the atmospheric landing tests using Enterprise in 1977 and the first four orbital test flights on Columbia, the two-person crews did have ejection seats for emergency escape during ascent or descent. Fortunately these were not required on the actual missions. On Columbia, they were deactivated for STS-5 and removed for STS-9. No other Shuttle orbiter carried the system, as two-person Shuttle crews ceased with STS-4. Escape capsules were considered, but were deemed impractical for the spacecraft's design.

Following the loss of Challenger in 1986, a slide pole was installed. Each of the crew wore escape pressure suits and had the capability to leave the vehicle to descend on their own parachutes—at least in theory. Crews trained for such evacuations as part of their mission preparation. Fortunately, this type of escape was never called upon. It would have required the vehicle to be in a relatively stable flight mode for the crew to avoid hitting somewhere on the orbiter as they evacuated through the side hatch. The Shuttle Orbiter was essentially a glider as it came home, capable of landing on the ground or even on the ocean, so evacuating the vehicle in stable flight seemed contrary to what it was designed to do—a controlled stable glide to an unpowered landing.

Conditions inside the orbiter as it fell through the atmosphere in an uncontrolled state would surely have made the slide pole an unlikely solution to crew escape. With the vehicle possibly breaking up in flight, the crew of up to seven would have had to leave their seats on the flight deck and mid-deck, moving around in bulky pressure suits to hook up to the slide pole. Then they would have had to hope to miss all the trailing debris as the vehicle dropped like a stone towards Earth.

In 2003, there was no time for the Columbia crew to react to the impending disaster. They were too high and traveling too fast to use the escape pole, even if they had had time to consider it.

During ascent, the Shuttle had a number of abort modes, giving the crew the option to return to the landing site, take it over the Atlantic to land at specific sites in Europe or Africa, to fly a single orbit and land on the next pass, or to abort the ascent into a low orbit. In the latter case, onboard systems would gradually have raised the orbit to at least an operational level, enabling the crew to conduct an alternative mission and probably return early. When the Shuttle launch abort modes were devised, each crew hoped they would not be called upon. They were there should the need arise and it did so during the 19th mission (STS-51F), in July 1985. On that mission, the loss of a main engine resulted in an abort to orbit and a revised, but still highly successful mission.

The abort modes were not new ideas. During Apollo, there were stages in the ascent where the mission could be aborted early to an emergency recovery in the Atlantic or to attain a lower-than-planned orbit to give the crew and Mission Control time to evaluate what to do next. The lunar missions featured points at which mission progress could be evaluated and the decision made whether to continue. For most of the Apollo missions, each of these decision points was passed to allow the missions to achieve most of what was planned. The exception, of course, was Apollo 13. Here, the redundancy built into the design of the program came to the forefront and contributed to the recovery of the crew. But there were still points at which the skills and endurance of the crew and ground controllers were pushed to the limit.

It is interesting to note that, apparently, when the Astronaut Office was approached to fly a test demonstration of the Shuttle return to launch site abort mode, their response was that it could be tested when it was needed. Clearly, turning the stack around in flight and heading back to the launch site minutes after leaving it was not a favored option for the astronauts, even given their varied and very capable flying experiences. Thankfully it was never put into practice.

A calculated but accepted risk

Timing is crucial in order to escape from pending disaster, but so is design. On Apollo 1, the 100% oxygen environment, bare electrical wires, poor communications, and a complicated hatch opening system all contributed to the loss of the crew on the pad during a demonstration test prior to launch. The first Soyuz seems to have been launched with inherent problems and the design of the parachute deployment system was at fault. On Soyuz 11, the fact that the crew did not have separate pressure suits for launch and entry meant that they lost consciousness and died when the atmosphere escaped rapidly from the descending crew module with the landing following the preplanned automated sequence.

No matter the precautions, training, and practice, there is always the potential for the unexpected or "bad luck" to affect a mission. To date (2012), no crew has actually been lost in space, all crews having begun the recovery process, although three crews (Soyuz 1, Soyuz 11, and Columbia STS-107) have not survived it. Loss in space may well happen, but if and when it does, is the program mature enough to handle that type of tragedy? Are the politicians or public willing to accept that type of sacrifice? Only time will tell.

In reviewing 50 years of human space endeavor, what continues to shine through are the outstanding technology, skill, and professionalism that has safely carried most of the crews from launch to landing. With the development of new spacecraft, it will be interesting to see what escape options the crews have. When so-called private, commercial, and "tourist" flights arrive, as they surely will, how will the participants approach the fact that space flight is, and always will be, inherently dangerous?

2

In the steps of Gagarin

No matter how many humans venture into space, on which vehicle, or for how long, they will always be following the trail of Yuri Gagarin. He was the first, the pioneer of a mode of transport untried before 1961. After centuries of dreams, decades of theory, and years of development, all the preparation had to lead to the point when the first mission could be launched with a man aboard. Unmanned missions had paved the way for over three years but on that spring day in 1961, a brave young man was about to put theory to the test. No one really knew if he would return alive, or with his faculties intact. Even with all the data collected from the unmanned flights of Vostok, Gagarin really was taking a leap of faith into the unknown.

For every mission since, though the risks and dangers can never be totally eliminated, none of them will ever have to face the uncertainty and risk of the first step off the planet that confronted Gagarin.

HUMAN SPACE FLIGHT OPERATIONS

The first half century of human space operations has, by any measure, been spectacular and rapid. It had taken hundreds of years to progress to this point, where human space exploration was more than just a dream. After mastering the techniques of balloon flight, gliders, and finally powered flight, it took another half a century to devise systems, procedures, and infrastructure to place man-made objects into space. In the half a century since then, we have created a huge space complex based on the experiences of at least nine earlier space stations, explored the Moon, and launched pathfinder probes to the farthest planets in the solar system, pioneering the way for humans to follow at some point in the future. Human endurance on space flight has increased from minutes to months and the number of crew rose from single-seat space flights to successive international

Sergei Korolev and Yuri Gagarin, heroes of the Soviet Space Program.

expeditions of up to six on the ISS. The station itself has operated with a crew continuously, 24/7 and 365 days a year, for over a decade.

Plans for large space complexes, bases on the Moon and colonies on Mars, the exploitation of asteroid minerals, and planetary journeys have been suggested for decades. They will surely occur, perhaps not in our lifetime, but not so far in the future to think that it is totally impossible from the standpoint of current technology. Interstellar travel may remain within the realms of science fiction for some considerable time to come, but who knows for sure?

It has taken humans thousands of years to expand across our planet and develop the knowledge, skills, and experience to "live" here, and we are still learning and exploring. It took over 300 years to explore the Pacific Ocean and its environs and we are still investigating the deep jungles, high mountains, and frozen polar caps. We have only touched upon the vast expanses of the ocean beds. All of this could be classed as planetary exploration, but of the planet we all live on. With all this covering such a passage of time in the history of human exploration of Earth, why should we expect so much, so quickly from our explorations in space after only half a century?

Launch systems

Active participation in a flight into space starts, logically, with the flight. This is a short, 8–10-minute, exciting, explosive, but always interesting, trip from the launchpad to low Earth orbit. Getting off the planet is always the first hurdle and, as the German rocket engineer Werner von Braun once explained, "Once you have left Earth, you are halfway to anywhere."

For the first journeys into space, adapted ballistic missiles were used to carry a human crew, riding on Vostok, Voskhod, Mercury, and Gemini spacecraft. For Apollo, a new family of "space boosters" was developed—the Saturn rockets— which were powerful enough to take the first men to the Moon and to launch America's only space station. Unfortunately, the rocket developed for the Soviet manned lunar program did not perform as planned and cosmonauts never rode the goliath off the pad. It was the smaller, ballistic missile, designated the R-7, which became the workhorse for the Soviet and subsequent Russian space program. More recently, it has also given international crew members access to space when the Shuttle was unavailable and following its retirement in 2011.

For over 50 years, the R-7 in its various guises has propelled cosmonauts from the national launch site in Baikonur to orbit on over 100 missions. The reliability and ruggedness of the design and the foresight of the decision to go with the launch vehicle in the 1950s are remarkable. Despite all the international technology, advanced designs, and countless proposals and plans, the most reliable

A Soyuz is rolled out to the launchpad on a rail transporter.

and long-lasting launch system to Earth orbit is one that has been around for as long as we have ventured into space. In 50 years, it has endured very few launch-pad aborts and only two launch incidents; the 1975 third-stage separation failure and 1983 pad abort both resulted in the safe recovery of the crew. It stands as a remarkable credit to Russian engineering that it is this R-7 which has outlasted all the other launch vehicles that have taken humans into space.

With the retirement of the Shuttle, there remain (2012) only two operational manned (orbital) space flight launch systems available in the global program, the Russian R-7 and the Chinese Long March 2F. There are plans and designs in development for carrying human crews on board, but these are still some years away from operational use.

The Mercury-Redstone, Mercury-Atlas, Gemini-Titan II, Apollo-Saturn 1B and Saturn V were very successful, with no launch failures and very few "near misses" considering the pioneering nature of their use in the early years of manned space flight. Over the 135 launches of the Space Shuttle system, there was only one tragic launch accident (Challenger) resulting in the loss of life and one launch abort to orbit (STS-51F). Though termed operational from its fifth launch, the Shuttle system in hindsight could only be termed a research vehicle rather than totally "operational" in the true sense of the word, due to the changes in launch manifests and delays in ground processing. Indeed, it would be difficult to term any manned launch system, apart from perhaps the R-7, as operational, due to the length of time in service and the number of launches completed.

FIVE DECADES OF OPERATIONS

In reviewing five decades of manned space flight operations, it is difficult to define any specific decade as a singular era, but in general the 1960s could be termed the pioneering decade; the 1970s the decade of ascending learning curve; the 1980s the reality decade; the 1990s the decade of application; and the 2000s perhaps a decade of expansion. What lies ahead is for the pages of history.

Connections

The physical quest for space flight can be traced back to the era of stratospheric balloon ascents during the early decades of the 20th century. This was followed by aviation pioneers in their quest for speed, height, duration, and eventually the international drive to break the sound barrier and pushing the limits of rocket aircraft propulsion. All of these linked steps led up to the dawn of human space flight operations. What is less often considered are the significant, but connected developments in other areas of science, technology, medicine, and human endeavor, including of course military advancements, which have all contributed to applications now used in space exploration. There is often discussion about the benefits of space flight and the spin-offs from the investment and technology developed, but this works both ways. There are technologies and procedures which have been incorporated into the space program which have filtered down to improve aspects of life here on the ground.

Lessons learned from other endeavors are crucial to developing the next steps in space. For example, underwater exploration is currently being used to prepare astronauts for flights on the Space Station and in supporting simulations related to future explorations of the asteroids, the Moon, and Mars. Other extreme environment operations are to be found in the Arctic and Antarctic regions of Earth, in long-duration isolation chamber experiments (such as the recent Mars 500 experiment) and even experience from what are now termed extreme sports.

The history of polar exploration has analogues in long-duration space flight, and studies of the close confines of living and working in nuclear submarines, submerged for days or weeks in isolated environments under operational and stressful situations, have also been used to evaluate crew behavior and performance on programs such as the Space Station. This work, including that being conducted by expedition crews on the ISS, will have direct application for our eventual return to the Moon and out to Mars, where long-term research bases will have to be staffed and operated remotely from our planet by self-reliant crews, with support from Earth coming in an advisory or backup role.

We are at a key point in human space flight history. After 50 years, we can no longer consider ourselves to be pioneering. It is now time to homestead space and to expand our horizons, creating a reliable, economical, and sustainable infrastructure to move away from Earth, not only to explore new planets, moons, and asteroids but also to safely exploit their resources. We must still monitor our own world to ensure its survival and get the best use from its finite resources and we must discover how to balance our need for those resources with protection of the natural environment to ensure we can continue to live here. If we learn these lessons here on Earth, we can apply them to other worlds with confidence and perhaps a clear conscience.

Apollo 8 astronaut Bill Anders once said that the most valuable return from going to the Moon was to discover Earth. In expanding our knowledge and understanding of our own planet, we can put our best efforts into exploring new worlds. The last five decades have created the foundations for a concerted international effort to move out into the cosmos. Never again can we look up at the night sky and wonder what it would actually be like to go there, because we have done that. We just need to keep going a little farther.

Hindsight is a wonderful way to interpret past events and experiences and to think how things could have been done better or differently. It is quite easy to look back and wonder what might have been if certain events had or had not happened or if fate had intervened a different way. In this context, you could ponder endlessly what would have happened if the Americans had launched the first satellite; if the Soviets had landed on the Moon before the Americans; if Apollo had not been canceled in 1972; if the Shuttle had been authorized with its liquid-fueled manned booster; if Buran had become operational; if a Moon base and 50-man space station had been authorized; and so on. That small word "if" could lead to countless such speculations but can never affect what actually happened.

In reality, as humans we can only do our best and hope we get it right. In space flight, "our best" has yielded some spectacular achievements over the past half a century. Whether the decisions made were the right decisions is irrelevant and unchangeable, but they can be learned from for such decisions in the future. Here, we can only briefly summarize the achievements and decisions of these first five decades, to provide an awareness of how we arrived at this point in space exploration and allow us to decide where to go next.

1961–1970: THE PIONEERING YEARS

In an era of strained East/West relations following WWII, the growth in communism and the fear of a new, potentially nuclear, global conflict helped to create what has been termed the "Cold War". Amid these fears of annihilation emerged a dramatic increase in the military arsenals of both the United States and the Soviet Union, creating the "Arms Race". For military planners, gaining the high ground meant mastering a strategic advantage over the enemy. Attaining space flight capability in orbit around the Earth, or perhaps as far as the Moon, was about as high as you could hope to get in the 1950s. Of course, the only countries capable of achieving such a feat in the late 1950s were the communist Soviet Union and the capitalist United States. The "Space Race" which evolved between these two powerful nations and their ideologies was a competition to place the first object (satellite) in Earth orbit, send payloads to the Moon, and put the first man in space. The overriding goal during the first years of human space exploration was not scientific in purpose, but simply to beat the other superpower to the line. Doing so would be a clear demonstration, it was thought, of superior technology, implying strong military capability and making a great political and ideological statement.

It soon became obvious that enormous financial investment and infrastructure would be required to develop, mount, and sustain manned space flight operations. Military involvement was critical in the early years, if not in operational activities, then certainly to support launch, tracking, or recovery, or to provide the hardware and infrastructure to launch the payloads. As the program developed, so the balance between military and civilian participation shifted accordingly, with the support of the scientific community varying as required. For the Soviets, the complicated structure of the design bureaus and the supersecret nature of the civilian and military programs frequently served only to delay the development of advanced programs, such as the manned lunar landing, space shuttle, or space station efforts.

In contrast, the creation of the U.S. government space agency NASA (National Aeronautics and Space Administration) in 1958 helped focus the "civilian" U.S. space program, but it did create the potential for competition and confrontation with the U.S. military—primarily the USAF. This friction was to

The original seven Mercury astronauts (left to right): Cooper, Schirra, Shepard, Grissom, Glenn, Slayton and Carpenter.

surface in congressional support for the development of the USAF Manned Orbiting Laboratory and NASA's Apollo Applications (Skylab) space station programs and, much later, in convincing the Pentagon of the value of the NASA Space Shuttle for Department of Defense needs.

When President Kennedy committed America to reach the Moon by 1970, ideally before the Soviets, the Arms Race that had become the Space Race now evolved into a Moon Race. Even though it turned out to be purely one sided, at the time this was not clear, even to the Americans. When the U.S. committed to the Moon, only Yuri Gagarin had orbited the Earth, and then only once, during a 108-minute mission. America's own space explorer, Alan Shepard, had experienced space but had not entered orbit, completing a 15-minute suborbital space "hop". A second such flight was completed by Virgil Grissom in July 1961, but in August this was overshadowed by a mammoth 24-hour flight by Soviet cosmonaut Gherman Titov in Vostok 2. The bold commitment to place Americans on the Moon came as a surprise even to those involved in determining whether it could actually be done, so there was much work ahead. The clock was ticking and the Americans were clearly well behind in the race ... for now.

Three for space, but which one? (left to right): Glenn, Grissom, and Shepard.

The first tentative steps

What followed was a series of one-person flights that pushed the boundaries of human space flight endeavor from a few minutes to up to a few days in the next two years. It was a rapid advancement. The four U.S. Mercury orbital missions during 1962 and 1963 increased U.S. durations from 4 hours to 8, then 22 hours. A planned three-day Mercury mission was canceled, as the agency wanted to move to the more advanced two-man Gemini missions in preparation for the Apollo series.

In 1962, and again in 1963, the Soviets flew two dual flights, increasing their mission durations to between 3 and 5 days and once again stealing the headlines by flying the first female space explorer, Valentina Tereshkova, in June 1963 (Vostok 6) and the longest solo space flight (5 days), by Valeri Bykovsky. The one-person Vostok was then modified to carry additional crew members (all without spacesuits). It reappeared as Voskhod 1 in October 1964, flying a three-person crew (a pilot, doctor, and scientist), before the two-person Voskhod 2 in

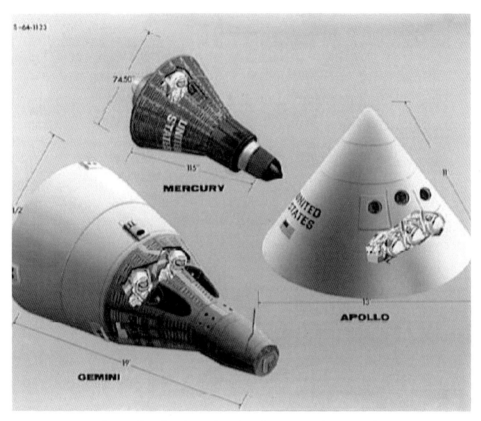

Comparing the pioneering American manned spacecraft.

March 1965 during which Alexei Leonov performed the world's first space walk. These missions were politically motivated, inserted as "space spectaculars" designed to upstage the American Gemini flights, as well as filling time until the more advanced three-person Soyuz was ready. The Voskhod flights were very risky and achieved very little apart from Leonov's space walk. They diverted resources away from the Soyuz program, the real Soviet competition to Gemini and Apollo.

The multi-purpose Soyuz was designed to perform a variety of tasks for the Soviets. These included rendezvous and docking missions, as a space station crew ferry, as a solo scientific research platform, some with clearly military objectives and, in its guise as Zond, as a two-person lunar transport craft with separate one-person lunar lander. Developed by the OKB-1 design bureau under the powerful leadership of Chief Designer Sergei Korolev, the driving force of the Soviet space program since its inception, Soyuz was hoped to be the salvation of the Soviet effort in space, trying to capture the headlines from the Americans. Unfortunately, the program was plagued with problems from the start. In January 1966 Korolev died on the operating table during surgery and with his loss came a power struggle

Gaining the experience—Gemini Agena rendezvous.

not only at OKB but with other design bureaus for government-supported space programs. Then in April 1967 Cosmonaut Vladimir Komarov was lost in a landing accident on Soyuz 1, and over the next few years difficulties in man-rating the lunar launch vehicle the N1 saw the Soviets abandon reaching the Moon in favor of the creation of large space stations. They issued statements to the effect that the Moon was never actually a target for their cosmonauts at all, which is now known not to be the case. It was a difficult time for the Soviet program.

The Americans, on the other hand, were very successful with their second-generation spacecraft, Gemini. Between March 1965 and November 1966, 10 two-man missions were flown with remarkable consistency. Gemini was planned as a stepping stone between the one-man Mercury and three-man Apollo and was designed to extend the experience of American missions from 1 to 14 days, which was the expected average duration of the Apollo lunar missions. Another objective of Gemini was to perfect the skills of rendezvous and docking and close quarters formation flying, known as Proximity Operations, or "Prox-Ops". The program also provided the opportunity to conduct far more extensive space walks than had been possible on Voskhod. Incorporating a crew compartment hatch that could be opened in space allowed one astronaut on each of Gemini flights 4, 9, 10, 11, and 12 to conduct pioneering American EVAs. Gemini also afforded the opportunity to fly a number of small experiments and conduct scientific observations on its longer missions and also to perfect the techniques of precision splashdown. On the whole, NASA gained a significant amount of experience with Gemini, taking the long-duration record from the Soviets, increasing their EVA experience, and perfecting docking with unmanned targets. These were important skills to be mastered in advance of the Apollo series of missions and beyond. Indeed, the experience gained during Gemini makes it one of the most important, if still most overlooked, programs in human space flight history.

Off to the Moon

Sadly, the start of Apollo manned operations was marked by a tragedy, not in space but on the ground, with the loss of the Apollo 1 crew in a pad fire a couple of weeks before the planned Earth-orbiting mission. This set back the program almost two years, with the first crew not flying in an Apollo spacecraft until October 1968. It is also important to point out that there were other issues (with the qualification of the launch vehicles and reducing the weight of the Lunar Module) that further delayed the manned missions. It is therefore reasonable to conclude that the Apollo landing attempts would probably not have occurred before 1969, even without the Apollo 1 fire.

In December 1968, the Apollo 8 mission became the first to carry astronauts around the Moon. Occurring at Christmas time, it gave the chance for a significant worldwide audience to watch the TV transmissions from lunar orbit. Then, in March 1969, the Apollo 9 crew tested the bug-like Lunar Module (LM) in Earth orbit and evaluated the Apollo lunar suit in a short EVA. Two months later in May, Apollo 10 astronauts took the LM to nine miles above the lunar surface, clearing the way for Apollo 11 to make the first landing attempt. In July 1969 millions saw Neil Armstrong step into history, followed by Buzz Aldrin as the first humans to land, walk, and live on the Moon, if only for a few short hours.

Once Apollo started flying with astronauts on board, the missions for the rest of the decade progressed remarkably smoothly, given the complexity of the missions and what they were trying to achieve. This apparent ease of success contributed to a general impression in both the politicians and public that space flight was becoming commonplace and that flying to the Moon was routine. One mission would soon demonstrate how wrong that impression was.

Meanwhile, the Soviets, while watching Apollo grab the headlines, quietly resumed the Soyuz missions at about the same time that Apollo returned to flight after Apollo 1, prompting Western observers to erroneously suggest that the race to the Moon was back on. The primary goal of the first Soyuz missions was for the cosmonauts to gain experience of manned rendezvous and docking, something they had lost ground with to the Americans. After the failure of Soyuz 3 to dock with the unmanned Soyuz 2, there was a concerted effort to achieve manned ren-

The Apollo Lunar Module sits on the barren lunar surface.

dezvous and docking with another crew in a second spacecraft. This had been the original objective of Soyuz 1 and the canceled Soyuz 2 in 1967 and was finally achieved with Soyuz 5 linking up to Soyuz 4 in January 1969. Two of the cosmonauts also completed an EVA from Soyuz 5 to 4, returning in the second craft. At the time, this was promoted as the world's first space station, but as more details emerged this bold claim was shown to be stretching the point a little. But it was still a significant achievement and a remarkable step forward for the Soviets.

Though not clear at the time, this was also a demonstration of the technique planned (but never demonstrated) for the Soviet lunar program, in which a lone cosmonaut would have spacewalked from the main craft to the lander and later, after returning from the Moon, would have completed a second EVA to enter the return craft for the trip home. Unfortunately, when the feat was tried again (this time without an EVA planned), Soyuz 8 could not dock with Soyuz 7. Both vehicles did compete a group (*troika*) flight with Soyuz 6 in which the first space welding experiments were performed, but it was another bitter blow to the Soviets in the wake of the success of Apollo 11 and the failure of their unmanned Luna 15 sample return craft. It lent credence to the argument for abandoning the Moon to the Americans and pressing on with creating a space station in Earth orbit instead, some years before the planned U.S. Skylab station was launched.

Shortly after the Soyuz *troika* flight came Apollo 12, which repeated the success of the previous Apollo missions by landing on the Ocean of Storms. Pete Conrad and Al Bean conducted two Moon walks to deploy a suite of surface experiments and then visited the unmanned Surveyor III which had landed nearby some 30 months earlier. Unfortunately, a failed TV camera did little to help viewer ratings back home with the audience having to hear what the two astronauts were doing instead of watching their activities.

Plans for Apollo originally included at least 10 landings, followed by the creation of a rudimentary space station, cleverly constructed from elements of Apollo/Saturn hardware. It was hoped that more extensive lunar exploration missions and further Saturn workshops would be launched, leading to far larger space stations by the 1980s. These were expected to be crewed by up to 50 astronauts and supplied by a reusable space ferry called a "space shuttle". By the 1980s, Apollo-derived hardware could be used to send the first humans to the planet Mars. This was the grand plan in 1969.

One of the major stumbling blocks in securing this grand vision was the April 1970 mission of Apollo 13. The explosion suffered on the way to the Moon aborted the planned landing and almost claimed the lives of the crew. The dramatic recovery of the three astronauts after such a perilous journey around the Moon and back home passed into NASA lore. It showed the agency at its very best at a time of great difficulty. Unfortunately, the seeds of success with Apollo were also maturing to throttle its future at the height of its accomplishments. NASA astronauts had reached the Moon within the timescale that President Kennedy had proclaimed and had achieved the feat twice. But now the American public was questioning why there was any need to keep going back when there was no sign of competition from the Soviets or anyone else, there were so many difficulties at home, and there was a very costly conflict on the other side of the world draining American resources.

In the firing line of all this was Apollo and the grand plan for what was to follow. Budgets had been tight for a while and, with new President Richard M. Nixon in office, were about to become much tighter. The first casualty was Apollo 20, which was canceled in January 1970, with the remaining seven flights stretched out over the next four years. In September 1970, five months after nearly losing Apollo 13, two more flights were canceled. Apollo would now end with flight 17 and, following the lunar flights, only one Saturn workshop (now called Skylab) would fly instead of the planned two or three. There would be no series of extended lunar missions or Apollo's flying in Earth orbit to utilize the skills gained and hardware proven for other objectives. On a more positive note, although they had lost the so-called "race" to the Moon, relations between the Soviets and the U.S. had improved and plans were being developed to fly a joint docking mission with cosmonauts during the mid-1970s. There were also signs that the Space Shuttle might still be authorized, although the large space stations it was originally planned to service were struggling to find support and funding. Any mention of manned missions to Mars was quietly dropped.

By the time the Soyuz troika missions flew in October 1969, Apollo 11 had won the race to the Moon for the Americans, while Soviet manned lunar hardware had still to leave the ground. The final blow for Soviet manned lunar exploration had been dealt and the leadership was planning a shift towards mastering long-duration space flight. They still held out hope for lunar success until 1974, when the lunar effort was finally abandoned. A major stepping stone in support of the space station goal, however, was the highly successful 18-day Soyuz 9 mission, flown in June 1970, the final mission flown in the first decade of manned space flight. It was an indication of things to come, looking to extend the duration of human flights in space rather than sending them out to explore distant worlds, at least for the near future.

Reflecting on the first decade

With 10 years of flight operations in the logbook from 39 manned space missions (16 Soviet and 23 American), significant lessons had been learned. Foremost of these, of course, was the simple fact that it was possible to send humans into space, keep them alive while performing useful work there, and bring them home safely. This was of great significance to the spacecraft engineers and designers, who for years had predicted the feasibility of human space flight. The Vostok and Mercury missions had proven them correct.

These early flights also established confidence in the launching systems, flight support infrastructure (such as tracking and communications), and the recovery process. Flight endurance was extended from a few minutes up to 5 days and a number of basic experiments were flown.

While the Voskhod missions did not extend the experience envelope much farther, apart from demonstrating that EVA was possible, the American Gemini series certainly did so. Gemini astronauts quickly discovered that performing suited operations in open space was not as easy as first thought, or indeed trained for. Their experiences led to the utilization of a large water immersion tank for EVA training, with the astronauts and test subjects wearing weighted suits to reproduce some of the conditions likely to be encountered during a space walk. This is one of the often overlooked lessons learned from the Gemini era and initiated an EVA training system that is still utilized in current space walk preparation programs around the world.

Gemini also provided significant experience of extended duration space flight and some of the psychology of flying a long mission in the small confines of a spacecraft. Perhaps the most important lessons learned from Gemini were in rendezvous, docking, and proximity operations, which had direct bearing on Apollo and still remain applicable to present day operations with space stations. The Gemini program also provided valuable experience of planning, preparing, launching, and controlling both manned and unmanned spacecraft in a tight manifest, and of controlled reentry and pinpoint recovery.

Clearly the experience gained from Gemini in preparing crews for tightly packed timelines was of benefit to the follow-on program, as many of the key crew positions on Apollo were filled by former Mercury and Gemini astronauts. An important decision taken during the development of the Apollo program was to cancel several missions which were repeats of the testing performed on an earlier successful flight, adopting a compressed and stepped approach to qualifying the complicated hardware called "all-up testing". The American missions flown between October 1968 and November 1969 had demonstrated that the LOR profile was indeed the most suitable to reach the Moon by the end of the decade using Apollo hardware. Even the aborted Apollo 13 in April 1970 had been a successful failure, by showing that NASA could adapt its mission profile after such an extensive, and fortunately not fatal, mishap. Sadly, the Russians never got the chance to demonstrate their techniques and hardware on a similar journey.

The seven manned Soyuz missions flown between October 1968 and June 1970 were of mixed success, which was very frustrating for the Soviets already suffering from the poor reliability of their own lunar hardware. These missions did achieve the first docking between two manned spacecraft and partial crew transfer by EVA, the first three-spacecraft rendezvous missions, and set a new endurance record. Although unclear at the time, the docking exercises were connected to the delayed lunar program as much as the planned space station program but, despite the *troika* flight falling short of the lunar objectives, the launch of three spacecraft from the same pad on consecutive days (and their recovery on consecutive days) seemed an impressive demonstration of the flexibility and ruggedness of the Soviet hardware, launch, control, and recovery teams, and systems for expanding flight operations in the future. But was it? It now seems that it was a belated attempt to detract attention from the success of Apollo 11 and give the Soviets much needed headlines and completed with no firm plans for follow-on extensive launch operations.

For the Apollo landings, NASA proved that, within the Apollo profile, landing on and exploration of the Moon on foot was possible. The short traverse by the Apollo 12 astronauts gave confidence to more extensive surface excursions planned from Apollo 13. Unfortunately, the landing was not to be on that mission, but Apollo 13 provided a set of valuable lessons of its own. Together with those of Apollo 1, these were some of the hardest lessons experienced by NASA to date.

Other lessons and experiences gained during Apollo 13 were equally valuable, but on a more positive side. NASA came to learn significantly about system reliability, backup and redundancy hardware and procedures, the provision of alternative mission plans, and simply never giving up on getting the crew home safely. All this came to the forefront during what has been described as NASA's greatest moment, and within a year of realizing what would be considered its greatest achievement—the first manned lunar landing. Despite the safe return of the Apollo 13 astronauts, the writing was already on the wall for Apollo, with little to follow of the grand plans of less than 12 months earlier.

The long-duration flight by two cosmonauts on Soyuz 9 was another mission which quietly achieved its goal and remained somewhat overlooked. But it provided important baseline data for medical and operational considerations to build upon for the first space stations, which were planned for the following year. Soyuz 9 was another milestone, like the 14-day Gemini 7, the eight-day Gemini 5, the 5-day Vostok 5, and the 4-day Vostok 3 before it, which pushed the boundaries of human space flight experience. This final mission of the decade, like the first, was a Soviet flight and it would provide them a head start in the race to master long-duration space operations, a path the Soviets had chosen to focus upon and one at which they would excel over the following three decades.

In the background of the headline-grabbing missions in Earth orbit and out at the Moon was the X-15 rocket plane program which explored the challenges of flying at speeds up to Mach 6 and to the fringes of the atmosphere. Supplementing this research was the series of flights of blunt (lifting) body vehicles, designed to provide valuable flight data to support the concept of flying and returning winged vehicles (such as the proposed Space Shuttle) from space. But this was for the future. For now, capsules would remain the primary method of sending men into space but, significantly in this pioneering decade (with one exception), not women. They were barred from flying jets in the U.S., one of the requirements for astronaut selection by NASA, and were mainly overlooked (after Tereshkova's propaganda-inspired mission) for selection to the cosmonaut team.

1971–1980: THE LEARNING CURVE YEARS

The new decade opened with a year of great success and shocking tragedy. The first mission, Apollo 14, continued the success of the series with a third lunar landing, this time at Fra Mauro, the intended site of Apollo 13. The mission confirmed that walking on the Moon for short distances was possible but could be confusing and disorientating without adequate reference points or navigational aids. The astronauts were aided by a two-wheeled rickshaw-style equipment transport to ease the load they had to carry, but often found themselves carrying the device when it became stuck in the lunar regolith, which tired them even more.

Six months later, the Apollo 15 astronauts benefited from the first electrically powered manned lunar roving vehicle. This remarkable device significantly increased the capabilities and productivity of the two astronauts on the surface. Meanwhile, in lunar orbit the third crew member participated in an expanded orbital program of science and observations, using equipment and experiments housed in a special instrument bay in the Service Module and from within the Command Module. In the closing months of the Apollo program, NASA was trying to make the most of the three remaining missions to the Moon by flying as much science as possible.

A salute to Gagarin

In contrast, the Soviet program was not going well. All their effort for manned operations was now diverted to creating a space station. The objectives of this program were, typically, not forthcoming from the secretive Soviets, and would not be for some years. However, it was subsequently revealed by Soviet space watchers that there would be two distinctly separate space station programs, either civilian and scientific in nature or more military in intent, which would both be run under one heading, called "Salyut" to help mask the clandestine nature of the military stations.

In April 1971, as part of the celebrations for the 10th anniversary of the Gagarin flight, the first Salyut was launched. Tragically, the first man in space had not lived to see the tribute, having been killed in a plane crash in March 1968 at the age of 34. He had never had the opportunity to return to orbit. The first Salyut was a compromise between the DOS scientific stations (identified later as "civilian" in the West) planned by OKB-1 and the Almaz ("Diamond") military-focused stations designed by OKB-52. It was plagued by difficulties from the start. The first crew (Soyuz 10), launched a few days after the Salyut, could not complete a hard docking and returned after only two days in space. Then the second crew was grounded when one of them failed a medical, so their intended mission was flown by the backup crew. Launched as Soyuz 11, this replacement crew completed a challenging 21 days on board the station but tragically died during the recovery phase due to a faulty air equalization valve.

As a result, the Soviet manned program was grounded and a planned third visit to the Salyut canceled. It would be two years before the next Soyuz crew entered orbit to test the improvements to the ferry craft and three years before a

The ill-fated Soyuz cosmonauts (left to right): Dobrovolsky, Patsayev, and Volkov.

cosmonaut crew would enter another Salyut in orbit. During this period, the Americans completed both the Apollo program and the three Skylab program manned missions, forging ahead in both lunar and space station experiences much to the chagrin of the Soviets.

The final Apollo missions to the Moon were flown in 1972. Apollo 16 and 17 completed the J-series of scientific missions to more geologically challenging sites, maximizing the variety of samples retrieved from the six landing sites. The Apollo program was a highly successful series and, even with the Apollo 13 aborted landing, significant experience and confidence was gained in sending crews out to the Moon and from performing the first EVAs on the surface of another celestial body. Unfortunately, a number of factors contributed to the closure of the program, which prevented the Americans from expanding upon this experience and contributed to the diversion of hardware, funds, and investment elsewhere. Most notably, this was to the planned Space Shuttle program, which had been authorized in January 1972 and approved in April that year while the Apollo 16 astronauts were exploring the Moon.

Just over a decade after manned space flight had become possible, the emphasis was already changing. No longer was it a race to achieve leadership at the Moon. Now, a concerted effort to look back at Earth began to develop and with it, hopefully, the creation of economical access to and from Earth orbit, with additional emphasis on sustained and extended operations in low Earth orbit. When Apollo stopped flying to the Moon in December 1972, no one really thought it would be over 35 years before we would even consider going back there seriously with a new program, and probably over 50 years before we finally make it.

Application by design

Using redirected lunar hardware, the Skylab space station became the first (and, to date, only) American domestic space station, another example of the long, complicated, and troubled American space station history within both NASA and the USAF. The military-orientated Manned Orbiting Laboratory program (utilizing a variant of the NASA Gemini spacecraft for crew transport) was canceled in 1969 after six expensive years of development and only one unmanned test flight. The NASA "civilian" space station program began quietly in the early 1960s, discarding the various grand plans revealed in the 1950s for giant stations circling the Earth and instead focusing upon creating payloads of scientific hardware flown on adapted Apollo spacecraft. These would supplement, but not replace, the lunar effort. Preliminary studies both within NASA and at contractor level revealed that the Apollo Command and Service Module, the Lunar Module, the Saturn family of launch vehicles, and the supportive hardware had the flexibility and potential to complete far wider missions than just sending men to the Moon for short visits to set up scientific instruments, return a few boxes of Moon rock, and plant a flag.

These ideas were soon identified as Apollo Extension (or Apollo X) missions. They were an obvious continuation to the basic lunar profile missions, which

An artist's impression of Skylab in orbit.

could be flown throughout the late 1960s and into the 1970s or beyond. However, as the effort intensified to develop the hardware required to reach the Moon, hopefully before the Soviets, so too did the administrative and political pressure not to divert substantial funds away from the main Apollo lunar program. Any new program suggestions suffered accordingly. Indeed, there was so much focus on simply getting the men to the Moon and safely home that even the science intended to be conducted on the Moon was reduced to almost nothing for the first landing. This was of course partly because of the immense challenge in simply achieving the landing in the first place and a desire not to overcomplicate the first attempt. Any expansion of science could be deferred to later missions once the commitment to reach the Moon by 1970 (and beat the Soviets) had been achieved.

The studies continued, however, and in an attempt to mask their appearance as a "new start" the programs were restructured. In 1966, Apollo X and all its connotations were gathered under a new program branch, identified as the Apollo Applications Program (AAP) Office. The primary focus of this effort centered upon using spent Saturn launch vehicle stages fitted out in space as rudimentary space laboratories. There were other missions proposed, but the Orbital Workshop (OWS) concept was the flagship of the AAP program. The rocket stage intended

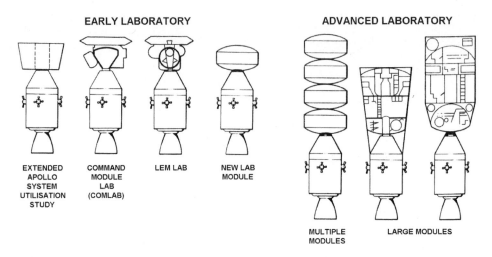

Early designs from the Apollo X studies.

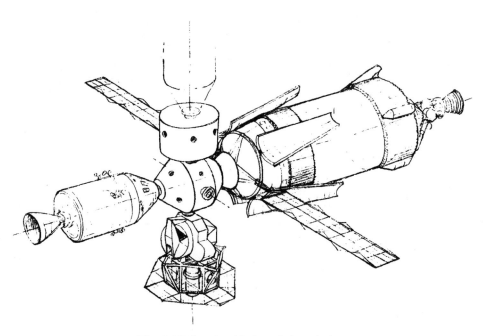

The AAP "wet" orbital workshop design.

for use as the station would be fueled and used during the launch. Once in orbit and empty, it would be decontaminated by the crew, who would arrive in a CSM ferry craft. This became known as the "wet" workshop design. NASA had to be careful to avoid promoting the space station program as a new start because of the possible threat to the budget allocation for the mainline Apollo effort. The way round this was to identify the "new program" (that was not officially new) as one that was simply applying the hardware and skills of Apollo to a range of further missions beyond the original remit.

Most of these missions were expected to fly in Earth orbit in between the lunar missions. After the initial Apollo landing missions (probably four or five), AAP missions to lunar orbit would follow, creating an OWS there followed by 14-day missions to the surface, hopefully setting up a small research base and extended surface expeditions. Unfortunately, concerns over the cost of the main-line Apollo and the decline in both interest and support for going to the Moon by both politicians and the public signaled the end of extensive AAP operations.

By the end of 1969, the lunar landing by Apollo had indeed been accomplished, twice, but only one AAP space station remained on the manifest. Three manned missions were planned, but it would not become the primary focus until after the Apollo missions to the Moon were completed. By early 1970, AAP had been renamed Skylab and the "wet" workshop design had also changed. Now, the laboratory would be launched on a two-stage Saturn V, with the third stage pre-fitted out as a fully equipped "dry", or unfueled, workshop. The three teams of astronauts were planned to fly 28, 56, and 56-day missions during 1973.

Skylab 1, the unmanned laboratory, was launched in May 1973 and was almost lost during the attempt, with one of the solar power arrays and micro-meteorite shields being ripped off by aerodynamic stress. Almost limping to orbit, the problems in cooling and powering up the station caused the first mission to be delayed to give NASA time to develop plans and hardware to recover the station to as near planned operating levels as possible. The success of the first and second crews in deploying the remaining array and installing protective solar shades saved Skylab, allowing a full three-mission program to be completed. The missions set new endurance records of 28 days, 59 days, and an impressive 84 days (both instead of the planned 56), rendering a potential 21-day fourth visiting mission unnecessary. The Skylab teams gained significant experience and gathered impor-tant results from Earth observations, solar studies, medical investigations, and a wide range of other experiments and research studies in materials processing, astronomy, and education.

Unfortunately, the fully functional backup laboratory, "Skylab B", was not launched due to budget restrictions. Instead, the flight-ready module was sent to the National Air and Space Museum in Washington, D.C. for public display. To this day, former astronauts who could have flown to and lived in the station are reluctant to visit the display, recalling lost opportunities. Other unflown Apollo Saturn hardware from the canceled lunar missions was allocated to the Johnson, Marshall, and Kennedy Space Centers as museum pieces. This was very disap-pointing, not only for those who had built the vehicles and those who had hoped to fly on them, but also for those who had negotiated the funding to pay for them under the Apollo program. Instead they remained on the ground, stark reminders of what might have been.

Following Skylab, the only remaining American mission firmly on the launch manifest was the docking mission with a Soviet Soyuz, which was designated the

Skylab 4 commander Jerry Carr enjoys microgravity inside the orbital workshop.

Apollo–Soyuz Test Project. This one-off mission occurred in the summer of 1975, when Apollo "18" rendezvoused and docked with Soyuz 19. The ensuing handshake in space between astronauts and cosmonauts as they opened the connecting hatches for the first time featured in the headlines around the world; a demonstration of growing detente between the two superpowers both on Earth and in space.

The ASTP had evolved from talks between representatives of the American and Soviet space programs in the late 1960s and early 1970s, with agreements on the exchange of data on both manned and unmanned missions. In early plans, it was suggested that an American Apollo might dock with a Soviet Salyut space station crewed by Soyuz cosmonauts. This was soon dismissed as impractical by the Soviets, since Salyut did not then have two docking ports. What was not revealed at the time was that a second-generation Salyut station was in development which did indeed feature two docking ports, a design which would not be revealed until later in the decade. Following the ASTP mission, talks continued for some years and included the possibility of docking a Shuttle orbiter to one of these, now revealed, second-generation Salyuts. Unfortunately, the political climate worsened, so talks on joint manned missions were abandoned for over a decade.

Publicity shot for Apollo–Soyuz crews (left to right): Slayton, Brand, Stafford, Kubasov, and Leonov.

Salyut takes over

While the Americans were completing their final excursions on the Moon, setting records on Skylab, and preparing to dock with Soyuz, the Soviets were recovering from the setback not only of Salyut 1, but also the officially unannounced launch failure of the second Salyut in July 1972. The in-orbit failure of the first Almaz station in April 1973 (which had been designated Salyut 2 to disguise its military objectives) was followed by the loss of a third Salyut just a month later. The latter one failed so soon after entering orbit that it was not assigned a Salyut designation but instead was identified as Cosmos 557 to once again mask its true, failed mission. These frustrating setbacks were balanced, to a degree, by the successful solo flights of two manned Soyuz missions. In September 1973, Soyuz 12 evaluated the new improvements to the basic ferry design in a pre-announced and planned two-day test flight. This was followed by the week-long Soyuz 13 astronomical research mission in December 1973.

Things began to look up for the Soviets from 1974, with the launch of another Almaz (designated Salyut 3). Then came a civilian station (Salyut 4) in 1975 and another Almaz (Salyut 5) in 1976. A series of eight Soyuz ferry craft, each carrying a two-man crew, supported operations with these stations. The reduction from

Soyuz, workhorse of the space program.

the previous three-man team was a direct result of the findings of the Soyuz 11 disaster, which required the cosmonauts to be protected in pressure suits. The additional support equipment replaced the mass of a third crew member, until an improved Soyuz variant could be introduced.

Two missions were flown to Salyut 3 in 1974, with the Soyuz 14 crew completing a successful 14-day residency and the first totally successful Soviet space station mission. But Soyuz 15 failed to dock with the station and came home after just two days, while Soyuz 16 was a dress rehearsal for the ASTP Soyuz mission and not associated with any Salyut. In early February 1975, the Soyuz 17 crew completed a 30-day mission to Salyut 4, followed just over three months later by the Soyuz 18 crew with a 63-day residency. This made Salyut 4 the first Soviet space station to host two resident crews. In between, though, came the first recorded launch abort of a manned mission, on April 5, 1975, when the original

"Soyuz 18" suffered separation failure of a spent rocket stage. The ascent had to be aborted and the crew endured a rather hair-raising 20-minute ballistic trajectory flight and recovery near the border with China. The crew survived the ordeal but it was their backup crew who eventually flew the 63-day replacement mission. Their landing, just after the return of the Soyuz 19 ASTP crew, demonstrated Soviet ability to handle two separate manned space missions at the same time. The final mission to Salyut 4 was the unmanned Soyuz 20, which further tested the proposed robotic resupply missions being planned for the next-generation Salyut station.

The next Soviet station (Salyut 5) was a military Almaz version, with Soyuz 21 completing a 49-day visit in 1976, followed by an 18-day visit by Soyuz 24 early the following year. As with all the previous Salyuts, there were setbacks as well as successes. The Soyuz 21 crew were forced to return earlier than planned, due to "sensory deprivation" according to the official line, and reports that the crew encountered an "acrid odor" inside the station. This must have been resolved because two months after the crew came home a second Soyuz was launched to the station. Unfortunately, Soyuz 23 only completed a two-day flight after failing to dock with the station. This mission is remembered more for its hazardous landing and recovery than its achievements in orbit. The landing occurred during a snowstorm and, to make things worse, on a frozen lake. The recovery effort was one of the most challenging ever encountered during a manned space flight and, although both cosmonauts survived their ordeal, neither ever flew in space again.

Partially in preparation for the next Salyut and also to fly a backup spacecraft that had previously been assigned to the ASTP, another "solo" Soyuz was flown in September 1976 to test new equipment and procedures. During the Soyuz 22 mission, the two cosmonauts evaluated a new Earth terrain camera intended for the forthcoming second-generation Salyut station. At the time, reports indicated that the flight was part of a planned series of "solo" scientific Soyuz missions flown independently of space station operations, but this proved to be the final "solo" Soyuz mission. Perhaps this was misunderstood information, or a move by the Soviets to mask the fact that the two solo Soyuz missions of Soyuz 13 (astrophysical) and Soyuz 22 (Earth resources) were flown because of the absence of a civilian Salyut (and to utilize available hardware approaching the end of its operational lifetime). In any event these missions, along with Soyuz 6 (space welding) and Soyuz 9 (biomedical), provided the Soviets with the chance to conduct research relevant to their space station program in more depth.

Another interesting development during Soyuz 22 was an official release revealing that candidates from Eastern Bloc countries would soon be selected as cosmonauts (within a program known as Interkosmos), to fly on future Salyut missions with Soviet commanders. This was the first time a cosmonaut selection process, albeit international, had been announced ahead of time. It appeared to be in direct response to the news that American and international candidates would be selected for dedicated science missions, or to accompany specific payloads, flown on the Space Shuttle. These "part-time" astronauts would become known as payload specialists, while their cosmonaut equivalent would be known as cosmonaut researcher. With new NASA selections for career astronauts for the Shuttle pending, a new era of space exploration was rapidly approaching.

In September 1977, one of the most successful space stations, Salyut 6, was launched. Over the next four years, the program would include 18 Soyuz missions and the first flights of the new unmanned resupply craft—Progress—based on the Soyuz design. To accommodate this increase of traffic, the new Salyut featured two docking ports, in theory to allow crews and vehicles to be exchanged for continuous manning of the station. On Salyut 6, however, although some Soyuz vehicles were replaced for fresh vehicles on orbit as their operational life came to an end, leaving a new vehicle with the resident crew; the expected exchange of resident crews did not occur.

What *was* introduced on this station was EVA capability, with the completion of the first space walks by cosmonauts since 1969. There was a series of missions flown by representatives of the Interkosmos countries between 1978 and 1981 and new world endurance records were set of 96, 140, 175, and then 185 days. Late in 1980, a short evaluation mission was flown to confirm that the station could support one final residence of 75 days which would allow the crew to host the final two Interkosmos missions. The Salyut 6 program also included an unmanned test flight of the new Soyuz variant (Soyuz T) and three test missions (Soyuz T2, T3, and T4) to confirm its operational integrity, prior to full operations with the next Salyut. Towards the end of the station's operational life, it acquired the first add-on module (Kosmos 1267), which was developed from an intended military manned spacecraft and ferry vehicle but now designed to test the potential for adding scientific modules to future space stations and to evaluate the structural integrity between two such vehicles.

Looking back on the second decade

There were significant developments during the second decade of manned space flight operations, which progressed the program forward as the emphasis changed from pioneering missions and lunar exploration to extended duration space flight and international cooperation.

During the final Apollo missions (14–17), the emphasis switched to more extensive surface activities and orbital science operations. The use of the Lunar Roving Vehicle and more mobility in the pressure suits helped the efficiency of the astronauts but one thing that became abundantly clear from the surface activities was that the disturbed lunar material would be a significant factor in planning any future lunar excursions (although at the time no one really thought this would be over 50 years in the future). The lunar dust found its way into everything, covering the suits, the equipment, and cameras. It was carried inside the LM at the end of the moonwalks and, in the one-sixth gravity, lingered in the environment inside the LM. When the next-generation lunar spacecraft or scientific research base appears, it will likely include an airlock-type facility, or at least an airflow barrier, to isolate the outside environment and EVA equipment area from the living quarters. Another important lesson learned from the later Apollo missions was that back-to-back EVA operations were tiring for the crew concerned, something that would have to be factored into planning for extensive EVA operations from the Shuttle.

From Skylab, the Americans experienced a totally new learning curve. Prior to the space station missions, the longest U.S. flight had been the 14-day Gemini 7 mission of 1965, with little mobility available in the close confines of the crew compartment. Even the three final Apollo landing missions with a packed timeline only lasted 11 to 12 days. There had been a gradual buildup of U.S. duration records over the first decade of operations, but Skylab extended the experience significantly over a period of just nine months. The Skylab missions set the achievement bar high for the rest of that decade and beyond.

Skylab has been an often overlooked program, in the shadow of Apollo, but like Gemini before it Skylab established some of the most important and influential experiences and achievements in U.S. space flight history. The program has more in common with today's space station program than with the historic Apollo lunar missions. In some respects, Apollo could be considered a diversion from the logical progression of early manned space flight activities, from the first attempts through to experience of extended space flight operations in low Earth orbit, prior to the expansion of human exploration away from Earth. It could be argued that, like the Concorde supersonic passenger plane, and perhaps even the Space Shuttle, the Apollo missions were ahead of their time and suffered accordingly. Mastering operations in low Earth orbit and establishing a firm foothold there before moving outwards seems to be the way the global program is being directed for the 2020s. Perhaps without the distraction of the Space Race, we may have already gone farther along this path. But, then again, without that background of competition, we may not have gone very far at all. Once again, future history will reveal just how important these early programs were in establishing permanent human presence in space and far from the Earth.

One of the key lessons learned from the Skylab missions was the importance of scheduling the crew's time and workload. There was an eventual realization that introducing new activities or objectives for which the crew had little or no prior experience would be less productive than allowing the crew to have the choice to follow a basic flight plan, with a "shopping list" of priorities. Tasks needed to be flexible, so that they could be completed on the day, added to a list of things that would be desirable to complete as soon as possible, or which could be slipped into the schedule as and when time allowed. Trying to micromanage the timeline, as was the case on Apollo, was not the best way to plan longer missions on a space station.

Skylab also highlighted the need for the crews flying the missions to be capable generalists rather than necessarily dedicated specialists. Each of the three missions included a scientist astronaut who had worked on the program for some years, but few of the pilot astronauts had been on the program for that long, many of them having moved over from the Apollo program. Skylab 4 Commander Jerry Carr, a Marine pilot on his first space flight, soon realized that, while learning to operate and monitor the Apollo Telescope Mount and its suite of solar observation experiments, he became a far better solar observer when he stopped trying to become a solar scientist.

It remains a bitter disappointment to many, both inside and outside of the program that, following the glowing success of Skylab A (especially after recovering the station from the brink of failure), the backup OWS could not be launched as Skylab B in the second half of the 1970s. It would have been a golden opportunity to capitalize on the experiences of the first workshop and to correct the mistakes made first time around, as the Soviets were beginning to learn from their Salyut series of stations.

Selecting new astronauts

For NASA, a second Skylab could also have helped the transition from the Apollo era to the Space Shuttle era and given some of the veteran astronauts remaining in the office an opportunity to fly and to pass on their skills and experiences to the Shuttle era selections. Unfortunately, this was not to be. Some members of the original NASA astronaut selections faced a wait of over seven years from the end of Skylab to the first Shuttle mission, with only the 1975 ASTP mission flown in that period. For some, this was much too long to wait and they left the program to pursue other goals, taking their experience with them.

But in 1978 the first NASA astronaut selections in a decade did bring in the first female and monitory astronauts to the team. Things were certainly changing as the decade drew to a close. The "original" four NASA pilot astronaut selec-

The first scientist–astronauts (left to right): Kerwin, Gibson, Michel, Graveline, Schmitt, and Garriott, in 1965.

The changing era, Skylab scientist–astronaut Joe Kerwin with Shuttle mission specialist candidate Anna Fisher.

tions, and former MOL astronaut transfers, in the 10 years between 1959 and 1969 were drawn from those with experience of military or civilian jet and test pilot roles. This reflected the need for the "flying" skills thought to be beneficial to the Mercury–Gemini–Apollo series of missions. As the nature of the missions evolved from the pioneering steps, so NASA brought in two groups of "scientist astronauts" to train for later missions on Apollo and AAP (Skylab). Their backgrounds were more academic than operational flying, but they still all had to qualify from a military jet pilot course to be assigned to Apollo era missions. For some this was a qualification too far and they left the program without flying in space, while others adapted well in gaining new flying skills. Four of the scientists flew between 1972 and 1974, while others performed backup and support roles on Apollo and Skylab. But they still had a long wait to fly on the Shuttle.

By the late 1970s, the scientist astronaut designation had changed briefly to senior scientist astronaut before eventually becoming the mission specialist designation that would become familiar in the new Shuttle program crewing policy. But the name wasn't the only change, as the role was now widened to encompass a broader range of skills and education. A greater diversity of specialists and qualifications were now considered acceptable for astronaut selection, encompassing engineering, pure and applied sciences, alongside operational accomplishments and new technologies.

Jet pilot training was now no longer a prerequisite. In fact, NASA reasoned that past experiences and qualifications served to demonstrate a candidate's ability to learn, so they assigned each new astronaut selection into a basic astronaut training program. It soon became evident, in the ongoing mystery of NASA flight crew selection, that being a professional astronomer did not lead to a flight on Shuttle astronomy missions; neither did qualification as a medical doctor guarantee automatic assignment to a medical mission. Such qualifications did increase the chances of such an assignment as the program unfolded, however. Unlike the early selections to NASA's astronaut program, who were designated "astronaut" from the first day, those selected from 1978 only received the designation after completing the Astronaut Candidate (ASCAN) training program.

Second-generation Salyut

Across the globe in the Soviet Union, the cosmonauts remained focused upon crewing a series of Salyut (or Almaz) space stations, flying to and from the station in Soyuz. From 1978, the mission durations began to increase markedly, supported by the regular resupply flights of Progress vehicles. These "space freighters" delivered fresh supplies of fuel, air, water, food, equipment, and other small items of hardware. Once emptied by the crew, they could be filled with trash and unwanted material for a destructive burn-up in the atmosphere, thus freeing up valuable room on board the station.

During the Salyut missions, each two-man crew had their hands full completing all the assigned science objectives while maintaining the onboard systems and keeping the facility clean and habitable. Generally, the crewing on most missions included a military pilot cosmonaut as commander and either a design bureau flight engineer on the civilian Salyut or a military engineer on the Almaz missions. There were very few equivalent scientist astronauts in the cosmonaut team and those who were selected, even with medical background, had little opportunity to fly on a mission. When the new variant Soyuz (Soyuz T) was introduced, it was once again possible to plan three-person crewing on the stations. However, when a third seat was available, it was normally filled by a second engineer from a design bureau (mostly from OKB-1, the Korolev design bureau), guest cosmonauts, or physician-cosmonauts.

From 1978, a change occurred for the visiting missions to a Salyut. The first civilian cosmonaut commanders (again from OKB-1) were accompanied by a representative from the East European/Interkosmos countries for short, week-long missions. The Interkosmos cosmonauts were certainly not of the "mission specialist" class, and were mostly military officers who were given a short course of space training for a one-flight opportunity, mainly for political–propaganda reasons (and to install foreign equipment on the Salyut). Essentially it was a Soviet way of combining the roles of the Shuttle payload specialists and manned space flight engineers that would soon be seen on the Shuttle. The Interkosmos program evolved into a series of commercial agreements with other countries, which flew in the 1980s on Salyut 7 and Mir and later developed into the so-called "tourist flights" of "space flight participants" seen on the ISS in recent years.

In September 1977, the Soviets launched the second-generation Salyut 6 station, of which much was expected. Reports indicated that the first Soyuz mission to the station would be, in part, a proud celebration of the 20th anniversary of Sputnik 1. So when Soyuz 25 failed to dock successfully, it came as a bitter blow and cast a shadow over the all-rookie crew. Though they were later exonerated of all blame, the die was cast and significant changes were implemented for future crewing policy. By October 1977 there had been 14 Soyuz manned dockings attempted with either another Soyuz or a Salyut/Almaz station since October 1968. Six of these had failed. As a direct result of the Soyuz 25 failure, it was decided that no all-rookie crew would be flown again, especially not for such an important, high-profile mission. Eventually, the criteria were relaxed, but it would be another 17 years before the next all-rookie Russian Soyuz crew would launch (Soyuz TM-19 in April 1994 with Yuri Malenchenko and Talgat Musabayev aboard).

One of the problems was that there was no leeway in the docking attempt. The stripped-down battery version of Soyuz, used on Salyut station missions since 1974, had a limited independent orbital life of just two days, barely enough to get to and from the station in the first place. Unlike Apollo 14, which took six attempts to extract the LM from the top of the S-IVB stage en route to the Moon, repeated attempts at docking for the Soyuz were out of the question. It was an expensive lesson to learn, from the point of view of wasted resources and hardware. Improvements made for the Soyuz T helped to resolve the orbital flexibility of the spacecraft, but not before another failed docking had occurred in 1979 due to a malfunctioning main engine on the Soyuz.

When the missions to the stations were successful, it added to an ever growing database of long-duration information that would enable the Soviets/Russians to develop their space station operations with greater confidence. Round-the-clock ground support for months on end; experiences of small crews working together in restricted confines of the orbital laboratory; masses of biomedical information and psychological data (including the stresses of command, work over long periods of orbital flight, and the difficult decision of whether to tell a crewman in flight of a family bereavement or major incident back home). All of this was essential information to the growing program and those missions yet to come.

There were also plenty of challenges to overcome and learn from. The Soviet stations were limited in air-to-ground communications coverage, due to the lack of a global tracking network. Maintenance and housekeeping chores increased as the stations got older, making it difficult to strike a balance with important and often time-critical research objectives. From Salyut 6, there was the added dimension of disruptions to the routine from visiting crews, both at arrival and after departure. Operationally, the program had to learn about the challenges (and consequences) of docking more than one spacecraft to the station core module at the same time and the dynamic stresses on the whole structure this entailed. Postflight recovery techniques and protocols following such long missions also had to be improved—valuable lessons for even longer expeditions that were already being planned.

Salyut operations during the 1970s were also evolving the cosmonaut mission training cycle. For the Vostok missions, the Soviets had created a small training group of cosmonauts taken from the larger corps, from which they would select the prime and reserve crews. This method had been successful and continued into the Salyut program. This experience, of having several crews going through the preparation cycle for assignment as reserve, backup, or prime crew, would prove highly successful and flexible. Separate training groups were formed for visiting crews, or to evaluate new versions of the Soyuz. These experiences would be adopted over 20 years later as the International Space Station evolved—a lasting tribute to the Soviet crewing policy devised in the Gagarin era.

1981–1990: THE REALITY YEARS

At the start of the new decade, Salyut 6 operations were winding down. Its last operations included the demonstration of the first add-on module docked to the core, in preparation for the launch of its replacement, Salyut 7, in 1982. Similar in appearance to the previous station, the new facility would continue where Salyut 6 left off, flying increasingly longer missions, with further visiting crews (this time more international than Interkosmos in nature). It would also see the first on-orbit partial crew exchanges and introduce a further add-on module to the main core station. Salyut 7 would also feature new endurance records for two expedition crews (212 then 237 days), with a further residency of 150 days by a third crew in addition to partial crew residences of 112, 168, and 64 days.

In 1984, and again during 1985, cosmonauts demonstrated the value of having a crew in space when things go wrong by overcoming serious problems to keep the station operating successfully. These activities safely extended its working life until 1986 when a new, improved station appeared. The final crew was able to visit Salyut 7 in 1986 to complete the planned program.

Salyut operations were expected to be the mainstay of orbital operations for the Soviets for the rest of the decade. It was not exactly clear what form the much anticipated Salyut 8 would take, nor was it evident how things were about to dramatically change, not just in the national space program but also across the Soviet Union itself. This would have considerable global consequences as the decade closed.

As the Soviets transitioned from Salyut 6 to Salyut 7, the headlines were being generated by the return of American astronauts to orbit after a gap of six years. It was the start of the Space Shuttle era.

The dawn of the Shuttle era

The Shuttle program had evolved over the previous two decades and had gone through a multitude of designs and formats before the final configuration was decided upon in the early 1970s. The winged spacecraft, with a huge cargo bay and two-deck crew module, would be launched like a rocket, powered by two solid rocket boosters and three liquid-fueled main engines. These would be fed with fuel from a giant external tank attached to the SRBs, under the belly of the

Testing Shuttle Enterprise from the back of a jumbo jet (1977).

Shuttle orbiter. The spent boosters jettisoned after two minutes and completed a parachute landing before being retrieved from the Atlantic and towed back to the Cape for refurbishment and reuse. After fueling the main engines up to orbital velocity, the external tank separated after 8 minutes and burnt up in the atmosphere.

Once in space, the orbiter's own maneuvering engines could place it into its required orbit. Now, the Shuttle would become a spacecraft, flying a range of missions of up to two weeks, but nominally between 7–10 days. These missions would feature a changing configuration of payloads, spacecraft, and hardware in the huge payload bay, demonstrating the flexibility and variety the Shuttle could offer to customers and investigators. The payloads would include commercial satellite deployments and retrievals, the development of on-orbit satellite servicing, and a range of science missions with a variety of payloads, carried in European-developed Spacelab pressurized laboratory modules or on unpressurized pallets. There were also planetary probes and large observatory deployments planned and

Columbia OV-102 lands at the end of the first Shuttle mission.

a number of classified military missions manifested. At the end of the mission, the orbiter would reenter and complete an aerodynamic, unpowered, glider-style approach, landing on a runway near to its launch site to enable a quick turnaround for its next flight. That was the plan.

The composition of Shuttle crews would be a mix of pilots to fly and command the mission and mission specialists to handle payloads, the robotic arm, and space walks. There would also be occasional payload specialists, chosen for specific or one-off missions. The orbiter had the capability to support multiple space walks from an integral airlock using a wide range of support equipment, including (in the early years) a manned maneuvering unit for untethered operations close to the orbiter (for safety reasons). Another innovation was the Canadian-built Remote Manipulator System (RMS), or robotic arm, which could lift large items out of and back into the payload bay, or support EVA astronauts in their spacewalking tasks.

The original plan was to fly one mission approximately every two weeks from converted Apollo launchpads in Florida and then introduce a series of high-inclination (polar) military launches from the USAF complex at Vandenberg AFB in California. To meet this expected demand, a fleet of orbiters would be required. In 1977, the orbiter Enterprise (OV-101) had been flown on the back of a converted Boeing 747 (which would also serve as a carrier aircraft for the orbiter

Challenger OV-099 launches on her maiden mission.

when required) to evaluate the atmospheric qualities of the design on a series of approach and landing tests. These were supplemented by a series of ground tests and launchpad evaluations prior to the first manned launch. That historic launch occurred on April 12, 1981, the 20th anniversary of Gagarin's flight. STS-1 was a stunning success and signaled the start of an impressive series of missions which would stretch across the next three decades. STS-1 was also the first U.S. manned space flight which had not been preceded by an unmanned space flight test prior to committing a crew to a new vehicle. This was a huge gamble, but it paid off.

Despite the success and proof of concept of the Space Transportation System over the next five years, the demand for commercial launches was not as great as expected and, consequently, the predicted reduction of launch and operating costs did not materialize. It was not all down to marketing the Shuttle as an all-encompassing national launch system. There were ongoing difficulties in preparing the vehicle for launch because the process was complex and not as routine as expected. This affected launch manifests and thus "selling" the Shuttle to fare-paying customers, who were looking for an affordable, dependable, and reliable system free from delays and mishaps. On top of this, the military never fully adopted the system for its own needs and the expected corporate commercialization of space by flying groundbreaking research equipment in orbit never really evolved beyond preliminary experiments.

Nevertheless, four more Shuttle orbiters were built and delivered, all of which flew during the first half of the decade. Options for a fifth were agreed, with a full set of spares being fabricated in case of the loss of one of the vehicles and the need to build a replacement orbiter. Columbia (OV-102) flew the first four orbital flight tests (between 1981 and 1982) and the first operational mission on the fifth mission (also in 1982). She then flew the first Spacelab mission in 1983 and, after upgrades, the 24th mission in 1986. Former ground test vehicle Challenger (OV-099) was upgraded to replace Enterprise as an orbital vehicle, as it was found

Discovery OV-103 completes its first mission in space.

to be too expensive to convert the latter vehicle after its ground tests. Between 1983 and 1985 Challenger completed nine missions, including three Spacelab missions, a number of satellite deployments, the first tests of the MMU, and satellite servicing of Solar Max. During 1983, the first American female (Sally Ride) and ethnic minority (Guion Bluford) astronauts also flew their first missions on Challenger.

In 1984, a new orbiter, Discovery (OV-103), was commissioned, followed the next year by Atlantis (OV-104). In a little over a year, Discovery supported a number of satellite deployment missions, satellite servicing–related EVAs, and the first dedicated DoD classified military mission. Atlantis was introduced in the fall of 1985 and completed a DoD mission on its first flight. Its second supported EVAs devoted to space construction demonstrations. By the end of 1985 not only had NASA astronauts flown on the Shuttle, but also representatives of the U.S. military, scientific and commercial payload specialists, political observers, and a number of astronauts from other nations, including Canada, Germany, The Netherlands, France, Mexico, and Saudi Arabia.

From 1986, there were plans to fly even more foreign payloads and crew members, to deploy space probes and space telescopes in the second half of the decade, and begin the initial flights in support of the creation of a large space station, called Freedom, which had finally been authorized by President Ronald Reagan in 1984 after years of debate, redesign, and deliberation. Space Station Freedom (SSF) would be created and operated by international collaboration between the U.S., Europe, Canada, and Japan and would be assembled by a series of Space Shuttle missions over several years.

January 1986

To make the Shuttle more commercially viable, and to address the lack of room in the lower deck of the orbiter when not flying Spacelab, a commercial augmentation module called Spacehab had been designed. This would add over 50 new lockers to the capacity, partly in an attempt to promote the commercial potential for small experiments and payloads to the wider market. The reduced launch loads on the Shuttle (compared to the earlier ballistic capsule launches) allowed NASA to relax the selection criteria for passengers. Following politicians and a Saudi royal prince, a U.S. schoolteacher was to be launched on the 25th mission and a U.S. journalist was being selected to follow on a later mission. There were unofficial reports of potentially flying artists, singers, poets, and other celebrities; perhaps even actors to film scenes for a "space movie". The Shuttle offered greater potential to fly in space than any other program before it, if it could be proved that it was a safe and reliable system for those who wanted to step aboard.

All these hopes for what Shuttle might have delivered tragically ended with the loss of Challenger and her crew of seven, including schoolteacher Christa McAuliffe, on January 28, 1986 (19 years and one day after the loss of the Apollo 1 crew). The tragedy occurred in full view of the TV cameras and disbelieving

Atlantis OV-104 performs a flight readiness firing (FRF) of its main engines prior to its first launch.

onlookers at the Cape. It was one of those horrible moments in history that those who witnessed the events, or watched the news, would never forget. On that day, the hopes of the American space program fell from the sky along with the wreckage of Challenger. As President Reagan observed during the nation's mourning, the brave crew of Challenger had "slipped the surly bonds of Earth and touched the face of God." In the quest for space, the reality of the dangers that each crew face was clearly revealed in the tragic events on that cold day in January.

As part of the inquiry into the tragedy, the Shuttle fleet was immediately grounded and all mission training halted. For the next two and a half years, while the accident was investigated and recommendations instigated, payloads were canceled, delayed, or reassigned to expendable unmanned launch vehicles. NASA was closely scrutinized. The agency would never be the same again. Many employees had left after the end of the Apollo program, taking with them the skills and experience that took America to the Moon. Now, another serious blow to the program would lead to further changes to the agency, not only for the Shuttle program but for NASA itself. It was another dark time for American manned space flight.

Crew of STS-51L Challenger during a mission briefing (left to right): Onizuka, McNair, Jarvis, Resnik, McAuliffe (teacher in space), and backup Barbara Morgan.

A new space complex

While America was reeling from the loss of Challenger and her crew, the Soviets were about to launch their next space station. Twenty-two days after the Challenger accident, it was not Salyut 8 that was placed in orbit but a new station called Mir ("Peace"). This new station, it was explained, was just the core module of a planned larger complex, incorporating six docking ports (five around a forward-located "node" and the sixth at the rear) to accommodate visiting Soyuz and Progress spacecraft and additional modules being launched over the next few years. There would also be provision to receive the yet-to-appear Buran space shuttle. With more than one docking facility, planned crew rotation and resupply could be completed without the need to vacate the station, thus maximizing the efficiency of the crew and eliminating the need to power down or reactivate station systems between expeditions.

In 1987, Kvant 1 became the first module attached to the Mir core. This was packed with astrophysics instruments and was originally intended for Salyut 7. It housed the initial set of gyrodynes, which enabled the complex to maintain its attitude without firing its thrusters and using up precious onboard propellants. The next module, Kvant 2, arrived in 1989. Kvant 2 became an extension facility

Challenger's replacement, Endeavour OV-105 stops off at Ellington Air Force Base near Houston on its way to Florida to begin its career in space.

to the main core module and was used for scientific research and EVA operations using an integrated airlock section for the space walks. In 1990, the Kristall space processing facility (also known as the technological module) arrived at Mir, bringing experiment furnaces and other equipment to expand the scientific research program on the station. There was an additional docking facility on Kristall originally intended for Buran. Although this was never used by the Soviet shuttle, the American shuttle docked to this module via a Russian-built docking module added in 1995. Delays in delivering these modules stretched the Mir manifest and caused the station to be temporarily abandoned for a few months during 1989.

In 1988, the much anticipated Soviet Buran space shuttle completed its maiden, unmanned flight. Despite this impressive achievement Buran never flew again due to budget restrictions. After years of development, construction of facilities, and training of crew members, the program disappeared in the wake of changes to the former Soviet Union. The Buran cosmonaut team finally disbanded in the mid-1990s, with the residual hardware abandoned, and the program providing a new chapter in space history of lost opportunities and wasted resources.

Following the loss of Challenger, the Americans reevaluated the Shuttle program and, as a result, the majority of commercial satellite deployments were shifted to expendable launch vehicles. The emphasis of the Shuttle manifest for the rest of the decade was on the recertification of both the Shuttle system and

The first U.S./Russian Space Station crew (rear left to right): Onufriyenko, Solovyov, Dunbar, Budarin, Poleshchuk; (front row): Thagard, Dezhurov. and Strekalov.

each individual orbiter, following the recommendations of the Challenger inquiry and to catch up with several important and long-delayed payloads which had to be launched on the Shuttle. Discovery was the first to return to flight, in September 1988, followed by Atlantis in December that year and finally Columbia in August 1989. The primary payloads launched in this period (1988–1990) were two Tracking and Data Relay Satellites, which would provide increased orbital communication coverage for future Shuttle missions. Other missions flown included the delayed deployment of the planetary probes Magellan (STS-30) to Venus, Galileo (STS-34) to Jupiter, and the solar polar observer Ulysses (STS-41). There were five fully classified DoD missions and the retrieval (STS-32) of the Long Duration Exposure Facility (LDEF), which had been deployed in 1984 and was originally planned for retrieval in 1985. The Hubble Space Telescope was also finally deployed (STS-30) and the first dedicated Shuttle pallet science mission, Astro-1 (STS-35), was flown to great success.

A difficult third decade

The 1980s were termed the reality years for good reason, for both the Soviets and Americans. The Soviets, despite great success with their Salyut and Mir programs,

Inside a clutter—Mir during the 1990s.

were suffering from lack of funding to expand the program as planned. An improved station, Mir 2, was struggling for funding even to be built; Buran had flown successfully, proving the concept, but was also unable to attract funds to continue. On top of all this, the country itself was under extreme pressure both internally and externally to change and reform. The surprising and sudden demise of communism towards the end of the decade in Eastern Europe—and the eventual breakup of the U.S.S.R.—was a major catalyst with global repercussions, and the pride of Soviet achievement—the Soviet space program—suffered. If it was to survive, the program would have to find other avenues of funding, much to the frustration of the older leadership in the design bureaus, the corridors of the Kremlin, and the top brass in the military. For a while, the program struggled on, but talks were in hand to help restore its pride and ensure its survival through the final decade of the century and beyond.

For NASA, the Shuttle was not delivering what was promised. The launch costs could not be lowered and the USAF had backed out of ordering their own orbiters and operating them from California. Commercial applications that took advantage of the unique characteristics of both the microgravity environment and the Shuttle transportation system were few and far between. It was becoming harder to keep the Shuttle flying and far from routine to launch them on time. The Challenger accident remained a painful memory.

On the positive side, as well as some remarkable flight operations once the vehicle attained orbit, significant lessons were learned and experience gained from

Working inside the Spacelab science module during a Shuttle mission.

repeated ground turnaround, from planning to processing, launching, and recovery to postflight analysis of the fleet of vehicles. A similar learning curve came from flying the dozens of payloads and experiments that the fleet did carry, from the smallest school experiment to complex Spacelab payloads.

Hovering in the background of all this was the Freedom space station program, which was slowly gaining momentum following its 1984 authorization. Unfortunately, it was also gaining in size, mass, complexity, and cost. The new decade would once again signal changes, both on Earth and for operations in its orbit.

1991–2000: THE APPLICATION YEARS

As the new millennium approached, work on Mir continued, but only just. The demise of the Soviet Union, the creation of a new Russia, and the independent development of former communist states left much uncertainty. Russia now had little funding available, even for the bare essentials, so there was precious little available for space exploration. The program was rapidly losing its popularity in the new Russia, with former Soviet space museums becoming nightclubs and unused hardware left to rust in unused playgrounds. The severe reduction in funding meant Mir was in serious trouble.

Farewell to Mir, a parting shot from the final Shuttle mission to the Russian Space Station.

In the short term, foreign investment helped, with a series of commercially supported visiting missions supplementing (and, on occasion, joining) long-duration expeditions. But, in the long term, something had to occur to keep the program going. That "something" would come from across the Atlantic.

Shuttle–Mir

The American Freedom space station program was running high over budget, exceeding its workable size and becoming too complex and too expensive. With new President Bill Clinton in the White House, the word came down to trim the program or it would be canceled. With so much having already been invested, the desire was to redesign the program, looking at new and alternative options. One of these, developed through talks with the Russians, was to use elements of the planned but grounded Mir 2 as a starting block for a stripped-down station. A further cost saving would be to use the well-proven Soyuz as a crew rescue vehicle until a dedicated vehicle could be developed. America would help fund the completion of the former Mir 2 elements and agreed to fly cosmonauts on the Shuttle in return for sending astronauts to Mir.

Identified as Phase 1 of the international program, a series of Shuttle–Mir docking missions were created to give America some long overdue docking practice, something they had not conducted since 1975. It would also give them the chance to gain their first long-duration experience since Skylab almost 20 years before, with a series of American astronaut residencies on Mir. As a result of these plans, several Shuttle flights were canceled and an assembly sequence worked out for what was now termed the International Space Station.

After further political and financial wrangling, as well as some doubts raised on both sides, the Americans and Russians agreed, with the other international partners, to cooperate on the ISS. Eventually, the new plans were sanctioned by the White House and Congress. At last, the Shuttle had a firm objective in building the ISS and the Russians had funds to keep Mir flying. It seems somewhat ironic given their past differences that the American space station Freedom would use pieces of the former Soviet Mir 2 to develop the multinational ISS. Times were certainly changing.

With Mir reprieved, a new cooperative partnership had to be forged, with a NASA team at the Gagarin training center near Moscow and a Russian team at JSC in Houston. The Shuttle–Mir program envisaged a series of flights of Russian cosmonauts as mission specialists on Shuttle missions, for which they only required to complete basic MS training in Houston since they were already part of the cosmonaut team. What is clear, but quietly overlooked, was that for those cosmonauts chosen to cooperate with the Americans, most had far more flight experience than their American colleagues. For the Americans, training for Mir would be significantly different than preparing for a Shuttle mission. Despite some of those selected for flights to Mir having previous flight experience on the Shuttle, they would all have to face an extensive training program on the Russian Soyuz spacecraft and Mir systems. Although it was agreed to use English as the default language for ISS, this did not apply to Mir, so for the NASA astronauts (and their support teams), the program of training in Russia included mastering the Russian language and moving to Star City near Moscow for months on end.

Mainstream Shuttle operations

The main Shuttle program changed emphasis again to more scientific missions, many of which were linked to the forthcoming ISS program. Within the decade, there were 63 Shuttle missions, including the 10 missions associated with Mir (the "near Mir" rendezvous mission of STS-63 and 9 docking missions STS-71, 74, 76, 79, 81, 84, 86, 89, and 91), and the first 6 ISS assembly missions (STS-88, 96, 101, 106, 92, and 97). The remaining missions were aimed at catching up with the delayed manifest and providing information useful to the proposed research programs on the ISS.

The classified military Shuttle missions quickly came to an end. In fact, the three which were flown (STS-39, STS-44, and STS-53) were only partially classified. The Shuttle also continued its program of deployments of NASA's Great Observatories, as well as larger payloads such as the Compton Gamma Ray Observatory (STS-37), the Upper Atmosphere Research Satellite (STS-48), the Advanced Communications Technology Satellite (STS-51), and the Chandra X-Ray Observatory (STS-93).

After the deployment of Hubble in 1990, problems were discovered with its optical clarity. Corrective optics were designed and these had to be installed on the first of a series of planned servicing missions. Within this decade of operations, there were three such missions to Hubble (STS-61, 82, and 103), which featured a total of 13 EVAs working at the telescope.

The Shuttle program of the 1990s also included a number of Spacelab module or pallet missions, which utilized the Shuttle's unique capabilities for science in low Earth orbit. Between 1991 and 2000, these missions and payloads included: Space Life Science 1 (STS-40) and 2 (STS-58) and the advanced Neurolab (STS-90); the International Microgravity Laboratory 1 (STS-42) and 2 (STS-65); Atmospheric Laboratory for Applications and Science 1 (STS-45), 2 (STS -56), and Atlas 3 (STS-66); U.S. Microgravity Laboratory 1 (STS-50) and 2 (STS-73); the U.S. Microgravity Payload 1 (STS-52), 2 (STS-62), 3 (STS-75), and 4 (STS-87); Space Radar Laboratory 1 (STS-59), 2 (STS-68), and the advanced Shuttle Radar Topography Mission (STS-99); the Japanese Spacelab J (STS-47) and the German Spacelab D2 (STS-55); Astro-2 (STS-67); the Life and Micro-gravity Spacelab (STS-78); and the Material Sciences Laboratory 1 (STS-83) and its re-flight (STS-94).

Contingency spacewalking had been an option for emergency or unplanned situations since the start of the program, so each Shuttle crew featured an EVA-trained team, whether for planned or unplanned space walks. The first Shuttle EVA had occurred during STS-6 in 1983 and since then EVA had supported a number of satellite-servicing and recovery/repair operations. Now, additional EVAs were being added to the program to evaluate hardware, training, procedures, and operations planned for the ISS.

Several Shuttle flights included demonstrations and evaluations of techniques and equipment in preparation for the ISS assembly missions, which would begin in 1998. Once that huge construction program started, the Shuttle program shifted emphasis again, beginning in 1999 and for the rest of its operational service, from mainly science to mostly ISS assembly and resupply. In fact, during the period of station assembly (November 1998–July 2011), there were only six missions (STS-93, 103, 99, 109, 107, and 125) which were not directly related to the ISS out of 43 Shuttle missions completed.

The final Mir years

For the Russians, the additional funding from the Americans certainly helped their own program, though for some it was a bitter pill to swallow. After three decades of national achievement under the Soviets, suddenly everything had changed and future operations depended so much upon U.S. dollars. There was a drive to try to market Mir commercially to fare-paying passengers or investors and though this generated interest, some of which filtered over to the early stages of ISS, the realization that Mir was near the end of its life was growing.

However, the station still had an important role to play for the rest of the decade. The final Mir modules arrived, starting with Spektr in 1995. This geophysical module was originally intended for military-focused experiments and research, but had been grounded for years due to lack of funding. Helped by American funding, the old module was dusted off and reconfigured to perform civilian scientific research instead. Later, during 1995, the second American Shuttle mission to the station (STS-74) delivered a Russian-built docking module, which was attached to the end of the Kristall module. Incorporating an androgynous docking system, this would be able to receive the American Shuttle orbiter on the remaining missions.

Mir received its final module, Piroda ("Nature"), in 1996. This was an international ecology module designed to conduct remote sensing of the planet from orbit. When the cooperative U.S./Russian program was agreed, Piroda was another partially completed module in long-term storage, but the new cooperation finally allowed it to be completed and launched into space. Mir crews had, for a while, found their time being spent more on maintenance, repairs, housekeeping, and overcoming setbacks, which had a knock-on effect on the amount of "research time" available to them, even with the arrival of the long-awaited modules. These difficulties hit the headlines in 1997, with an onboard fire and the collision with a Progress resupply craft, which ended the operational usefulness of the Spektr module and almost ended Mir altogether.

While the program recovered from these incidents, there were growing concerns about the Russians' ability to keep Mir flying while putting their full effort into the ISS program. This was especially concerning because so much hinged on the launch of the first ISS core elements, Zarya and Zvezda. These two modules would form the core Russian segment, allowing a resident crew to live on board the embryonic station and forming a base from which to expand the station to its full potential. In June 1998, veteran Salyut 6 cosmonaut Valeri Ryumin, who had flown to Mir on STS-91 with the task, according to NASA, of determining the overall state of Mir in preparation for it to be decommissioned, instead indicated that it was still viable for future operations. This did not please the Americans.

After 10 years of permanent manning of Mir by a succession of expeditions, the 27th main crew came home on August 28, 1999. Despite rumors of possible continued occupation through to 2002, the promised funding never materialized and it appeared that the Mir program was finally at an end. However, in the

Beginning construction, the first ISS element Zarya in orbit with Node 1 Unity prior to docking during STS-88 in December 1998.

spring of 2000, a private company called MirCorp (part owned by RKK Energiya), supported another 72-day mission (Soyuz TM-30). During this 28th residency, the two cosmonauts again evaluated the station's potential future uses, potentially including leasing the station for space tourism to advertising companies, for industrial production, and even for film companies. Understandably, NASA was not happy with this proposal, but the cosmonauts reported that the station was in good shape and could see no reason not to continue operations, subject to available funding.

These future plans were affected by delays with the required foreign investment, given that the ISS was the new jewel in the space crown for which funding was the priority. As a result, plans for launches to Mir slipped into 2001 before the final realization that it was time to end the Mir program. The station was de-orbited on March 23, 2001, just over 15 years after the core module had been launched into space. It was a sad day for the Russians, who had hoped to maintain a national station and try to restore some of the glory of the "old days". But the work to assemble the ISS had begun and signaled that a new era was about to dawn.

A fourth decade of experience

There were three streams to manned space flight operations during the 1990s. First, the main Shuttle program completed a range of independent engineering, science, deployment, and servicing missions. The Russian national Mir program flew a series of long-duration expeditions and a number of shorter visiting missions, some of which were international cooperative ventures. The third element was the Shuttle missions associated with the ISS program, including the series of rendezvous and docking missions with the Mir station and the start of ISS assembly operations.

For the Shuttle program, an important learning curve was climbed by sending orbiters to the Russian Mir station. When Space Shuttle was originally proposed, one of its major objectives was to create and support space station operations. This objective was lost in the early years of Shuttle operations and almost forgotten well into the 1980s, but by the 1990s the Shuttle finally had an objective and the target it was designed for. As plans to send the Shuttle to a space station were finalized, the 1990s also saw a range of missions flown by the Shuttle which at times seemed very much like cleaning a house: flying long-delayed missions, adapting others to support the new Mission to Planet Earth program; expanding the scope of the Hubble Space Telescope and deploying other great observatories and research satellites; and preparing for the start of ISS construction.

It is amazing now to look back at this period. It was a time of great change in the Soviet Union and a struggle for funding for the new Russian space program. Across the Atlantic, there were similar difficulties in securing a future for the U.S. Freedom station and continuing with Shuttle operations. The unique circumstances of these events allowed both nations to come together and, with some effort, overcome their differences to launch a true partnership that was extended to include 16 nations, creating what we now know as the International Space Station. The program of cooperation has, in hindsight, allowed the creation of the ISS to run relatively smoothly during the last decade or so, even with further tragic setbacks and operational hurdles to overcome.

The Mir program was the cornerstone for both the Russians and, to an extent, the Americans to progress on to the ISS. One of the most important lessons learned from Mir was adaptability. Being able to adapt flight plans, overcome difficulties, and have the skills, alternative systems, and procedures to keep flying was a testament to the robust nature of the Soviet/Russian hardware, if not its refined technology. Being prepared for the unexpected was a lesson well learned by the Americans during the seven residencies on the station.

During Mir, something new was learned on every expedition; there were often unexpected lessons. It was surprising to some how long the program continued to be operational. Mir, as a space complex, had been continuously manned between September 7, 1989 and August 27, 1999 by a succession of crews for 3,640 days, 22 hours, and 52 minutes. That record has now been surpassed by the ISS, but at the time this was a huge achievement, especially with the difficulties in keeping the station flying at all in its latter years. Often overlooked was the significant amount of science research conducted on the station over a period of 14 years. By 2000, the station had over 240 scientific devices on board with an accumulated mass of over 14 tons. Over 20,000 experiments had been conducted (see Andrew Salmon's "Firefly" in *Mir: The Final Year*, edited by Rex Hall, p. 8, British Interplanetary Society, 2001).

It was not smooth sailing by any means. Simply maintaining Mir systems began to consume most of the main crews' time on board, as did trying to find places to store unwanted equipment. Once cosmonauts began to make return visits to the station, sometimes years apart, they reported that the time spent on repairs had increased significantly over their earlier tour. They found equipment already stored in locations meant for experiments to be set up, which caused valuable experiment time to be lost because of relocating the logistics. International crew members described finding equipment on board the station a nightmare, with conditions on the station reported as detrimental to an efficient working environment. But much of this would prove valuable for the new ISS program. In quantifying the value of their series of missions to Mir and supporting the seven NASA residencies, John Uri, the NASA mission scientist for Shuttle–Mir, commented in the fall of 1998 that NASA had developed a host of lessons learned from Phase 1 operations at Mir. Of these, he estimated that 80–90% would have useful and direct application for the planned science program on the ISS. The others were somewhat peculiar to performing science on Mir; yet there is always something to be gained. Clearly identifying what not to do can be an equally valuable lesson as learning how to approach and complete a task.

A view of the completed International Space Station from STS-135 in July 2011, the final Shuttle mission.

2001–2010: THE EXPANSION YEARS

From the start of the new decade and the new millennium, the emphasis of human space flight focused around the creation and operation of the International Space Station. All but 3 of the 31 Shuttle missions flown in this decade were associated with the assembly and supply of the station. The first non-ISS mission was the fourth Hubble Servicing Mission, SM-3B (STS-109) in 2002. The second conducted ISS-related science, but flew independently of the station. This was the research flight STS-107, which ended tragically on February 1, 2003, with the high-altitude destruction of Columbia just 16 minutes from the planned landing. Following the loss of a second orbiter and crew in 17 years, the decision was made to retire the fleet. Originally, this was expected to be in 2010 but this was subsequently delayed until 2011, after 30 years of flight operations. As a result of the Columbia tragedy and inquiry, there was a review of the remaining payloads and hardware still to be launched to the station and a revised launch manifest was released for the remainder of the program, now with some of the planned elements deleted from the schedule. Once the Shuttle completed its return to flight qualification in 2006, there were no further serious delays in completing ISS assembly and finally retiring the Shuttle fleet.

In addition to the effort to complete the ISS, there was a move to fly one more service mission (SM-4) to Hubble. This had originally been canceled following the Columbia loss but was reauthorized after lobbying from the scientific community and the public. This third non-ISS mission of the decade was flown with great success, as STS-125 in 2009. The flight rounded out an impressive series of six missions specifically associated with the telescope but, sadly, the end-of-mission return to Earth on board a shuttle for the Hubble, which was on the manifest for about 2013 prior to the loss of Columbia, had to be abandoned. Had this been possible, a significant amount of information could have been gained from studying the telescope's physical structure after over 20 years in orbit, prior to displaying the historic spacecraft in a museum for public viewing.

For the Russians during the first decade of the new millennium all effort was focused on supporting ISS operations, with the slow expansion of the Russian segment but regular resupply via Progress unmanned freighters. The venerable Soyuz was still in service at the end of the decade and in a new variant, the Soyuz TMA-M. This was the fifth major upgrade to warrant a separate designation. Though other vehicles have often been proposed by the Russians as Soyuz replacements, the almost 50-year-old design remains the primary crew ferry and rescue vehicle. In the period of 2003–2006, while the Shuttle was grounded or being requalified for operational flight, only the Soyuz was able to ferry crews to and from the station. It enabled the station to continue to be manned and operated, albeit with a reduced two-person caretaker crew. In hindsight the option taken in the early 1990s to include Soyuz in the program has been, with the grounding and eventual retirement of the Shuttle, a wise one and has saved the ISS program until something else replaces the Shuttle orbiter to take American astronauts into orbit.

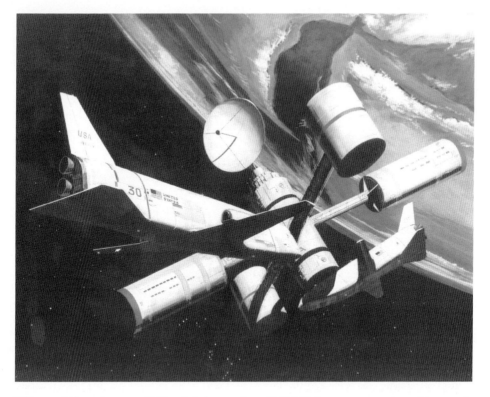

What could have been—a 1969 artist's impression of NASA's large space station with an early design of the Space Shuttle nearby.

A new player in orbit

This decade also brought a new, third player into the field of manned space flight operations—the Chinese. Long thought to have keen interest in developing human space flight capability, a planned program to place Chinese citizens in space in the 1970s was abandoned due to more pressing difficulties in the country. In 1992 a new manned space flight program was authorized, and from 1999 a series of unmanned test flights of the Shenzhou vehicle finally qualified the system to put a man into space.

In October 2003, taikonaut (or yuhangyuan) Liwei Yang flew Shenzhou 5 on a 21-hour mission, certifying the vehicle for human space flight operations. Two years later, in October 2005, a trio of taikonauts flew Shenzhou 6 on a five-day manned test flight, qualifying the vehicle for more extensive operations. It would be almost three years later, in September 2008, before the next Shenzhou crew entered orbit. This was also a three-man mission, but much shorter, with the specific objective of performing the first Chinese EVA. This was accomplished by mission commander Zhai Zhigang on September 27, 2008. All of these missions were explained by the Chinese authorities as planned steps toward the creation of a small space research laboratory, leading to larger space stations and eventually to Chinese manned expeditions to the Moon.

The first decade of the 21st century would put in place the infrastructure to expand the capabilities of the U.S., Russia, and the other partners in the ISS program, and to support the future direction that would be pursued over the coming two decades. The emergence of China added a whole new element to human space exploration and the appearance of their first space laboratory signals their intention to create a permanent presence in space, possibly far beyond low Earth orbit.

Five decades of experience

The retirement of the Shuttle in 2011 left something of a void. While any operational experience can be learned from, Shuttle operations will have little application directly to the proposed vehicles that will follow. The Shuttle system and hardware were unique and the new designs have more in common with the Apollo Command and Service Module than Shuttle orbiters. However, many of the lessons learned from the Shuttle–Mir program did have direct application to ISS operations.

For those vehicles which will eventually follow the Shuttle, a whole new learning curve might need to be scaled. By the time the new vehicles fly, those who were around for most of the Shuttle program will probably have retired, losing core experience that, like the Apollo era, will be hard to replace. There is, however, one significant difference between the transition from the Apollo era to the Shuttle era and the one from the Shuttle era to whatever replaces it.

In the 1970s, relatively few former astronauts moved to managerial roles in the space agency or the industry. Most of the engineers and managers who were at NASA during Apollo moved on to work on the Shuttle program, at least in its early years, bringing with them valued experience. A generation later, things were much different. The industry was much larger, most of the original employees at NASA had retired, and there were far more opportunities for former astronauts to move across to managerial roles, both inside and outside the agency, within the broader space program.

At the end of the Apollo era, many of the veteran astronauts, managers, and engineers decided to leave the program. In contrast, many of those who flew or managed the Shuttle are now in key positions within the space industry, working on a variety of new programs or projects including those contractors developing the Shuttle replacement. With the end of the Shuttle and some years before its replacement arrives, it is likely that more former Shuttle astronauts will retire. It will be interesting to monitor the career path of the ex-Shuttle pilots and mission specialists who climb the corporate ladder or advance in the administration of NASA and leading contractors in the coming decades. In Europe, former ESA astronauts are also beginning to move to higher administrative roles. In contrast, trying to track the progress of cosmonauts after they stop flying was always (and continues to be) difficult. Most of the military cosmonauts retire on a pension,

Table 2.1. New mission entries (September 2006–December 2012).

2006	
Soyuz TMA-9	ISS Expedition 14; Visiting Crew 11
STS-116	ISS 12A.1 P5 Truss

2007	
Soyuz TMA-10	ISS Expedition 15; Visiting Crew 12
STS-117	ISS 13A S3 Truss and S4 Truss
STS-118	ISS 13A.1 S5 Truss
Soyuz TMA-11	ISS Expedition 16; Visiting Crew 13
STS-120	ISS 10A Node 2 Harmony

2008	
STS-122	ISS 1E ESA Columbus Laboratory
STS-123	ISS 1J/A JAXA Kibo Laboratory and Dextre
Soyuz TMA-12	ISS Expedition 17; Visiting Crew 14
STS-124	ISS 1J JAXA Kibo Pressurized Module and RMS
Shenzhou 7	First Chinese EVA
Soyuz TMA-13	ISS Expedition 18; Visiting Crew 15
STS-126	ISS ULF2 Logistics

2009	
STS-119	ISS 15A S6 Truss
Soyuz TMA-14	ISS Expedition 19/20; Visiting Crew 16
STS-125	Hubble Service Mission 4 (last mission to the Hubble Telescope)
Soyuz TMA-15	ISS Expedition 20/21 (six-person capability)
STS-127	ISS 2J/A JAXA Kibo Research Module
STS-128	ISS 17A Final Solar Array Sections
Soyuz TMA-16	ISS Expedition 21/22; Visiting Crew 17
STS-129	ISS ULF 3 Logistics
Soyuz TMA-17	ISS Expedition 22/23

2010	
STS-130	ISS 20A Node 3 Tranquillity and Cupola Module
Soyuz TMA-18	ISS Expedition 23/24
STS-131	ISS 19A Logistics
STS-132	ISS ULF4 Logistics and Russian Mini Research Module Rassvet
Soyuz TMA-19	ISS Expedition 24/25
Soyuz TMA-M	ISS Expedition 25/26 (maiden flight of new Soyuz variant)
Soyuz TMA-20	ISS Expedition 26/27

2011	
STS-133	ISS ULF5 and Permanent Multipurpose Logistics Module Leonardo
Soyuz TMA-21	ISS Expedition 27/28
STS-134	ISS ULF6 and Enhanced ISS Boom Assembly (EIBA)
Soyuz TMA-02M	ISS Expedition 28/29
STS-135	ISS ULF7 and AMS-2 (the final Space Shuttle flight)
Soyuz TMA-22	ISS Expedition 29/30
Soyuz TMA-03M	ISS Expedition 30/31
2012	
Soyuz TMA-04M	ISS Expedition 31/32
Shenzhou 9	First Chinese space station crew; first Chinese female in space
Soyuz TMA-05M	ISS Expedition 32/33
Soyuz TMA-06M	ISS Expedition 33/34
Soyuz TMA-07M	ISS Expedition 34/35

while the civilian engineers resume work at Energiya until they retire. A few, like Alexei Leonov and Vladimir Titov, secured positions in leading Russian corporate businesses.

Former NASA astronauts, including Bob Crippen, Frank Culbertson, Bill Lenoir, Brian O'Connor, Loren Shriver, Dick Truly, and more recently Charles Bolden, Mike Coat, and others have made the transition from the astronaut office to the management side of the agency and then stepped across into industry. Their personal experiences from flying in space have been applied back into the program, but at a much higher level. It will be interesting to monitor whether such moves have a lasting legacy for the future space program.

2011-PRESENT: UTILIZING THE RESOURCES

In 2011, the 50th anniversary of manned space flight was celebrated. With it came the retirement of Shuttle and the transition of ISS from a construction site to a fully operational science research facility. It is too early to fully analyze what lessons have been learned from this decade as we are not yet a third of the way through it, but based on operations on the station during the previous decade, the difficulties of managing time on the facility remain. Even with a six person crew and full experiment facilities attached, there is still too much time spent in maintenance, repair and cleaning. Locating items and keeping track of the inventory only works, even with the help of a barcoding recording system, if the data is input correctly. This was a lesson learned back in the days of Skylab, but has it

The end of an era—Space Shuttle Atlantis on the ground at the end of the final Shuttle mission STS-135 July 2011.

fully translated to today's space station? Exercise and fitness programs in preparation for the return home also impact on crew time for science and research, and anything new can disrupt even the most trained and prepared crew. Visiting crews can be a welcome relief to the daily routine on station, but they also impede the natural flow of the main crew built up over time and it takes a while for a new crew to get up to speed.

Between the return of the American Skylab 4 astronauts in February 1974 and astronaut Norman Thagard launching to Mir in March 1995, the flow of experience of space station operations was lost by the Americans and the skills had to be learned all over again. Thagard commented, after coming home from a challenging mission in July 1995: "As we move into the era of long duration space flights, we need to consider that the problem of conducting complex experiments with little training is unavoidable." Thagard went on to explain that if a mission is long in duration, the first sessions of some experiments may not be conducted for some months after last training on them and if these experiments were in early development or preparation, the crews may not have even trained on them at all. This was first recognized by the Skylab astronauts and had long been known by the Soviets. It could be a new challenge as crews move out from Earth and do not conduct the experiments they trained for until some months after their mission began.

As the sixth decade continues, new lessons will be learned, continuing to expand the skills and knowledge of the first fifty years. These lessons can hopefully be applied not only to future ISS and Tiangong expeditions, but also to follow-on programs which take us deeper into space. Despite over 50 years of human exploration of space we have still only taken a few steps in the trail first pioneered by Gagarin. With each new space flight a further step forward is taken, but these missions are destined for future entries in the log book of manned space flight.

3

Completing the fifth decade: 2006–2010

SOYUZ TMA-9 (UPDATED)

International designator	2006-040A
Launched	September 18, 2006
Launch site	Pad 1, Site 5, Baikonur Cosmodrome, Republic of Kazakhstan
Landed	April 21, 2007
Landing site	Northeast of the town of Jezkazgan, Republic of Kazakhstan
Launch vehicle	Soyuz-FG (R7 serial number Ц15000-23), TMA-9 (serial number 219)
Duration	215 da 08 h 22 min 48 s (López-Alegría, Tyurin) 10 da 21 h 05 min (Ansari)
Call sign	Vostok ("East")
Objectives	ISS resident crew transport (13S), ISS-14 research program, visiting crew 11 research program

Flight crew

LOPEZ-ALEGRIA, Michael Eladio, 48, USN, NASA ISS-14 commander, Soyuz TMA flight engineer, fourth mission
Previous missions: STS-73 (1995), STS-92 (2000), STS-113 (2002)
TYURIN, Mikhail Vladislavovich, 46, civilian, RSA ISS-14 flight engineer 1, Soyuz TMA commander, second mission
Previous mission: ISS-3 (2001)
ANSARI, Anousheh, 40, civilian, U.S. space flight participant, visiting crew 11 (returned with ISS-13 crew on TMA-8)

Shuttle delivered ISS-14 crew members—STS-121 up (STS-116 down)
REITER, Thomas Arthur, 48, ESA (German) ISS-14 flight engineer 2, second mission
Previous missions: Soyuz TM-22 (1995)

STS-116 up (STS-117 down)
WILLIAMS, Sunita Lyn, 41, US Navy, NASA ISS-14 flight engineer 2

Flight log

The 14th resident crew boarded the International Space Station (ISS) two days after they had been launched from the Baikonur Cosmodrome. On board with López-Alegría and Tyurin was U.S. space flight participant Anousheh Ansari, who would spend just over a week aboard the station conducting a small research program. The day after docking, the crew were able to witness the reentry of

Returning the ISS resident crew to three: Expedition 14 crew members Mikhail Tyurin (left), Thomas Reiter (center), and Michael López-Alegría share a meal at the galley in the Zvezda Service Module.

Atlantis at the end of the STS-115 mission, four days after the Shuttle had undocked from the station.

The original intention was for Japanese businessmen Daisuke Emomato to fly to the ISS on TMA-9 with Ansari serving as his backup, but he failed his preflight medical on August 21. The following day, Ansari replaced him on the prime crew. Ansari is the Iranian-born naturalized U.S. citizen who cofounded Telecom Technologies Inc. in 1993, supplying softswitch technology to the telecommunications industry. With her brother-in-law, she made a multimillion dollar contribution to the X-Prize suborbital space flight record attempt foundation, which was officially named the Ansari X-Prize in recognition of the contributions by her family. (The X-prize was won by SpaceShipOne in 2005.)

With the formal handover to the ISS-14 crew completed on September 27, Ansari returned to Earth on September 29 aboard TMA-8 with the ISS-13 crew members NASA astronaut Jeffery Williams and Russian cosmonaut Pavel Vinogradov.

While Ansari's fellow Soyuz TMA-9 crew mates were undergoing their induction in ISS systems prior to assuming formal residency from the outgoing ISS-13 team, the latest space tourist completed her own program. During her nine days on board the station, she conducted three TV broadcasts, amateur radio broadcasts, and a series of photographic and video surveys in the Russian segment of the station for educational purposes. She also participated in bio-medical experiments, including researching the mechanisms behind anemia, muscle changes that influence back pain, and the consequences of radiation on crew members.

With the departure of the ISS-13 crew, López-Alegría and Tyurin began to work with ESA astronaut Thomas Reiter, who had arrived on July 6 aboard Discovery during the STS-121 mission. The German astronaut was working on the ESA Astrolab program and would leave the station on December 20 aboard Discovery (STS-116) after handing over his ISS-14 FE2 role to NASA astronaut Sunita Williams who arrived on the station on December 11. Williams would continue on board the station with the ISS-15 crew after the departure of López-Alegría and Tyurin. During their first days on the station, the new resident crew was occupied with troubleshooting the Russian Elektron oxygen system, which had been switched off due to overheating just before they arrived on the station. This was followed by the shutdown of Control Moment Gyroscope #3 (this would be replaced during STS-118) and the repair of the American Carbon Dioxide Removal Assembly (CDRA) system. Maintenance work takes up as much time as experimental work on most expeditions. Thomas Reiter's arrival saw the resumption of delivering a third permanent crew member via Shuttle, an important step towards resuming normal operations after the loss of Columbia in February 2003. Other "normal" operations included the relocation of Soyuz TMA-9 from the aft port of Zvezda to the nadir port on Zarya on October 10. The crew relocated the Soyuz a second time on March 31, moving the spacecraft from Zarya back over to Zvezda to clear the hatches for other operations. During their expedition, the crew would also receive the Progress M-58 and M-59 resupply vessels and host the STS-116 crew.

Throughout their expedition, the resident crew continued to work on the expanding experiment program in both the Russian and American segments. These were now being supplemented by the ESA program conducted by Reiter. There were 114 hours of crew time planned for American science operations and 266 sessions on 41 experiments in the Russian segment. The crew also conducted extensive robotics work with the Canadarm2 unit on the exterior of the station.

A total of 33 hours 42 minutes of EVA time was accumulated by the ISS-14 crew. López-Alegría conducted all five space walks, accompanied on the first and fifth by Tyurin from Pirs using Russian Orlan suits and by Sunita Williams for the other three, operating in U.S. EMUs from the Quest airlock.

The first EVA (November 23, 2006, 5 h 38 min) involved the repositioning, deployment, and relocation of equipment on the exterior of Zvezda, as well as the commercially sponsored Canadian "golf experiment" in which a golf ball was placed in orbit using a gold-plated club. The next three EVAs (January 31, 2007, 7 h 55 min; February 4, 7 h 11 min; and February 8, 6 h 40 min), from the U.S. segment, focused upon rerouting electrical and fluid quick disconnect lines from the soon-to-be-disconnected Early External Active Thermal Control System to a permanent cooling system in the Destiny Laboratory. The two astronauts also began work for the Station-to-Shuttle Power Transfer System (SSPTS), which would enable the Shuttle to draw electrical power from the station for extended visits to the facility. On their third EVA, the two astronauts jettisoned two large shrouds from solar array truss P3 Bays 18 and 20 and installed an attachment for cargo carriers. The final EVA of this expedition (February 22, 6 h 18 min) was back on the Russian segment, where the two astronauts retracted a stuck antenna on Progress M-58 and performed a series of equipment photography and similar lesser tasks. In a total of 10 EVAs, López-Alegría had accumulated an American astronaut record of 67 hours 40 minutes by the end of his residency.

On March 31 (March 30, GMT), the Soyuz TMA-9 spacecraft was relocated from the Zarya port to the rear Zvezda port. Although this operation took only 24 minutes, the operation to shut down the ISS and then restart it again took several hours each side of the relocation flight, during which the station remained unmanned.

A combination of the delay to the launch of STS-117 (due to hail damage on the External Tank on February 26) and the rescheduling of TMA-10 from March to April ensured that López-Alegría and Sunita Williams would both set new American astronaut endurance records for stays in space. The replacement resident crew arrived at the ISS aboard Soyuz TMA-10 on April 9, 2007. Traveling to the ISS with the ISS-15 cosmonauts was American businessman Charles Simonyi. Following the customary welcoming ceremonies, Simonyi exchanged his Soyuz seat liner with Williams, who officially joined the ISS-15 crew. The formal handover between ISS-14 and ISS-15 crews took place on April 17. After a joint program of 10 days, the ISS-14 crew loaded the Soyuz TMA-9 with items for their return, together with Simonyi, on April 21.

Milestones

250th manned space flight
102nd Russian manned space flight
 95th manned Soyuz flight
  9th manned Soyuz TMA mission
13th ISS Soyuz mission
11th ISS Soyuz visiting mission (VC11)
Longest flight by a Soyuz spacecraft (215 days)
Set new endurance record for ISS Expedition (215 days)
López-Alegría set career EVA record for an American at 67 h 40 min (10 EVAs)
He also set record for longest U.S. space flight (215 da 8 h 22 min 48 s)
Ansari was the 4th (1st female) space tourist and the 1st Iranian in space

STS-116

International designator	2006-055A
Launched	December 9, 2006, 20:47
Launch site	Pad 39B, KSC, Florida, U.S.A.
Landed	December 22, 2006
Landing site	Runway 15, Shuttle Landing Facility, KSC, Florida, U.S.A.
Launch vehicle	OV-103 Discovery/ET-123/SRB BI-128/SSME: #1 2050, #2 2054, #3 2058
Duration	12 da 20 h 45 min 16 s (STS-116 crew)
	171 da 03 h 54 min 05 s (Reiter who had been launched on STS-121)
Call sign	Discovery
Objectives	20th ISS Shuttle mission (12A.1), delivery and installation of ITS P5 truss assembly, Spacehab single-cargo module, ISS resident crew member exchange

Flight crew

POLANSKY, Mark Lewis, 50, civilian, NASA commander, second mission
Previous mission: STS-98 (2001)
OEFELEIN, William Anthony, 41, USN, NASA pilot
PATRICK, Nicholas James MacDonald, 42, civilian, NASA mission specialist 1
CURBEAM Jr. Robert Lee, 44, USN, NASA mission specialist 2, third mission
Previous missions: STS-85 (1997), STS-98 (2001)
FUGLESANG, Arne Christer, 49, ESA (Swedish) mission specialist 3
HIGGINBOTHAM, Joan Elizabeth, 42, civilian, NASA mission specialist 4

ISS-14 crew member up only
WILLIAMS, Sunita Lyn, 41, USN, NASA mission specialist 5, ISS-14 flight engineer 2

ISS-14 crew member down only
REITER, Thomas, 48, German Air Force, ESA (German) mission specialist 5, ISS-14 flight engineer 2, second mission
Previous mission: Soyuz TM22/Mir 20 (1995)

Flight log

This was one of the most challenging missions in the history of the program. During the 13-day mission, the crew rewired the ISS power system and continued

Installing the P5 truss assembly: NASA astronaut Robert L. Curbeam and Swedish ESA astronaut Christer Fuglesang participate in the first of three EVAs.

the construction phase by installing the P5 truss assembly. The flight also featured the exchange of ISS resident crew member Thomas Reiter with Sunita Williams.

Originally planned for a December 7 launch, the mission was delayed 48 hours due to low cloud cover, as no favorable conditions were expected to support a launch until December 9. The first nighttime launch in four years (due to post-Columbia safety limits for ascent photography), the ascent to orbit went without a problem. This was the first mission to feature the Advanced Heath Management System (AHMS) designed to improve the safety of the SSME. On this flight only performance data were collected, but on future flights the system would cut off the SSME if it detected a failure was about to occur.

The next day, the crew used the orbiter boom sensor system mounted on the end of the RMS to sweep the orbiter's surface carefully, surveying for any damage incurred during ascent. The survey revealed no significant problems and the 2-day approach to the ISS continued.

However, a wing inspection was subsequently called for after a minor vibration reading on a port wing sensor was recorded. Analysis of inspection imagery determined that the heat shield could support a safe reentry and no further inspection was required. After backflipping the orbiter to allow the ISS crew to visually and photographically inspect the heat shield, Discovery was docked with the station on December 12. Following integrity checks and hatch-opening ceremonies, Williams officially became a member of the ISS-14 crew and Reiter joined the STS-116 crew.

The installation of the P5 truss was supported by three EVAs. Shortly after docking, the truss was lifted out of the payload bay and passed to the station's robot arm, where it was suspended overnight. The following day (December 12) Curbeam and Fuglesang conducted a 6 h 36 min EVA to attach the P5 truss, as well as replacing a failed camera that would be required to support future EVA tasks. Launch locks were removed and the astronauts completed plugging in the new segment to allow the P6 segment to be attached to the end of the P5 unit in readiness for when it was moved from its temporary location. A number of get-ahead tasks were also completed. At the end of the EVA the backbone of the ISS had increased by a further 11 feet (3.35 m).

The second EVA on December 14 lasted 5 hours, with the two crew members continuing the rewiring to incorporate the new truss. Despite problems fully retracting the P6 solar array (with only 17 of the 231 bays or panels folded as designed), its retraction was sufficient to allow the P4 array to rotate and track the Sun, generating power to the station. The astronauts were also able to relocate the two main carts on the rails of the main truss, place a thermal covering over the station's RMS, and install bags of tools for future EVA support.

For the third EVA on December 16, Williams joined Curbeam to complete the rewiring operations. They also attached three bundles of Russian debris shield panels on the exterior of the Zvezda Service Module. These would be fully installed on future EVAs. After installing a robotic arm grapple fixture, the pair returned to the P6 array to continue its retraction, shaking it while it was reeled in one bay at a time. At the end of the attempt, 65% of the array had been retracted. This EVA lasted 7 hours 31 minutes.

The final EVA on December 18, by Curbeam and Fuglesang, was added to complete the P6 retraction. It would be relocated during a later 2007 mission. After securing insulation on the station arm, the EVA ended at 6 hours 38 minutes. At the completion of EVA activities, STS-116 had logged a total of 25 hours 45 minutes. In addition, Curbeam had accumulated a career total of 45 hours 34 minutes EVA time across two of his three Shuttle missions, setting a new record.

While the EVAs were being conducted, the crews transferred over two tons of food, water, and equipment to the station under the direction of Load Master Joan Higginbotham. They also transferred about two tons of unwanted equipment and samples from experiments into the Spacehab module for the return to Earth. Just two minutes short of a full eight days of joint operations, Discovery was undocked from ISS on December 19. The next day, the crew again inspected the orbiter heat shield for any damage incurred during orbital flight. They also deployed three small scientific satellites, checked out the landing systems, and completed stowage for reentry.

Originally scheduled for December 21, the landing was postponed due to the addition of the fourth EVA. Inclement weather at the Cape then forced a cancellation of the first attempt there and it was too windy to land at Edwards. However, conditions at the Cape turned dramatically to allow a landing on the second attempt on December 22. A total of 17,900 commands were sent on this mission setting a new record, with over 5,000 more commands from MCC-H than for any previous flight.

Milestones

251st manned space flight
147th U.S. manned space flight
117th Shuttle mission
 33rd flight of Discovery
 20th Shuttle ISS mission
  7th Discovery ISS mission
First Swedish citizen in space
First Swedish astronaut to perform EVA
First flight of Advanced Health Management System
New record of 17,900 commands sent from MCC-H
New Shuttle EVA record (accumulative) of 45 h 34 min set by Curbeam

SOYUZ TMA-10

International designator	2007-008A
Launched	April 7, 2007
Launch site	Pad 1, Site 5, Baikonur Cosmodrome, Republic of Kazakhstan
Landed	October 21, 2007
Landing site	10 km from the settlement of Tolybai, Republic of Kazakhstan
Launch vehicle	Soyuz-FG (serial number Ц15000-019), Soyuz TMA (serial number 220)
Duration	196 da 17 h 04 min 35 s (Yurchikhin, Kotov) 13 da 18 h 59 min 50 s (Simonyi down on TMA-9)
Call sign	Pulsar
Objectives	ISS resident crew transport (14S), ISS-15 resident crew program, visiting crew 12 program

Flight Crew

YURCHIKHIN, Fyodor Nikolayevich, 48, civilian, RSA ISS commander, Soyuz TMA flight engineer, second mission
Previous mission: STS-112 (2002)
KOTOV, Oleg Valerievich, 41, Russian Federation Air Force, RSA ISS flight engineer 1, Soyuz TMA commander
SIMONYI, Charles, 58, U.S.A., space flight participant, visiting crew 12

ISS-15 Shuttle-delivered crew members
WILLIAMS, Sunita Lyn, 41, USN, NASA ISS flight engineer 2
(up on STS-116, down on STS-117)
ANDERSON, Clayton Conrad, 48, civilian, NASA ISS flight engineer 2
(up on STS-117, down on STS-120)

Flight log

The 15th ISS resident crew launched along with the 12th visiting crew member, Charles Simonyi, aboard the Soyuz TMA-10 spacecraft. Two days later the Soyuz, under the command of rookie cosmonaut and medical doctor Oleg Kotov, docked with the Zarya FGB module. The commander of the expedition, Fyodor Yurchikhin (call sign Olympus), had previously visited the station five years before as a member of the American STS-112 mission and fulfilled the role of flight engineer for the Soyuz flight to and from the station. Kotov served as flight engineer 1 for the duration of the expedition, resuming the command of the Soyuz for the return home. Formal handover to the ISS-15 crew took place on April 17.

Soyuz TMA-10 refresher docking training is completed by flight engineer Oleg Kotov as commander Fyodor Yurchikhin (to Kotov's left) and space flight participant Charles Simonyi (at Kotov's right) look on.

Two American (NASA) astronauts would complete the resident crew complement. Sunita Williams was already on board the station having arrived on STS-116 (12A.1) and having worked as part of the ISS-14 crew. She would continue with the ISS-15 crew before returning on STS-117 (13A) in the spring. Aboard that flight was her replacement, Clayton Anderson, who was to continue with the ISS-15 crew and then transfer over to the ISS-16 expedition before his own return to Earth on STS-120 later in 2007. This was the second in a series of resident crew exchanges via the Space Shuttle over the next three years. Both NASA astronauts were assigned to the MS5 position for ascent and descent on the Shuttle, transferring to the station's resident crew in the ISS FE2 role.

The third TMA-10 crew member was space flight participant and naturalized U.S. citizen, Charles Simonyi, one of the founders of the Microsoft Corporation, who had developed Word and Excel computer software. Simonyi, who had been born in Hungary, became the fifth space tourist and completed the longest stay on the ISS by a tourist up to that point. The duration of his visit had been decided upon by the requirement to land TMA-9 in Kazakhstan before sunset. During his time aboard the station, as the two main crews completed a series of handover activities, Simonyi completed a program of experiments that included a number of ESA life science experiments. He also wrote a log (which was not posted on his website until well after his return) and narrated video tours of the station. He landed with López-Alegría and Tyurin in Kazakhstan aboard TMA-9 on April 21.

During the ISS-15 expedition, 272 sessions of 47 Russian experiments were planned, 43 of which were continuations of previous increments. The expedition also continued the program of experiments in the U.S. segment, totaling over 119 hours of planned operations. The ISS-15 increment received visits from three Shuttle crews: STS-117 (13A), STS-118 (13A.1), and STS-120 (10A). They also worked with the unmanned resupply vehicles Progress M-59, M-60, and M-61. In addition to their scientific activities and logistics transfers, the main crew completed a wide range of maintenance, repair, and housekeeping duties throughout their stay on the station.

On April 16, during the handover activities, Sunita Williams became the first person to run a marathon while participating in a space flight. Using the TVIS treadmill she "ran" 26.2 miles (42.16 km) as competitor 14,000 of the Boston Marathon in 4 hours 24 minutes. Ten days later, Williams was told she would return on STS-117 instead of the original STS-118, which had slipped in its launch cycle due to the hail damage and 3-month delay to STS-117. That mission arrived on June 10, carrying her replacement, Clayton Anderson. Sunita Williams returned on Atlantis after a mission of just under 195 days, surpassing the female space endurance record held by Shannon Lucid for the previous 11 years.

The ISS-15 residency would include three space walks totaling 18 hours 43 minutes: one from the U.S. segment and two from the Russian segment. Yurchikhin participated in all three excursions and was accompanied by Kotov on the first two space walks from the Pirs module at the Russian segment and by Anderson from the U.S. Quest module for the third.

The first EVA (May 30, 5 h 25 min) involved attaching further debris shields around the Zvezda module and rerouting a high-frequency cable for a Russian GPS system. The second EVA (June 6, 5 h 37 min) featured the installation of the Biorisk samples exposure experiment, which would be retrieved on a later EVA. More cables and debris panels were also installed. The third EVA (July 23, 7 h 41 min) featured the removal and jettison by hand of two large structures: the VSSA Flight Support Equipment and the Early Ammonia Servicer.

On September 27, TMA-10 was relocated from the nadir port on Zarya to the aft port of Zvezda in an operation that took less than 28 minutes, clearing the Zarya port for other spacecraft operations.

Yurchikhin and Kotov handed over formal command of the station on October 19 to the recently arrived ISS-16 crew, before returning to Earth aboard the TMA-10 spacecraft on October 21 2007.

During the descent, the landing computer in the Descent Module of the Soyuz switched unexpectedly to a ballistic return trajectory, which would result in a landing short of the planned recovery zone. Though not in increased danger, the crew did have to endure g-loads about 2g higher than the nominal 5g (up to 7g max) and undershot the landing zone by about 340 km. This was similar to the reentry flown during TMA-1 on May 4, 2003.

Yurchikhin and Kotov landed with VC-13 Malaysian space flight participant Sheik Muszaphar Shukor Al Masrie, who had arrived with the ISS-16 crew aboard Soyuz TMA-11. The landing of TMA-10 took place just 10 km from the settlement of Tolybai, Republic of Kazakhstan.

Milestones

252nd manned space flight
103rd Russian manned space flight
 96th manned Soyuz flight
 10th manned Soyuz TMA mission
 14th ISS Soyuz mission (14S)
 12th ISS Soyuz visiting mission
Kotov became Russia's 100th cosmonaut
Williams (April 16) "ran" the first marathon while in space

STS-117

International designator	2007-024A
Launched	June 8, 2007
Launch site	Pad 39A, KSC, Florida, U.S.A.
Landed	June 22, 2007
Landing site	Runway 22, EAFB, California, U.S.A.
Launch vehicle	OV-104 Atlantis/ET-124/SRB BI-129/SSME: #1 2059, #2 2052, #3 2057
Duration	13 da 20 h 12 min 44 s (STS-117 crew)
	194 da 18 h 02 min 01 s (Sunita Williams)
Call sign	Atlantis
Objectives	ISS assembly mission 13A (ITS S3 and S4), ISS resident crew member exchange

Flight crew

STURCKOW, Richard Frederick Wilford, 47, USMC, NASA commander, third mission
Previous missions: STS-88 (1998), STS-105 (2001)
ARCHAMBAULT, Lee Joseph, 46, USAF, NASA pilot
FORRESTER, Patrick Graham, 50, civilian, NASA mission specialist 1
SWANSON, Steven Roy, 46, civilian, NASA mission specialist 2
OLIVAS, John Daniel, 41, civilian, NASA mission specialist 3
REILLY II, James Francis, 53, civilian, NASA mission specialist 4, third mission
Previous missions: STS-89 (1998), STS-104 (2001)

ISS-15 crew member up only

ANDERSON, Clayton Conrad, 48, NASA mission specialist 5 and ISS-15 flight engineer 2

ISS-15 crew member down only

WILLIAMS, Sunita Lyn, 41, NASA mission specialist 5 and ISS-15 flight engineer 2

Flight log

This mission was manifested to deliver the S3/4 truss assembly and to exchange NASA resident station crew members Clayton Anderson and Sunita Williams.

Atlantis had arrived back at the Orbital Processing Facility on September 21, 2006 following the STS-115 mission. Processing for its next flight continued in the OPF until the orbiter was moved over to the Vehicle Assembly Building (VAB)

Cutting the cake marks the end of formal crew training.

for mating with the rest of the stack. The rollout to launchpad 39A occurred on February 15. However, on February 26, the thermal protection on both the orbiter and the External Tank suffered damage from hail. This was so severe that repairs could not be conducted on the pad. Atlantis had to be rolled back to the VAB on March 4 to complete the repairs and was returned to pad 39A until May 15.

The remainder of the processing and countdown proceeded as planned, with an on-time launch on June 8, 2007 and the 8 min ascent to orbit going according to plan. Initial orbit inspection of the left Orbital Maneuvering System (OMS) later that day revealed that part of the thermal protection system appeared to have pulled away from adjacent thermal tiles. The crew used the RMS for a closer inspection. The following day, Archambault (assisted by Forrester and Swanson) used the RMS with the boom-mounted extension system to inspect both the heat shield on the leading edge of the orbiter wing and the vehicle's nose cap.

Atlantis docked with the ISS at the PMA-2 port of the Destiny Laboratory on June 10. Following the traditional greeting between the Shuttle and station crews after hatch opening, the exchange between Anderson and Williams took place. Anderson became the new flight engineer on the ISS-15 resident crew, while Williams became a member of the STS-117 crew. The formal exchange included swapping the form-fitting Soyuz seat liners, which marked the point at which the two astronauts changed crews.

During the docked phase, four EVAs were completed by the Shuttle EVA crew from the Quest airlock, working in alternate teams of two. The initial EVA (6 h 15 min on June 11) by Reilly and Olivas focused upon completing the attachment of bolts, cables, and connectors to the S3/4 truss segment and preparing for the deployment of its solar arrays. The EVA had been delayed for an hour after the ISS had temporarily lost its altitude control. This was not entirely unexpected because the movement of the 17.8-ton mass of the S3/4 truss (equivalent to the size of a bus) skewed the symmetry of the station, causing the control moment gyro to go offline.

The second EVA (June 13) was conducted by Forrester and Swanson in 7 hours 16 minutes. The previous day, station controllers had unfurled the solar arrays on the recently attached truss to soak up the Sun's energy. The main task of this second EVA was to remove all of the launch locks which held the 3.4 m wide solar alpha rotary joint in place. During the EVA, the crew ran into difficulty when Forrester found that commands intended for a drive lock assembly they were installing were in fact being sent to a drive lock assembly installed during the first EVA. As a result, ground controllers checked and confirmed the safe configuration of the earlier installed assembly. The two astronauts also assisted with the retraction of an older solar array, which cleared the way to deploy the new array. A total of 13 of the 315 solar array bays were folded.

On June 15, Olivas and Reilly completed the third EVA of the mission, totaling 7 hours 58 minutes. During this EVA, they assisted with the retraction of the P6 truss, which required 28 commands over 45 minutes to complete the operation. The two astronauts also addressed separate tasks. Olivas spent 2 hours stapling and pinning down a thermal blanket on the orbiter's OMS pod, after a 10×25.5 cm corner had peeled up during ascent. Meanwhile, Reilly installed the hydrogen vent valve of a new oxygen generation system on the U.S. laboratory, Destiny.

Two days later (June 17), Forrester and Swanson ventured outside for the fourth and final EVA of this mission. During this 6 h 29 min EVA, the two men relocated a TV camera from a stowage platform on the Quest airlock to a new position on the S3 truss. They also verified the drive lock assembly #2 configurations, as well as removing the final six solar alpha rotary joint launch restraints. After clearing a path for the Mobile Base System (MBS) on the S3 truss, several get-ahead tasks were completed. The two astronauts installed a computer network cable on the Unity node, opened the hydrogen vent valve on Destiny, and tethered two orbital debris shield panels on the Zvezda Service Module. Activation of the rotary joint meant there were now four U.S. solar arrays tracking the Sun during each orbit, providing much needed additional power for station operations.

Over the four EVAs, the astronauts logged 27 hours 58 minutes in total. Reilly and Olivas logged 14 hours 13 minutes on the first and third EVAs, while Forrester and Swanson accumulated 13 hours 45 minutes during EVAs 2 and 4.

During the docked phase of the mission, the Shuttle's propulsion system was used to back up the station's control and orbital attitude adjustment after the Russian segment computers which normally handle this task experienced problems. Both Russian and U.S. teams on the ground worked on the problem, troubleshooting and then restoring the capabilities of the computer. On June 15, cosmonauts Yurchikhin and Kotov managed to get two of the three lines to each computer running after they bypassed an apparent faulty power switch with external cabling. This modification was repeated on the final two channels. Three days later the Russians demonstrated the ability to maintain station control using the computers, thus allowing Atlantis to depart.

After transferring 19 tons of food, water, and equipment across to the station, and now with Sunita Williams aboard, Atlantis undocked on June 19. Joint activities had logged 8 days 19 hours 6 minutes. Piloted by Archambault, Atlantis completed a fly-around of the station, offering a good inspection and photo-documentation opportunity of the reconfigured outpost. At a distance of 74 km from the station, the Atlantis crew then used the RMS and boom sensor system to inspect the orbiter's thermal protection system on both wing leading edges and the nose cap.

Three days later, after a 24 h delay due to adverse weather, Atlantis arrived safely on runway 21 at Edwards Air Force Base in California, after marginal weather had again prevented a landing at the Shuttle Landing Facility at the Cape in Florida. Setting a new record for female space flight endurance, Sunita Williams logged just short of 195 days in space during her residency.

The transfer of Atlantis back to the Cape did not begin immediately. After several days of preparation, the orbiter was bolted to the top of the Shuttle Carrier Aircraft (Boeing 747). Following several fuel stops and weather delays, the combination finally arrived back at the Cape on July 3.

Milestones

253rd manned space flight
148th U.S. manned space flight
118th Shuttle mission
 28th flight of Atlantis
 21st Shuttle ISS mission
Heaviest station payload carried by Shuttle to date (42,671 lb) (19,355.5 kg)
First launch from Pad 39A since STS-107 Columbia in January 2003
Suni Williams set a new endurance record of 195 days for the longest single space flight by a female, surpassing the 188-day record of Shannon Lucid set in 1996

STS-118

International designator	2007-035A
Launched	August 8, 2007
Launch site	Pad 39A, KSC, Florida, U.S.A.
Landed	August 21, 2007
Landing site	Runway 15, Shuttle Landing Facility, KSC, Florida, U.S.A.
Launch vehicle	OV-105 Endeavour/ET-117/SRB BI-130/SSME: #1 2047, #2 2051, #3 2045
Duration	12 da 17 h 55 min 34 s
Call sign	Endeavour
Objectives	ISS assembly flight 13A.1, ITS S5, External Stowage Platform 3, Spacehab single module

Flight Crew

KELLY, Scott Joseph, 43, USN, NASA commander, second mission
Previous mission: STS-103 (1999)
HOBAUGH, Charles Owen, 45, USMC, NASA pilot, second mission
Previous mission: STS-104 (2001)
CALDWELL, Tracy Ellen, 37, civilian, NASA mission specialist 1
MASTRACCHIO, Richard Alan, civilian, NASA mission specialist 2, second mission
Previous mission: STS-106 (2000)
WILLIAMS, Daffyd (David) Rhys, 53, civilian (Canadian), CSA mission specialist 3, second mission
Previous mission: STS-90 (1998)
MORGAN, Barbara Radding, 55, civilian, NASA mission specialist 4
DREW, Benjamin Alvin, 44, USAF, NASA mission specialist 5

Flight log

This mission continued the installation of the solar array truss segment and also featured the flight of the first educator astronaut—a teacher turned astronaut.

As a result of a major modification program for Endeavour, and the recovery from the loss of Columbia, launch preparations took some time. The orbiter had arrived at the OPF on December 7, 2002 following the landing of STS-113. Endeavour remained at the Cape for the next four years, being moved between facilities as required for both mission processing and upgrades. During the modification period, Endeavour, the newest of the orbiter fleet, received a number of upgrades including the installation of a "glass cockpit", a global positioning system to be used as an aid for landing and the Station-to-Shuttle Power Transfer

Education mission specialist Barbara Morgan floats on the middeck.

System (SSTPS). This would enable the orbiter to draw power from the station while docked with it, extending the time it could remain at the station.

On January 9, 2004 Endeavour was moved to the VAB for maintenance, returning to the OPF Bay 2 on January 21. On December 16 that year Endeavour was moved to High Bay 4 in the VAB for storage. Then, on January 12, 2005, the orbiter was relocated back to OPF Bay 2 once again. The move to the Shuttle Landing Facility (SLF) hangar on February 22 was to make room to conduct planned modifications to the OPF. Endeavour returned to the Processing Facility on March 18, where it remained for the next two years. Following mission processing for STS-118, Endeavour was rolled over to the VAB on July 2, 2007 for mating to the External Tank (ET) and two SRBs. The STS-118 stack was rolled out to Pad 39A on July 11, 2007.

Despite some problems with a stubborn side hatch on the Endeavour, the mission launched on time, heading into the early evening sky just before sunset. On board was the crew of seven, the SS truss, the Spacehab module, and the External Storage Platform #3, which included a Control Moment Gyroscope (CMG) to replace a faulty one on the station. The following day, the crew used the Shuttle's robotic arm and Orbiter Boom Sensor System (OBSS) to take a close look at the vehicle's heat shielding over the leading edge of the wings.

Shortly before docking with the station on August 10, commander Scott Kelly maneuvered Endeavour in a backflip, allowing the resident ISS crew to take digital photos of the underside of the vehicle and its upper surfaces to check the thermal protection system for possible damage. Subsequent analysis of this imagery revealed a $3''$ (8 cm) round dent on the starboard underside. Further inspection and analysis of the images and data showed that the damage had penetrated a tile down to the internal framework. Over several days, while the crew continued work on the station, the Mission Management Team (MMT) evaluated the evidence and authorized a further engineering analysis and series of tests. It was determined that direct tile repair by the crew on a contingency EVA was not required and that the damage would not pose a risk to the crew during reentry. It was therefore decided to leave the damaged tile alone until after landing. This decision revealed growing confidence in the changes introduced since the loss of Columbia, better capability in interpreting inspection imagery, and more understanding of the effects of minor damage on the TPS.

Endeavour docked to the station on August 10 while orbiting 214 miles (344.32 km) above the southern Pacific Ocean, northeast of Sydney, Australia. The MMT initially extended the flight to 14 days following a successful transfer of power from the station by means of the SSPTS. This enabled a fourth space walk to be added to the flight plan. However, later in the mission, with growing concern over the movement of Hurricane Dean towards the coast of Texas, the MMT decided to shorten the flight and end the mission one day early.

The mission's four space walks logged 23 hours 13 minutes in total. Rick Mastracchio and Canadian astronaut Dave Williams each completed three EVAs, with ISS-16 flight engineer Clayton Anderson joining Mastracchio for the third EVA and Williams for the fourth and final EVA.

During the first EVA (August 11, 6 h 17 min) installation of the 2-ton, 11 ft long spacer, Starboard 5 (S5) segment of the truss structure was completed. The astronauts also retracted the forward heat-rejecting radiator from the P6 truss assembly. This was planned for relocation to the end of the port truss during the following STS-120 mission. The mission's second EVA (August 13, 6 h 28 min), focused upon installation of the 600 lb (272.16 kg) CMG onto the Z1 segment of the station truss assembly. The failed unit was removed and stored outside the station pending its planned return to Earth on a later mission.

During the third excursion (August 15, 5 h 28 min), Anderson and Mastracchio relocated the S-band antenna subassembly from P6 to P1. They then installed a new transponder on P1 as well as retrieving the P6 transponder. Meanwhile, inside the station Pilot Charles Hobaugh and resident station flight engineer Oleg Kotov moved two CETA (Crew and Equipment Translation Aid) carts. It was during this EVA that Mastracchio noticed a hole in the thumb of his left pressure glove. As it was only through to the second of five layers it did not create a leak or endanger the astronaut. As a precaution, however, he returned to the airlock while Anderson completed the final tasks.

The fourth and final EVA of the mission (August 18, 5 h 2 min), conducted by Williams and Anderson, featured a number of lesser tasks, including installation of the External Wireless Instrumentation System (EWIS) antenna, installation of a stand for the temporary relocation of the Shuttle RMS extension boom, and retrieval of the two material experiment containers which were to be returned on Endeavour. The two other tasks planned for this EVA were deferred to a future space walk. These were relocating a toolbox to a more central location and cleaning up and securing the debris shielding. During each of the EVAs fellow crew members Caldwell and Morgan used the Shuttle and station RMS devices to move and locate the 7,000 lb (3175.2 kg) number 3 external storage platform that would be installed on the P3 truss.

On August 11, the primary command and control computer in the U.S. segment shut down unexpectedly. Fortunately this shutdown did not affect the EVA being conducted that day. The redundancy in the system worked as designed, with the secondary computer taking over and the third computer providing a backup role. Ground controllers brought up the third computer after determining that an errant software command was the cause of the shutdown.

Mission specialist Barbara Morgan was a professional teacher turned astronaut, who had originally been chosen as a backup payload specialist to fellow teacher Christa McAuliffe in 1985 under the Teacher in Space Program. Tragically, McAuliffe died in the January 1986 Challenger accident. Morgan continued her association with NASA and returned to teaching, but in 1998 she was chosen as an educator mission specialist in the 17th NASA astronaut group. During STS-118, Morgan conducted three educational events and on several occasions she and her colleagues answered questions from children at the Discovery Center in Boise, Idaho and the Challenger Centers for Space Science Education in Alexandria, Virginia and Saskatchewan, Canada.

Endeavour undocked from the Station on August 19 after 8 days 17 hours 54 minutes of joint operations. Ironically, the expected threat from Hurricane Dean never materialized but by then the orbiter was committed to an early return—weather permitting. There was indeed fine weather as the landing took place on Runway 15 at the Shuttle Landing Facility at KSC on the first opportunity.

Milestones

254th manned space flight
149th U.S. manned space flight
119th Shuttle mission
 20th flight of Endeavour
 22nd Shuttle ISS mission
Morgan became the first educator mission specialist to fly in space
The Zarya module completed its 50,000th orbit (August 14), as the oldest element of ISS
Installation of ESP3 using just the Space Station and Shuttle Remote Manipulator System (SSRMS) without help of EVA astronauts as on previous two installations
First use of the Station–Shuttle Power Transfer System
Caldwell celebrates her 38th birthday (August 14)

SOYUZ TMA-11

Internation designator	2007-045A
Launched	October 10, 2007
Launch site	Pad 1, Site 5, Baikonur Cosmodrome, Republic of Kazakhstan
Landed	April 19, 2008
Landing site	Republic of Kazakhstan
Launch vehicle	Soyuz-FG (serial number Ц15000-020), Soyuz TMA (serial number 221)
Duration	191 da 19 h 07 min 05 s (Malenchenko, Whitson) 10 da 21 h 13 min 21 s (Muszaphar down on TMA-10)
Call sign	Agat
Objectives	ISS resident crew transport (15S), ISS-16 research program, visiting crew 13 (Malaysian Angkasa MSM) research program

Flight crew

MALENCHENKO, Yuri Ivanovich, 45, Russian Federation Air Force, RSA Soyuz TMA commander, ISS flight engineer 1; fourth mission
Previous missions: Soyuz TM19/Mir EC-16(1994), STS-106 (2000), Soyuz TMA-2/ISS-7 (2003)
WHITSON, Peggy Annette, 47, NASA Soyuz TMA flight engineer, ISS-16 commander; second mission
Previous mission: STS-111/STS-113/ISS-5 (2002)
MUSZAPHAR, Shukor Al Masrie, 35, Malaysian space flight participant

ISS resident crew Shuttle transfers

ANDERSON, Clayton Conrad, 48, NASA ISS flight engineer 2
TANI, Daniel Michio, 46, NASA ISS flight engineer 2; second mission
Previous mission: STS-108 (2001)
EYHARTS, Leopold, 50, French Air Force, ESA (French) ISS flight engineer 2; second mission
Previous mission: Soyuz TM27/26/Mir (1998)
REISMAN, Garrett Erin, 40, NASA ISS flight engineer 2

Flight log

This was, by any account, a busy residency, which officially took over from the ISS-15 crew on October 19, 2007. During this 16th expedition to the ISS, the expansion of scientific facilities at the station finally resumed after the tragedy of Columbia. This included installation of an additional node (#2, named Harmony),

Expedition 16 crew poses for a Christmas photo. From left: Yuri Malenchenko, Peggy Whitson, and Dan Tani.

the European Laboratory (Columbus), and the first elements of the Japanese experiment facility (Kibo). The mission also featured the now familiar routine maintenance chores, expansion of the science program, and hosting the arrival of the first ESA Automated Transfer Vehicle (Jules Verne), laden with over 4.6 tons of cargo for the station.

There were three Shuttle assembly missions during this expedition, in addition to partial crew rotation of four members of the main expedition crew, docking of further Progress resupply craft, and four planned EVAs. Three of these would be from the U.S. segment and one from the Russian segment. During the three Shuttle missions, a further 12 EVAs were completed by Shuttle crew members.

The two ISS-16 main resident crew members docked with the Zarya Module on October 12. On board with them was Malaysian space flight participant Sheikh

Shukor Al Masrie Muszaphar, flying a 10-day mission. During the main residency, the core crew would be joined by three NASA astronauts and one ESA astronaut, all serving in sequence as ISS FE2 and all launched and returned via the U.S. Space Shuttle.

The Malaysian 'Angkasa MSM Project' was agreed with Russia under a contract signed on September 29, 2005. This involved a programme of science research activities and experiments to be conducted by a Malaysian citizen flying as a SFP on the thirteenth ISS visiting crew. The programme consisted of thirteen experiments; eight were Malaysian national experiments, while the other three were joint investigations with ESA. In addition, there was a range of public relations and symbolic activities planned. The experiment programme included 31 sessions for 8 experiments over five of the ten days in space. Two of the experiments were performed during the two-day flight of the Soyuz TMA spacecraft on the way to the station, with the remainder performed aboard the complex itself. These included four life science experiments, three biotechnology experiments and one education experiment. In total, approximately 18 hours were assigned for Shukor to complete the programme, occasionally assisted by Malenchenko. On October 21, 2007, after a flight of 11 days, Shukor returned to Earth aboard Soyuz TMA-10, together with the returning ISS-15 crew members Fydor Yurchikhin and Oleg Kotov after their six month mission.

When TMA-11 docked with the station, NASA astronaut Clayton Anderson was already aboard serving as FE2, having been delivered to the station by STS-117 (13A). He would remain aboard as FE2 with the ISS-16 crew until his replacement, Dan Tani, arrived on STS-120 (10A) and then come home with the STS-120 crew. Tani would return on STS-122 (IE), which delivered his replacement, French ESA astronaut Leopold Eyharts. Eyharts would support the installation and early setup activities of the European Columbus Experiment Module before himself being replaced by NASA astronaut Garrett Reisman. Reisman arrived with the first elements of the Japanese Laboratory Module on STS-123 (1J/A), on which Eyharts would return, and would remain onboard station with the ISS-17 crew when the ISS-16 main crew departed in the spring.

The Russian research program featured 251 sessions of 55 experiments, 44 of which were continuations of earlier studies, while 11 were brand new. To accomplish this objective, the crew had been allocated over 217 hours to operate the experiments during the expedition, mostly by Malenchenko. Across in the American segment, there were 38 experiments being conducted and, with the delivery of the Columbus module towards the end of the residency, an increasing amount of work on the European science program. Hardware and supplies delivered for the Japanese segment during the residency meant that the program of experiments planned for that facility would at last also be approaching fruition.

The ISS-16 crew would also work with the Progress M-61, M-62, and M-63 resupply vehicles during their time on the station. The major hardware delivered by the three Shuttle missions were the Harmony 2 Node (STS-120); the ESA Columbus Module (STS-122), the Pressurized Section of the Japanese Experimental Logistics Module (ELM-PS), and the Canadian Space Agency Special Purpose

Dexterous Manipulator, known as Dextre (STS-123). The 12 EVAs completed by the Shuttle crews were split between STS-120 (four EVAs), STS-122 (three EVAs), and STS-123 (a record five EVAs). Before formally joining the resident crew, Dan Tani assisted on the second EVA of STS-120 (6 h 33 min) on October 28 and Greg Reisman assisted on the STS-123 EVA 1 on March 13 (7 h 1 min) during installation of the Japanese elements (see STS-123).

The ISS-16 crew itself completed a program of five EVAs during their residency. The first three (November 9, 6 h 55 min; November 20, 7 h 16 min; November 24, 7 h 4 min) were associated with the relocation of the PMA-2 and Harmony Node 2 using the Canadarm2 during November 12–14, while the final two space walks saw the astronauts work on the starboard solar array truss.

In total, Whitson accumulated 35 hours 21 minutes across the five EVAs; Malenchenko logged 6 hours 55 minutes on the expedition's first EVA with Whitson, and Tani, who participated with Whitson for EVA 2 through 5, accumulated 28 hours 26 minutes on the ISS-16 crew in addition to that of his STS-120 EVA.

One of the hardest personal challenges to face is the loss of a family member. It is even harder when there is some distance involved and if you are off the planet, adding further barriers to overcoming the grief. Telling a crew member of a personal loss has always been a difficult decision for the ground support team, one that had to be addressed during this residency. Dan Tani was informed on December 20 that his mother, Rose, had died in an automobile accident. He was informed over the private communication loop by his wife, who was also a flight surgeon. Over the next few days, the astronaut was allowed to grieve and conducted a number of private calls to family members over the secure video channel. Tani was the first American crew member to lose a close relative while participating in a space flight.

The ISS-16 main crew of Whitson and Malenchenko returned to Earth on Soyuz TMA-11 on April 19, 2008, along with South Korean VC-14 crew member So Yeon Yi, who had arrived with the ISS-17 main crew aboard TMA-12. The formal change of command between the two resident crews had occurred on April 17. Reisman continued his residency with the ISS-17 crew, until he was replaced during STS-124 (1J).

The landing of TMA-11 in the Republic of Kazakhstan was not as smooth as planned. The descent module performed a ballistic reentry, with the crew enduring loads of 8.5g for a short time. The separation of the Orbital Module from the Descent Module occurred without incident, but when the Descent Module tried to separate from the Instrument Module a bolt remained attached, resulting in the configuration entering the atmosphere sideways and creating a raging sheath of flame outside the windows. It was the rigidity of the Soyuz design that protected the crew, until the two spacecraft elements finally (and thankfully) pulled apart, allowing the Descent Module to make a harrowing, but otherwise safe landing 420 km short of the planned recovery zone.

Following the mission, postflight debrief, and recovery, in October 2009 Peggy Whitson assumed the role of Chief NASA Astronaut. She was the first female and non-pilot to achieve this coveted role. Malenchenko, meanwhile, resumed ISS training for a return to station as a member of a new resident crew.

Milestones

255th manned space flight
104th Russian manned space flight
97th manned Soyuz flight
11th manned Soyuz TMA mission
15th ISS Soyuz mission (15S)
13th ISS Soyuz visiting mission
Whitson becomes 1st female ISS commander
New EVA record by a female in a career total of 39 hours 46 minutes across six EVAs

STS-120

International designator	2007-050A
Launched	October 23, 2007
Launch site	Pad 39A, KSC, Florida, U.S.A.
Landed	November 7, 2007
Landing site	Runway 33, Shuttle Landing Facility, KSC, Florida, U.S.A.
Launch vehicle	OV-103 Discovery/ET-120/SRB BI-131/SSME: #1 2050, #2 2048, #3 2058
Duration	15 da 02 h 24 min 02 s (STS-120 crew)
	151 da 18 h 24 min 09 s (Anderson)
Call sign	Discovery
Objectives	ISS assembly flight 10A, Node 2 (Harmony) connecting module, ISS resident crew exchange

Flight crew

MELROY, Pamela Ann, 46, USAF Ret., NASA commander, third mission
Previous missions: STS-92 (2000), STS-112 (2002)
ZAMKA, George David, 45, USMC, NASA pilot
PARAZYNSKI, Scott Edward, 46, civilian, NASA mission specialist 1, fifth mission
Previous missions: STS-66 (1994), STS-86 (1997), STS-95 (1998), STS-100 (2001)
WILSON, Stephanie Diana, 41, civilian, NASA mission specialist 2, second mission
Previous mission: STS-121 (2006)
WHEELOCK, Douglas Harry, 47, U.S. Army, NASA mission specialist 3
NESPOLI, Paolo, 50, civilian (Italian), ESA mission specialist 4

ISS resident crew members
TANI, Daniel Michio, 46, civilian, NASA mission specialist 5 (up), ISS-16 flight engineer, second mission
Previous mission: STS-108 (2001)
ANDERSON, Clayton Conrad, 48, civilian, NASA mission specialist 5 (down), ISS-16 flight engineer

Flight log

The primary objective of this mission was to deliver the Node 2 (Harmony) facility and relocate the P6 truss. This mission also featured the exchange of NASA astronaut Dan Tani (ISS-16 FE2) with ISS-15/16 flight engineer Clayton Anderson on the station.

Two female commanders greet each other in space. ISS commander Peggy Whitson (right) greets STS-120 commander Pamela Melroy following hatch opening.

Mission processing went relatively smoothly, with Discovery arriving at the OPR on December 22, 2006, following the STS-116 landing. Processing continued with the rollover of Discovery from the OPF to the VAB on September 23, 2007 for stacking with the ET and SRBs. A week later, on September 30, the STS-120 stack was rolled to Pad 39A.

The October 23 launch was on time and docking with the station was accomplished without incident. Discovery docked with the ISS at PMA-2 on Harmony on October 25, 2007 and later the same day Tani formally took over from Anderson as ISS-16 FE2. Anderson had spent a total of only seven days as a member of ISS-16, but had accumulated 131 days as a member of the ISS-15 crew. By the end of the STS-120 mission Anderson had logged a total of 148 days on the ISS and 152 days in space.

The completion of docking and hatch-opening operations on this mission would see the historic first greeting in space between a female commander of a Shuttle mission (Melroy) and a female commander of the Space Station (Peggy Whitson).

The main focus of STS-120 activity centered upon the installation of the new node (on October 26) and repair to the Solar Alpha Rotary Joint (SARJ) during the second EVA. In addition to the EVAs, the crew transferred over 2,020 lb (916.27 kg) of equipment and scientific samples to the Shuttle and delivered additional supplies to the station. Amongst the range of items being returned to

Earth for postflight analysis were metal shavings from the SARJ, to determine the probable cause of resistance in the starboard joint.

Paulo Nespoli, the third Italian to fly on the Shuttle, would serve as Inter Vehicular Activity (IVA) crew member for the planned EVAs. He would also complete a joint science program devised by the European and Italian space agencies under the label "Esperia", from the ancient Greek name for the Italian peninsula. This program included a range of human physiology and biology experiments as well as a number of educational activities.

The station management added a 360-degree visual inspection of the station starboard SARJ to the second EVA after it had shown increased friction for the past 30 days. Between the fourth and fifth EVAs, an extra day was added to the mission to allow the crew additional off-duty time and to prepare equipment for the fifth EVA. However, when a repair to a torn solar array was required on the fourth EVA the priorities changed, so the objectives of the fifth EVA would be completed by the station crew after the Shuttle had departed.

On the ground, teams worked around the clock to devise a workable plan for the repair. The crew fabricated a solar array hinge stabilizer from strips of aluminum, a hole punch, a bolt connector, and approximately 20 meters of wire. The stabilizer would work in a similar fashion to a cuff link on a shirt. The wire was fed through a hole on the array and was supported by the strip of aluminum. The astronauts also positioned the station's Robotic Arm and Mobile Transporter at the end of the truss to serve as a "cherry picker" and work platform. To protect against electrical currents while working, the astronauts insulated the tools with Kapton tape.

During the first EVA (October 26, 6 h 14 min), mission specialists Parazynski and Wheelock assisted with the installation of the Harmony Module (Node 2) to its temporary location and also readied the P6 for its planned relocation two days later. In addition, the two astronauts closed a window cover on Harmony that had inadvertently opened during the launch phase and retrieved a failed radio communication antenna. After their return to the station, Wheelock noticed a small hole in the outer layer of his right pressure glove thumb. This would be evaluated later, prior to his next EVA. Post-EVA analysis of the gloves revealed excessive wear, requiring a replacement glove for his next excursion on the mission's third EVA.

The next day, the hatches into the new module were opened and ISS-16 commander Peggy Whitson and ESA astronaut Nespoli were the first to float inside. As this was a new addition to the station, both crew members wore protective face masks and goggles in case of any floating debris not picked up in the rigorous ground processing. Air samples were taken and a process to refresh the air was run about five times inside the module as part of the unit's safety acceptance as a permanent element on the station.

For the second EVA (October 28, 6 h 33 min), Parazynski was accompanied by ISS-16 flight engineer Dan Tani. The main objective of this excursion was to disconnect cables from the P6 Truss, which would enable it to be removed from the ZI Truss. The two astronauts outfitted the Harmony Module, mated the power and data grapple feature, and reconfigured connections to the S1 Truss that would allow the radiator on S1 to be deployed by a ground command from the Control Center at a later date. In a busy EVA, Tani also inspected the SARJ and collected "shavings" he found under the joint's multilayer insulation cover. Mission managers authorized a limited use of the SARJ while the anomaly was assessed further and a repair plan formulated.

On October 30, the extra day added to the mission was announced. The originally planned 4 h 45 min EVA 4 would now be extended to a full duration of 6 hours 30 minutes and would be devoted to inspection of the starboard SARJ, instead of the previously planned demonstration of the Tile Repair Ablator Dispenser in the payload bay of Discovery. This would be deferred to a later mission. The fifth EVA would now be conducted by ISS-16 crew members Whitson and Malenchenko, completing further work on outfitting the exterior of the Harmony Module.

Parazynski and Wheelock paired up again for EVA 3 (October 30, 2 h 8 min), with Wheelock wearing one of the spare EVA gloves. During this space walk, the astronauts installed the P6 Truss segment (with its set of solar arrays) in its permanent position. In addition, they installed a spare main bus switching unit on a storage platform, for future use if required. Parazynski examined the port SARJ and compared it with the starboard one, finding it clear of debris. Towards the end of this EVA, when the P6 solar arrays were deployed, a tear appeared in one of the blankets. To allow analysis and prevent any further damage, the deployment was halted so that engineers on the ground could evaluate the situation and plan what to do next. Despite the 80% deployment, the array was still able to generate nearly normal power levels.

The fourth EVA was slipped 24 hours to study options for repairing the torn array. It was decided to concentrate primarily on repairing the array on EVA 4 and defer any work on the SARJ to later in the program. The planned EVA 5 would be completed by the ISS-16 crew after the departure of Discovery.

The fourth EVA (November 3, 7 h 19 min) was again completed by Parazynski and Wheelock. Before they began the EVA, the Orbiter Boom Sensor System (OBSS) was moved from the Shuttle RMS to the station arm. Over the next 90 minutes, the two astronauts rode the arm to work at the torn array area. This distance was 165 feet down the Truss and 90 feet up to the damaged area. Once there, Parazynski cut a snagged wire and installed homemade stabilizers to strengthen the array's structure and stability where the damage had occurred. Ground controllers were then able to complete the deployment. Deploying at one half bay at a time, this process took 15 minutes to complete.

The four space walks amassed a total of 27 hours 34 minutes. Individually, Parazynski had logged 27 hours 14 minutes (four EVAs); Wheelock 20 hours 41 minutes (three EVAs); and Tani 6 hours 33 minutes on a single excursion.

On November 5, Discovery undocked from the station after 9 days 21 hours of joint activities, completing a nominal landing on November 7. This followed a rare southbound trajectory which took the orbiter over the central states of continental America and which allowed a daylight landing at the Cape instead of the preplanned night landing.

Milestones

256th manned space flight
150th U.S. manned space flight
120th Shuttle mission
 34th flight of Discovery
 23rd Shuttle ISS mission
In honor of the 30th anniversary of the feature film *Star Wars* franchise, the Luke Skywalker light saber was flown on Discovery
The first time female commanders would lead Shuttle (Melroy) and station (Whitson) missions at the same time and meet in space
Use of OBSS during EVA 4 (November 3) was the first operational use of OBSS to reach a work site on the ISS

STS-122

International designator	2008-005A
Launched	February 7, 2008
Launch site	Pad 39A, KSC, Florida, U.S.A.
Landed	February 20, 2008
Landing site	Runway 15, Shuttle Landing Facility, KSC, Florida, U.S.A.
Launch vehicle	OV-104 Atlantis/ET-125/SRB BI-132/ SSME: #1 2059, #2 2052, #3 2057
Duration	12 da 18 h 21 min 50 s (STS-122 crew) 119 da 21 h 29 min 01 s (Tani)
Call sign	Atlantis
Objectives	ISS assembly flight 1E (Columbus Laboratory); ISS-16 residency partial crew exchange

Flight crew

FRICK, Stephen Nathaniel, 43, USN, NASA commander, second mission
Previous mission: STS-110 (2002)
POINDEXTER, Alan Goodwin, 46, USN, NASA pilot
MELVIN, Leland Deems, 43, civilian, NASA mission specialist 1
WALHEIM, Rex Joseph, 45, USAF, NASA mission specialist 2, second mission
Previous mission: STS-110 (2002)
SCHLEGEL, Hans Wilheim, 56, civilian (German), ESA mission specialist 3, second mission
Previous mission: STS-55/Spacelab D2 (1993)
LOVE, Stanley Glen, 42, civilian, NASA mission specialist 4

ISS resident crew members

EYHARTS, Leopold, 50, French Air Force, ESA (French) mission specialist 5 (up only); ISS-16 flight engineer 2, second mission
Previous mission: Soyuz TM-27 (1998)
TANI, Daniel Michio, 46, civilian, NASA mission specialist 5 (down only), ISS-16 flight engineer 2, second mission
Previous mission: STS-108 (2001)

Flight log

The 24th ISS assembly mission featured the delivery of the long-awaited European Science Laboratory called Columbus (named after the historic European explorer). The science payload for the European module would be managed by the

The ESA Columbus module is delivered to the space station.

Columbus Control Center, located in Oberpfaffenhofen, Germany. This would also be the center responsible for coordinating and managing the research and for collecting the results data. On board the station, experiment hardware would be operated mainly by European crew representatives, though not exclusively, as there would not always be an ESA representative on board as part of the main resident crew.

The orbiter Atlantis arrived back at the Orbiter Processing Facility at KSC on July 4, 2007 (America's 231st birthday) following a ferry flight from Dryden and arrival at the Cape the previous day. On November 3, 2007, Atlantis was moved from the OPF to the VAB for mating with the rest of the stack. It was then rolled out of the VAB on November 10, 2007 for the move to Pad 39A.

During December, the mission was twice delayed during the fueling of the ET due to false readings in the engine cutoff sensor systems. Tests subsequently revealed that the open circuits in the ET electrical feed through a connector were the most probable cause of the fault. One of many safety systems installed on the vehicle, this particular connector protected the SSME by initiating shutdown if fuel ran unexpectedly low. To resolve the fault, a modified connector (which had pins and sockets soldered together) was installed for the mission. As a result of these changes, the launch was rescheduled for February 7 and was achieved without further incident.

Docking with the ISS occurred on February 9. Earlier, the crew had completed the now customary backflip maneuver so that Atlantis could be photo-documented and laser-scanned from the ISS for analysis on the ground. The orbiter crew had previously used the RMS to scan the surfaces of Atlantis on February 8; this inspection by the resident station crew was an additional check into the integrity of the vehicle's heat shield.

Following the docking, ESA astronaut Leopold Eyharts officially joined the ISS-16 expedition, replacing NASA astronaut Dan Tani, who rejoined the Shuttle crew and ended his residency. Tani had spent 107 days aboard the station as a member of the resident crew. His stay on the station had been extended two months due to difficulties in getting the Shuttle off the launchpad.

Following closer inspection of the tile data, minor damage was discovered on a thermal blanket over the right OMS pod. Further inspections were made by the crew but the Mission Management Team eventually cleared the TPS for reentry. They also extended the mission an extra day to continue activation of the European Laboratory.

There were three EVAs conducted during the mission, totaling 22 hours 8 minutes. The first EVA had to be postponed a day due to a medical issue with Schlegel. It was later revealed that Love would replace the German astronaut on the first space walk.

That first EVA (February 11, 7 h 58 min) by Love and Walheim mainly focused on installation of the Columbus Laboratory. The astronauts installed a grapple fixture on Columbia while in the payload bay and prepared electrical and data connections on the module. Inside the station, astronauts Melvin, Tani, and Eyharts used the robotic systems to grab Columbus, lift it out of the orbiter payload bay and relocate it over to the starboard side of Harmony (Node 2). The EVA continued with the crew beginning work on replacing a large nitrous tank, which is used for pressurizing the station's ammonia cooling systems.

Schlegel was well enough to participate in the mission's second EVA with Wheelock (February 13, 6 h 45 min). The two astronauts replaced the nitrous tank and used the station's RMS to move the spent tank back into the orbiter payload bay. Minor repairs were also undertaken on the debris shield on the Destiny lab and several get-ahead tasks completed in preparation for the third and final EVA.

The third EVA of the mission (February 15, 7 h 25 min) was conducted by Walker and Love. Its first objective was to relocate one of two external experiment facilities (called SOLAR) to the Columbus module for installation. The EVA crew was guided by Poindexter, while Melvin used the station RMS for the transfer. The EVA crew then retrieved a stored, failed gyroscope and secured it in Atlantis' payload bay for return to Earth. Next, they installed the second experiment facility, called the European Technology Exposure Facility (EuTEF), on to Columbus. Their final task was to examine a damaged handrail on the exterior of the Quest airlock. The deterioration of the handrail was thought to be caused by years of repetitive glove abrasion. To check this, the astronauts rubbed it with a tool covered in EVA over-glove material to see if it left any new damage.

In total, Walheim logged 22 hours 8 minutes on three EVAs; Love amassed 15 hours 23 minutes on his two space walks; and Schlegel 6 hours 45 minutes on his single excursion.

All crew members worked throughout the docked period to activate the Columbus Laboratory, which included outfitting it with several experiment racks. Both the Shuttle and station crews spoke with German Chancellor Angela Merkel, ESA Director Jean-Jacques Dordain, and former ESA astronaut Thomas Reiter, now a member of the German space agency (DLR).

Prior to the departure of Atlantis, its orbital maneuvering propulsion system was used to reboost the station's altitude by about 2.25 km (1.4 miles), to achieve a proper alignment of the station in advance of the planned arrival of Endeavour on STS-123 in March. This was the first time since 2002 that an orbiter had been used for a reboost maneuver. Hatches were closed for a final time on February 17 and, early the following morning, Atlantis undocked from the ISS after 8 days 16 hours 7 minutes of joint operations.

The landing, on February 20, 2008, happened to coincide with the 46th anniversary (1962) of John Glenn's historic first U.S. manned orbital flight of three orbits (4 h 55 min) aboard Friendship 7 (Mercury-Atlas 6).

Milestones

257th manned space flight
151st U.S. manned space flight
121st Shuttle flight
 29th Atlantis flight
 24th Shuttle ISS mission
 12th Atlantis ISS mission
First EVA was the 100th devoted to the assembly of the ISS
Whitson's 48th birthday (February 9)
Melvin's 44th birthday (February 15)

STS-123

International designator	2008-009A
Launched	March 11, 2008
Launch site	Pad 39A, KSC, Florida, U.S.A.
Landed	March 26, 2008
Landing site	Runway 15, Shuttle Landing Facility, KSC, Florida, U.S.A.
Launch vehicle	OV-105 Endeavour/ET-126/SRB BI-126/SSME: #1 2047, #2 2044, #3 2054
Duration	15 da 18 h 10 min 54 s (STS-123 crew)
	48 da 04 h 53 min 38 s (Eyharts)
Call sign	Endeavour
Objectives	ISS assembly mission 1 J/A, delivery and installation of the Japanese Kibo Experiment Logistics Module-Pressurized Section (ELM-PS), Canadian Special Purpose Dexterous Manipulator (Dextre), ISS-16/17 residency partial crew exchange

Flight crew

GORIE, Dominic Lee, 50, USN retired, NASA commander, fourth mission
Previous missions: STS-91 (1988), STS-99 (2000), STS-108 (2001)
JOHNSON, Gregory Harold, 45, USAF, NASA pilot
BEHNKEN, Robert Louis, 37, USAF, NASA mission specialist 1
FOREMAN, Michael James, 50, USN, NASA mission specialist 2
DOI, Takao, 53, civilian (Japanese), JAXA, mission specialist 3, second mission
Previous mission: STS-87 (1997)
LINNEHAN, Richard Michael, 50, civilian, NASA mission specialist 4, fourth mission
Previous missions: STS-78 (1996), STS-90 (1998), STS-109 (2002)

ISS resident crew members

REISMAN, Garrett Erin, 40, civilian, NASA mission specialist 5 (up only), ISS-16/17 flight engineer
EYHARTS, Léopold, 50, French Air Force, ESA (French) mission specialist 5 (down only), ISS-16 flight engineer, second mission
Previous mission: Soyuz TM27/26 (1998)

Flight log

Following several years of delays, this mission saw the start of construction of the main Japanese element at the ISS. The Kibo ("Hope") Module was too massive to

Dextre arrives for operational assignment.

be launched in one go and would therefore be delivered over three Shuttle flights. This first mission carried the Equipment Logistics Module-Pressurized Section (ELM-PS) which would be attached temporarily to Harmony (Node 2). The more advanced Canadian Special Purpose Dexterous Manipulator, called Dextre, was also delivered on this flight. The new unit would supplement the Canadarm2 unit delivered in 2001.

Final processing for the mission began with the arrival of OV-105 (Endeavour) at the OPF on August 21, 2007. On February 11, 2008, the orbiter was transferred over to the VAB for final mating with the twin SRBs and ET. A week later, on February 18, the STS-123 stack was rolled to Pad 39A. Following a smooth countdown with no concerns over the weather, everything progressed as planned towards an on-time night launch. A low cloud bank meant that Endeavour disappeared from view from the ground soon after it began its journey to orbit.

A 5 h inspection of the orbiter's thermal protection system by the RMS was conducted by the crew the day before docking. The standard rendezvous pitch maneuver (backflipping the orbiter) for the resident ISS crew to inspect the underside was also completed successfully. Subsequent analysis of these data on the ground revealed no damage, allowing the Mission Management Team to clear the vehicle's thermal protection system for reentry.

Docking with the station occurred on March 12, but the hatches were not opened until the early hours of the following day. Shortly after entering the station, Reisman exchanged places with outgoing ISS-16 resident flight engineer Eyharts (France), who had logged 33 days as a member of ISS-16.

The station's Canadarm2 removed the Spacelab pallet containing the Dextre hardware from Endeavour on March 13, relocating and attaching it to the station's Mobile Base System. The station arm was also used later to relocate Dextre to a position on the Destiny Laboratory, attaching it to one of the laboratory's power and data grapple fixtures.

A record five EVAs were completed during the mission, totaling 33 hours 28 minutes. A trio of astronauts worked in pairs to complete the EVAs. To support this work, ISS-16 crew member Reisman also participated in the first EVA.

This first EVA (March 13, 7 h 1 min) saw Linnehan and Reisman remove a cover from the centerline berthing camera system on the top of the Harmony Module. This system had provided a live video link as an additional visual asset in the docking of spacecraft and modules. They then removed the contamination covers from the Japanese module's docking mechanism and disconnected other power and heater connections. Next, the two astronauts installed the "hands" of Dextre to its arms, and the Orbital Replacement Unit (ORU) tool change-out mechanism. Initial attempts to route power to Dextre during the EVA failed, but Canadian engineers were able to develop a bypass software patch to try at a later date.

The next EVA (March 15, 7 h 8 min) saw Linnehan and Mike Foreman attach the two arms to Dextre. This would allow the device to conduct installation and maintenance tasks controlled from inside the station. The astronauts also removed previously set up thermal covers from the robotic arm device.

During the third EVA (March 18, 6 h 53 min), Linnehan and Behnken continued work on Dextre. They installed the unit's toolholder assembly (which also serves as the "eyes" of the unit) and then the Spacelab logistics pallet was prepared for its return to Earth on Endeavour. The two astronauts next installed spare equipment for the station, as well as an external platform on the Quest airlock. This equipment included a spare yaw joint for the Canadarm2 and two spare direct current switching units. The crew also attempted to install the MISSE 6 experiment on the Columbus laboratory, but they were unable to engage the latching pins so this task was unavoidably deferred to a later EVA.

During EVA 4 (March 20, 6 h 24 min), Behnken and Foreman replaced an electrical circuit box, known as the Remote Power Control Module, on the station's truss structure. A major focus on this EVA was demonstration of a tile repair ablator dispenser (resembling a caulking gun), which was used to apply a sample material (Shuttle Tile Ablator-54, or STA-54) to samples of Shuttle heat shield tiles which had been deliberately damaged prior to the mission. The test samples were returned to Earth for more extensive testing to determine how STA-54 performed under microgravity and vacuum environments. Towards the end of the space walk, the astronauts removed a cover from Dextre and several launch locks that were still attached to the Harmony Node.

For the final EVA (March 22, 6 h 2 min), Behnken and Foreman stowed the Orbiter Boom Sensor System (OBSS) on the station's truss. This was a temporary move to make room in the payload bay of Discovery, which was currently being prepared to deliver the large Kibo science laboratory on the next mission (STS-124). This would take up most of the payload capacity of the orbiter. The OBSS would be returned on Discovery once the Japanese science laboratory had been delivered. After evaluating various methods of troubleshooting the latching pin problem, ground-based engineers advised Behnken of the best way to install MISSE-6 on the exterior of Columbus on this EVA. Meanwhile, Foreman inspected the SARJ to evaluate apparent damage, which had been revealed from photographs.

This was the first time a Shuttle flight had supported five EVAs and, across the series of space walks, three astronauts had logged three excursions each. Linnehan had accumulated 21 hours 2 minutes; Foreman 19 hours 34 minutes; and Behnken 19 hours 19 minutes. In his single excursion, Reisman logged 7 hours 1 minute.

Between the EVAs, Doi configured experiments and storage racks on the newly installed ELM-PS. Prior to the installation of Dextre, Reisman and Behnken had tested the joint bracket. Gorie examined minor condensation on a cooling line under the middeck flooring of the orbiter. This was later deemed not to impact orbiter operations, but was inspected periodically for the rest of the mission.

On March 19, between EVA 3 and EVA 4, Doi, Gorie, and station commander Peggy Whitson talked to Japanese Prime Minister Yasuo Fukuda, who congratulated the crew on their success in installing the first Kibo element. Later that day, the hatches were finally closed between Endeavour and the ISS, followed a few hours later by the orbiter undocking after a total of 11 days 20 hours 36 minutes of joint operations.

The first landing attempt was waived off due to unsuitable weather at KSC, but just one orbit later the weather cleared sufficiently to allow the landing there. Leopold Eyharts had logged 44 days on the space station during his 48-day mission.

Milestones

258th manned space flight
152nd US manned space flight
122nd Shuttle flight
 21st flight of Endeavour
 25th Shuttle ISS mission
  8th Endeavour ISS mission
First mission to fully utilize the Station-to-Shuttle Power Transfer System (SSPTS)
First Shuttle mission to feature 5 EVAs
Set Shuttle record for longest stay at the ISS (11 da 20 h)

SOYUZ TMA-12

International designator	2008-015A
Launched	April 8, 2008
Launch site	Pad 1, Site 5, Baikonur Cosmodrome, Republic of Kazakhstan
Landed	October 24, 2008
Landing site	89 km north of Arkalyk, Republic of Kazakhstan
Launch vehicle	Soyuz-FG (serial number ЦШ15000-024), Soyuz TMA (serial number 222)
Duration	198 da 16 h 20 min 31 s (Volkov, Kononenko) 10 da 21 h 19 min 21 s (Yi)
Call sign	Eridanus
Objectives	Resident crew transport (16S), ISS-17 research program, visiting crew 14 (Korean Astronaut Program, KAP) research program.

Flight crew

VOLKOV, Sergei Alexandrovich, 35, Russian Federation Air Force, RSA Soyuz TMA commander, ISS-17 commander
KONONENKO, Oleg Dmitryevich, 43, Civilian, RSA Soyuz TMA flight engineer, ISS flight engineer 1
YI, So Yeon, 29, civilian, Soyuz TMA research cosmonaut, South Korean space flight participant

ISS resident crew transfers
REISMAN, Garrett Erin, 40, NASA ISS flight engineer 2 (up STS-123/down STS-124)
CHAMITOFF, Gregory Errol, 45, NASA ISS flight engineer 2 (up STS-124/down STS-126)

Flight log

This mission featured the first space flight by a son of a cosmonaut. Sergei Volkov's father was veteran cosmonaut Alexander Volkov, who had flown missions to Salyut 7 and Mir. With first-timer Kononenko also on board, this was the first all-rookie Russian crew since the July 1994 flight of Soyuz TM-19 to the Mir space station with Yuri Malenchenko and Talgat Musabayev. The last all-rookie U.S. crew *in orbit* had been Joe Engle and Richard Truly on STS-2 in 1981, although Engle had flown on three X-15 astro-flights. The last completely rookie flight crew from the United States had been the Skylab 4 trio in November 1974.

Two second-generation space explorers. The crews of Expeditions 17 and 18 include Sergei Volkov (third from left) and Richard Garriott (fifth from left). Also in the frame are (from left) Oleg Kononenko, Yuri Lonchakov, (Volkov), Michael Fincke, (Garriott), and Greg Chamitoff.

Flying with the ISS-17 crew to the ISS and returning with the ISS-16 crew in TMA-11 was South Korean Space Flight Participant So Yeon Yi, on an 11-day mission. The South Korean Astronaut Program (KAP) encompassed 52 investigations across 15 experiments under a contract dated December 7, 2006 between Roscosmos and the South Korean Aerospace Research Institute. This program included a simple geophysical research experiment, five life science, six technological, one biotechnology, and one educational experiment, along with a single Earth resource experiment. This research was planned to take 53 hours 15 minutes of her time on station, with over 6 hours support from a Russian cosmonaut. The usual public affairs and symbolic activities would also be conducted aboard the station. Originally, Ko San was to have made the flight, but he was replaced shortly before the flight due to a breach of security related to official documentation. He took Yi's place as backup crew member.

The new expedition crew took over formal command of the station from the ISS-16 crew on April 17. During the main residency, the Russian cosmonauts were supported by two NASA astronauts who were delivered and returned by the Shuttle. When TMA-12 docked with the station, Garrett Reisman was already on board (having arrived aboard STS-123) and serving as an ISS-16 crew member. He subsequently transferred to ISS-17 operations and in turn was replaced by Greg Chamitoff (who arrived on the STS-124 mission which took Reisman home).

Chamitoff remained on board at the end of the ISS-17 residency and then transferred to the ISS-18 crew until the mission which would take him home arrived (STS-126).

The Russian ISS-17 research program encompassed 182 sessions of 34 experiments. Only two of these were totally new studies, but this research program was described by Energiya as of the highest priority. In order to accomplish this, over 229 hours of crew activity were planned; split between Kononenko (139 h) and Volkov (90 h). Within this program, there were nine human life research studies, six geophysical research experiments, two experiments devoted to Earth resources, and two each for technical research, contract activities, and the study of cosmic rays. Additionally, there were single experiments in education and space technology. On top of this, there were an additional 26 NASA managed experiments in human research, new technologies, biological and physical science, and education, with another 20 experiments planned by ESA and JAXA.

This residency had included plans for a Russian segment space walk, hosting two Shuttle flights and receiving three Progress cargo flights (including the first of the new series—Progress M-01M—which was subsequently delayed until the next expedition), as well as loading and unloading of the ESA ATV cargo spacecraft that arrived during ISS-16. Routine maintenance and housekeeping would also be a feature of this tour of duty. Less than a month after the departure of the ISS-16 crew with Yi on board, the new crew received the Progress M-64, which included a new Sokol pressure suit for Volkov. A slight bladder bulge on his original suit was noted prior to launch after a zip failed, although it passed pressure integrity checks and was cleared for launch. The replacement was loaded on the next available resupply craft as a precaution.

Following the departure of the STS-124 mission, which delivered the pressurized module of the Japanese Kibo laboratory and exchanged Anderson for Chamitoff, the ISS-17 cosmonauts prepared for their first EVA. This was not the one originally planned. The two previous unplanned ballistic reentries of the Soyuz Descent Module had raised concerns that a similar event may occur with TMA-12. The Russians determined that faults with the explosive bolts that separated the modules were the likely cause. On July 10, with Chamitoff safely located in the Soyuz Descent Module, with books and a laptop to keep him occupied, his two Russian colleagues completed their contingency EVA. Neither cosmonaut had performed EVA before so this was a challenging operation for the pair. They were also working in an area not normally equipped for EVA, so they had to install restraints and handholds and then cut into the Soyuz insulation to access the area, as well as installing protection for vulnerable propellant lines. The suspect bolt (one of two) from one of the five locks which secured the Instrument Module to the Descent Module was removed, stowed in a blast-proof container, and returned to the station, much to the relief of the flight controllers watching from Mission Control in Moscow. After reinstalling the protective insulation, the 6 h 18 min EVA ended.

Five days later the two cosmonauts, now veteran spacewalkers, conducted the EVA they originally planned for. This included installing a new docking target on the Zvezda transfer compartment in advance of the arrival of the Mini Research Module 2 in 2009. The pair then completed some inspection photography, installed a science experiment to study bursts of cosmic radiation on a handrail on Zvezda, and straightened out a bent ham radio antenna, before ending the EVA after 5 hours 54 minutes, giving both men a total of 12 hours 12 minutes of EVA experience. Volkov now had approximately two hours more EVA experience than his father had logged in his cosmonaut career.

With the EVAs completed, the crew resumed science and maintenance work and witnessed the departure of Progress M-64 and the undocking of ATV-1. The powerful Hurricane Ike temporarily closed down the Mission Control Center in Houston, with a temporary MCC being set up in Austin, Texas. The result of all this was that Progress M-65 docking was delayed a few days until the hurricane had passed over Houston. As a precaution, until the main MCC could be returned to full operating status, the starboard truss radiator on the station was repositioned, internal systems reconfigured, the Columbus Module placed in safe mode, and Kibo shut down. This episode once again demonstrated the importance of and reliance upon ground support facilities for ISS operations and the need for alternative communication centers in the event of terrestrial natural disasters or phenomena. Truly independent control of a manned spacecraft, from the vehicle itself, is still a long way off, but has to be a consideration, at least in part, for future deep-space missions to the asteroids and Mars.

The arrival of the TMA-13 spacecraft signaled the approaching end of the ISS-17 residency, as the ISS-18 crew arrived with space flight participant Richard Garriott, another son of a former astronaut. He would return with Volkov and Kononenko in TMA-12, while Chamitoff remained on board with the new expedition crew. This was the first time that two second-generation space explorers were in space at the same time on the same vehicle. Richard Garriott's father Owen had been one of NASA's original scientist–astronauts, selected in 1965 and flown on a 59-day mission to Skylab in 1973 and a 10-day Shuttle flight (Spacelab 1) a decade later.

The formal change-of-command ceremony occurred on October 22, officially ending the ISS-17 residency after 188 days. On October 24, Volkov, Kononenko, and Garriott boarded Soyuz TMA-12 and undocked from the station, landing safely in Kazakhstan later the same day after the nominal landing following the planned entry profile.

Milestones

259th manned space flight
105th Russian manned space flight
 98th manned Soyuz flight
 12th manned Soyuz TMA mission
 16th ISS Soyuz mission (S16)
 14th ISS Soyuz visiting mission
First South Korean in space
First flight of a cosmonaut's son (Sergei Volkov/Alexander Volkov)
Sergei Volkov becomes the youngest ISS commander (aged 35)
Kononenko celebrates his 44th birthday in space (June 21)

STS-124

International designator	2008-027A
Launched	May 31, 2008
Launch site	Pad 39A, KSC, Florida, U.S.A.
Landed	June 14, 2008
Landing site	Runway 15, Shuttle Landing Facility, KSC, Florida, U.S.A.
Launch vehicle	OV-103 Discovery/ET-126/SRB BI-133/SSME: #1 2047, #2 2044, #3 2054
Duration	13 da 18 h 13 min 7 s (STS-124 crew)
	95 da 08 h 47 min 5 s (Reisman)
Call sign	Discovery
Objectives	ISS assembly flight (1J), delivery and installation of Japanese Kibo pressurized module (JEM-PM) and Kibo RMS, ISS resident crew member exchange

Flight crew

KELLY, Mark Edward, 44, USN, NASA commander, third mission
Previous missions: STS-108 (2001), STS-121 (2006)
HAM, Kenneth Todd, 43, USN, NASA pilot
NYBERG, Karen LuJean, 38, civilian, NASA mission specialist 1
GARAN Jr., Ronald John, 46, USAF, NASA mission specialist 2
FOSSUM, Michael Edward, 50, USAF Reserve, NASA mission specialist 3, second mission
Previous mission: STS-121 (2006)
HOSHIDE, Akihiko, 39, civilian (Japanese), JAXA, mission specialist 4

ISS resident crew members
CHAMITOFF, Gregory Errol, 45, civilian, NASA mission specialist 5 (up only), ISS-17 flight engineer
REISMAN, Garrett Erin, 40, civilian, NASA mission specialist 5 (down only), ISS-17 flight engineer

Flight log

This was the second of three missions related to the installation of the Japanese Science Module Kibo and its associated facilities. Discovery was moved into the OPF on November 8, 2007 for processing, and then transferred to the VAB on April 26, 2008 for mating with the rest of the stack. The completed STS-124 stack was rolled out to Pad 39A on May 3, 2008.

Kibo expands. The Japanese Pressurized Module (foreground) and Logistics Module (top right) and a portion of the Harmony Node and Canadarm2 are visible during this image taken during one of the STS-124 space walks.

After a smooth countdown, STS-124 launched on time and reached orbit without incident. During the scheduled postlaunch walk-around of the pad, however, severe damage was discovered. A 75 ft. (22.86 m) × 20 ft. (6.096 m) section of the east wall of the north flame trench was affected. An investigation was begun immediately with the aim of determining the probable cause of the damage.

Up on orbit, the usual inspection of the orbiter's thermal protection system was limited to the RMS, as the boom attachment had been temporarily stored on the station during the previous (STS-126) mission in order to make room on this flight for the large Japanese payload. Prior to docking with the station on June 2, Discovery was inverted to allow the station crew to document the orbiter's surfaces for analysis. On June 3, the Japanese Pressurized Module was relocated to the Harmony Node using the station RMS, operated by Hoshide and Nyberg.

Greg Chamitoff exchanged places with Garrett Reisman on the ISS-17 resident crew four and a half hours after docking. Reisman had spent 81 days as a resident crew member. Due to changes in the scheduling that required an ISS crew exchange seat to be available on this mission, Chamitoff joined the STS-124 crew in late 2007, replacing the original mission specialist Steve Bowen who was reassigned to STS-126. Chamitoff's cousin ran a famous Fairmount bagel establishment in Montreal, Canada and as part of his personal mission allowance,

Chamitoff brought three bags of sesame seed bagels (six in each) from his cousin's bakery to the ISS, adding to the culinary delights on orbit.

There were three EVAs on this mission, all completed by Garan and Fossum and totaling 20 hours 32 minutes. The first EVA (June 3, 6 h 48 min) was dedicated to the very first U.S. EVA by Ed White from Gemini 4 exactly 43 years before. This latest American space walk saw the two astronauts disconnect cables and remove covers from the Kibo Japanese Pressurized Module (JPM) while in Discovery's payload bay. They also assisted in relocating the OBSS back to the orbiter payload bay, attaching it to the Shuttle RMS for its move. The EVA crew then demonstrated a technique to clean debris from the SARJ and, while Garan installed a new bearing on the joint, Fossum confirmed that damage noted previously was indeed a divot.

During the second EVA (June 5, 7 h 11 min), Fossum and Garan continued to outfit the exterior of the new Japanese module, installing the forward and rear TV cameras on the outside of the Kibo JPM. The astronauts also removed thermal covers from the Kibo RMS and prepared the JPM upper docking port for later attachment of the Kibo Logistics Module. They also prepared an External Stowage Platform (ESP) for the removal and replacement of a nitrogen tank assembly which was planned for the next EVA.

The third EVA (June 8, 6 h 33 min) focused upon the removal and replacement of this tank on the starboard truss. Fossum then returned to the port SARJ to take samples of particulate matter from inside the joint (using a strip of tape) for engineers to analyze back on Earth. He then removed thermal insulation from the Kibo robotic arm's wrist as well as elbow cameras and launch locks from one of the Kibo windows. He also deployed debris shields on Kibo and tightened a bolt holding a TV camera in place. Garan retrieved the video camera removed during the second EVA and reinstalled it. Some of the additional tasks completed on this EVA included the installation of a thermal cover on Harmony's outside connectors, relocating a foot restraint aid, and the removal of a launch lock on the starboard SARJ.

Nyberg, assisted by Chamitoff, later used the station RMS to reposition the Japanese Logistics Module from the Harmony Module to its permanent location on top of the Kibo laboratory. Following the installation of Kibo's Pressurized Module, Hoshide and Nyberg opened the hatch and were the first inside the newest ISS module, which still had to be outfitted. Hoshide held a sign up to the TV camera stating, in Japanese, "experiments and astronauts wanted". Following a Japanese tradition, he hung a small door curtain above the module's entry hatch. Shortly afterwards, the other eight crew members entered the new module to appreciate the added volume Kibo gave to the station. The Japanese astronaut noted that, although the new module was empty, it was full of dreams and that it gave new "hope" to the station. Hoshide and Nyberg later operated the RMS for its final deployment maneuver and then stowed the arm and checked out the brakes within its joint. The astronauts also opened the hatches between the Pressurized Module and Logistics Module for the first time to inspect the added storage volume.

Other activities completed during the docked phase included Kononenko installing the spare gas liquid separator pump in the station's toilet to return it to useful service. Reisman and Chamitoff replaced a bed in the carbon dioxide removal assembly that decreased air contamination aboard station. commander Kelly and mission specialist Hoshide spoke with Japanese Prime Minister Yasuo Fukuda on June 6, who congratulated them on the success of the mission. Reisman commented, shortly before leaving the station, on his relief that during his 91 days on board, he had not broken anything really expensive.

Hatches between Discovery and the station were closed four days later on June 10. Undocking took place the following morning, after 8 days 17 hours 39 minutes of joint operations. Pilot Ken Ham circled Discovery around the station for video and photo-documentation of the now 330-ton mass complex by the rest of the crew.

With the OBSS now relocated to Discovery's payload bay, the crew completed a late inspection of the vehicle's heat shield. Analysis of the images by experts on the ground determined that the heat shield was safe for entry and landing. On June 13, during day-before-landing system checks designed to verify entry and landing systems, the Mission Management Team revealed that an object was seen floating away from the vehicle. Engineers had concluded that the object in question was most probably a heat shield clip from the rudder/speed brake area on the tail of the vehicle. As this was used as a heat barrier during launch only it was not a concern for entry or landing.

Discovery swooped to a successful landing on Runway 15 at KSC on June 14, 2008. Reisman adapted to 1g quite quickly after three months in space and at the post-landing crew press conference, he stated that after greeting his wife his thoughts were focused on a pizza or T-bone steak. The orbiter, meanwhile, was towed across to the OPF the same day to begin the de-processing cycle prior to preparations for its next mission.

Milestones

260th manned space flight
153rd U.S. manned space flight
123rd Shuttle mission
 35th flight of Discovery
 26th Shuttle ISS mission
  9th Discovery ISS mission
Nyberg becomes 50th woman in space
Nyberg becomes 1st person to operate three RMSs: the station, the Shuttle, and Kibo's arms.
First time JAXA flight control team activated and controlled a module from Kibo MCC, Tsukuba, Japan
At 15 tons, Kibo was the largest and heaviest space station module lifted to orbit

SHENZHOU 7

International designator	2008-047A
Launched	September 25, 2008
Launch site	Pad 921, South Launch Site #1, Jinquan Satellite Launch Center, China
Landed	September 27, 2008
Landing site	Siziwang Banner, central Inner Mongolia
Launch vehicle	Long March 2F (CZ-2F)
Duration	2 da 2 h 27 min 35 s
Call sign	Unknown
Objectives	First Chinese EVA mission, first three-person Chinese space flight

Flight Crew

ZHAI, Zhigang, 42, PLA Air Force, commander
LIU, Boming, 42, PLA Air Force, orbital module monitor
JING, Haipeng, 42, PLA Air Force, descent module monitor

Flight log

On September 27, 2008, Zhai Zhigang became the first Chinese taikonaut to perform an EVA. This historic event, 43 years after the first EVAs had been performed, meant that China had become the third nation, after the Soviet Union (Russia) and the U.S.A., to achieve EVA capability. The buildup to China's third manned space flight had for some time indicated that a short EVA was part of the program, and that it would be a further step towards the creation of a small space station.

The flight into Earth orbit lasted approximately 583 seconds. In the central couch flew mission commander Zhai Zhigang, with Liu Boming to his right and Jing Haipeng to his left. They would occupy the same positions for landing. Five hours into the mission, the onboard engines were used to refine the orbit, allowing for a flight time of up to five days. The duration of this mission had previously been announced at only three days, to prevent Western speculation of an unplanned early return.

This was the first Chinese mission to fly three crew members, with a further three as backup—Chen Quan (commander), Fei Junlong (OM monitor), and Nei Haisheng (DM monitor)—making a training group of six. The prime crew for the flight had been announced on September 17 and of the backup crew only Chen Quan had not flown in space. The other two were the crew of Shenzhou 6, lending experience to preparations for this technically challenging mission.

A display model of the Chinese Feitian ("Flying Sky") EVA suit of the type used on Shenzhou 7. Photo copyright: MarkWade/Astronautix.com, used with permission.

The space walk was planned to last no more than 30 minutes, although the crew's EVA preparations took about 15 hours. This was mainly taken up with assembling the suit, putting it on, and then checking it out for use outside the orbital module. It was reported that there were 30 contingency plans developed for the EVA to ensure the safety of the crew and the integrity of both the suit and spacecraft during the operation. It was also revealed that a fire alarm at the Mission Control Center shortly before the start of the EVA was in fact a false alarm, presumably adding to the tension at the time.

The EVA actually lasted 22 minutes and was conducted by Zhai Zhigang, wearing the Chinese-developed Feitian space pressure suit which is similar to the Russian Orlan ("Bald Eagle") EVA suit used since 1977 for Russian space station-based EVAs. Supporting his commander from inside the depressurized Orbital Module of Shenzhou 7, Liu Boming wore a Russian Orlan M suit (No. 42) for comparison with the national suit. The name Feitian comes from the Mandarin *fei* (flying) and *tian* (sky).

Zhai Zhigang's exit from the module was hindered slightly by difficulties in opening the hatch. Once these were resolved and in full view of the exterior TV camera, Zhai Zhigang floated head first outside the module, and then used handrails mounted outside the module to perform a wandering excursion to retrieve experiment samples and wave a Chinese flag for the benefit of the camera. The experiment was a solid lubricant exposure device, about the size of a book. It was installed on the outer wall of the Orbital Module during launch preparation and retrieved during the EVA after about 40 h exposure to the space environment. The aim of the experiment was to study the characteristics of lubricants designed for future space-based "moving components in space facilities".

Limited during the EVA by restraint cables, the taikonaut remained close to the EVA hatch, his movements captured by two cameras providing spectacular panoramic images of the historic event. Conducting a short (approximately 4 min) stand-up EVA, Liu Boming handed his commander a national flag. Though ready to assist if necessary, he did not fully exit the module. The third crew member, Jing Haipeng, remained inside the Descent Module monitoring the EVA and the general condition of the spacecraft.

Though EVAs of up to 7 hours are reportedly possible in the Feitian suit, this initial excursion was limited to just over 20 minutes to evaluate the design and procedures, providing baseline data in order to plan longer excursions on future missions. No serious difficulties were reported, although it was clear that at one point the spacewalker became entangled in his tether. At the end of the EVA period, both men returned to the Orbital Module of the spacecraft, sealed the hatch and repressurized the module before opening the internal hatch and rejoining their colleague. Apart from the gloves, the suit units were not returned to Earth as there was not enough room inside the Descent Module to stow them. The success of the first Chinese space walk was feted across the world as a major milestone in the development of Chinese manned space flight. The total hatch open time was 22 minutes, with Zhai Zhigang actually outside for only 10 minutes, with about 4 minutes taken for exit and entry into the module.

Two hours after the end of the EVA, a small, 40 cm long, 40 kg monitoring satellite called "BanXing" was deployed from the nose of the Shenzhou 7 Orbital Module. It carried a liquid ammonia "boost device" maneuvering engine and a pair of 150-megapixel stereo cameras. Its objectives were to evaluate minisatellite technology, to observe and monitor the Shenzhou spacecraft and to test the tracking and approach technology being developed for space rendezvous and docking. After taking video and still images of the spacecraft, the satellite was maneuvered to approximately 100–200 km away. Following the return of the crew to Earth, the small satellite re-rendezvoused with the abandoned Shenzhou 7 Service Module and then orbited around it, imaging the module. The operational lifetime of the minisatellite was reportedly three months. It finally reentered on October 29, 2009, about 13 months after its deployment from Shenzhou 7.

As a further development in support technologies for improved coverage of manned missions the Chinese had launched their first data relay satellite Tianlian 1 ("Skylink 1") on a Long March 3C carrier rocket from the Xichang Satellite

Launch Center on April 25, 2008. The relay satellite was used to improve orbital communications with Shenzhou to approximately 60%. Ground stations and ocean tracking ships can only cover about 12% of each orbit, so the data relay satellite improved this by covering 50% of the orbit. Two new tracking ships, *Yuanewang 5* and *6*, were also commissioned in time to support the Shenzhou 7 mission.

The Chinese reported that 220 technical modifications had been implemented to the Shenzhou 7 mission after the flight of Shenzhou 6. Most notable were the removal of the solar panels on the Orbital Module to allow for the EVA, the installation of EVA handrails, and an additional camera installed to support the space walk. The removal of the solar arrays meant that the OM would not remain in space as on earlier missions, but would reenter after its separation at the end of the mission. There were also more than 30 other upgrades and improvements to the carrier rocket since the last mission, notably to the pipes inside the second stage.

Life on board the spacecraft was also improved for the crew, with a custom-made compact and storable toilet to allow collected urine to be recycled for drinking water. There were 80 food varieties available to them, compared with the 50 available for Shenzhou 6. The choices included spicy kung-pao chicken, de-shelled shrimp, and a selection of dry fruits.

On September 27, Shenzhou 7 reportedly passed within 45 km of the International Space Station and, although no statements were released by Chinese authorities, the American media reported that China had deployed its companion satellite BX-1 only four hours earlier. They suggested that it was a dual military–civilian mission, speculating that it may have been a test of orbital antisatellite inception technology and space station observation.

On September 28, with the primary objective of the mission completed, the de-orbit and entry maneuvers were followed by a landing in Siziwang Banner in central Inner Mongolia, between Hohhot and Erenhot. The success of this mission led to some speculation that the next Shenzhou missions would include docking with a rudimentary space laboratory.

Milestones

261st manned space flight
3rd manned Chinese spaceflight
1st Chinese three-person space flight
3rd manned Shenzhou flight
1st three-person Shenzhou flight
1st Chinese EVA
1st Chinese stand-up EVA

SOYUZ TMA-13

International designator	2008-050A
Launched	October 12, 2008
Launch site	Pad 1, Site 5, Baikonur Cosmodrome, Republic of Kazakhstan
Landed	April 8, 2009
Landing site	151 km northeast of Dzhezkazgan, Republic of Kazakhstan
Launch vehicle	Soyuz-FG (serial number Ill15000-026), Soyuz TMA (serial number 223)/17S
Duration	178 da 00 h 13 min 38 s (Lonchakov, Fincke) 11 da 20 h 35 min 37 s (Garriott)
Call sign	Titan
Objectives	ISS resident crew transport (17S), ISS-18 resident crew; visiting crew 15 (Generation II Astronaut, GTA) research program

Flight crew

LONCHAKOV, Yuri Valentinovich, 43, Russian Federation Air Force, RSA
Soyuz TMA commander, ISS flight engineer 1, third mission
Previous missions: STS-100 (2001), Soyuz TMA-1 (2002)
FINCKE, Edward Michael, 41, USAF, NASA Soyuz TMA flight engineer,
NASA ISS commander, second mission
Previous mission: Soyuz TMA-4/ISS-9 (2004)
GARRIOTT, Richard Allen, 46, civilian, American space flight participant

ISS resident crew exchanges

CHAMITOFF, Gregory Errol, 45, NASA ISS flight engineer 2
(up STS-124, down STS-126)
MAGNUS, Sandra Hall, 44, NASA ISS flight engineer 2
(up STS-126, down STS-119), second mission
Previous mission: STS-112 (2002)
WAKATA, Koichi, 45, JAXA (Japanese) ISS flight engineer 2
(up STS-119, down STS-127), third mission
Previous missions: STS-72 (1996), STS-92 (2000)

Flight log

The 18th resident crew arrived at the docking port of Zarya on October 14, 2008, two days after leaving the Baikonur Cosmodrome. They assumed formal residency from the ISS-17 crew on October 22.

Former Skylab and Space Shuttle astronaut Owen Garriott and his son space flight participant Richard Garriott talk with reporters outside the Cosmonaut Hotel crew quarters, Baikonur Cosmodrome. Photo credit: NASA/Viktor Zelentsov.

In command of the Soyuz, but serving as flight engineer on the station was veteran cosmonaut Yuri Lonchakov. He had previously flown on two visiting missions to the station on Shuttle in (2001) and on the maiden flight of TMA (2002). The flight engineer on Soyuz and commander of the residency was veteran NASA astronaut Michael Fincke, who had previously logged 187 days aboard the station on ISS-8. Arriving with the ISS-18 crew was Richard Garriott, the latest

SFP and son of Skylab and Spacelab astronaut Owen Garriott. He would return with the outgoing ISS-17 cosmonauts in TMA-12 after a flight of 10 days on station.

The experiences of his father were mirrored in Garriott's own program of research on the station 35 years later. The Generation II Astronaut (GTA) project featured nine experiments. There were two more under the ESA program and three under the U.S. medical program. The nine experiments featured research in life science, biotechnology, technical research, education, and humanities, as well as a range of public affairs and outreach programs. Six experiments were performed in the Russian segment, the other three in the U.S. segment. Garriott was assigned 33 hours of experiment time, supported by one of the Russian cosmonauts for photo-documentation. The memory of his father's Skylab mission was also recalled with an updated Leonardo da Vinci figure used as the mission emblem as was the case with his father's Skylab 3 emblem. Garriott's return on TMA 12 completed a highly successful 12-day mission.

For the main crew, the Russian research program consisted of 46 experiments, only 5 of which were new investigations. The remainder were continuations of previous investigations. There were 6 experiments in human life research, 7 in geophysical research, and 3 in Earth research. A further 14 experiments were under the heading of space technology, with 6 technical research investigations, 2 contracted activities, and 2 on the study of cosmic rays. There were also 3 educational experiments and 3 more in space technology and material sciences. A total of 161 hours were allocated for Russian science across the mission.

Over in the U.S. segment, the research to be completed during this residency included 40 NASA-managed experiments in human research, exploration technology testing, biological and physical sciences, and education. In addition, there were 33 experiments planned by the European and Japanese space agencies.

During this expedition, three other crew members worked with the Russians. When the main crew arrived, Chamitoff was already aboard. He was replaced by Sandra Magnus on STS-126 in November 2008 and she in turn was replaced in March 2009 by Japanese astronaut Koichi Wakata on STS-119. The two Shuttle missions delivered further logistics to the station, with STS-126 being the second dedicated Utilization and Logistics Flight and STS-119 delivering the long-delayed final set of solar arrays and truss element S6. Both flights included several EVAs in support of the activities by Shuttle crew members.

ISS-18 crew members Fincke and Lonchakov conducted two EVAs totaling 10 hours 27 minutes from Pirs, wearing Orlan-M suits for the final time before the improved Orlan MK suits were introduced. The first EVA (December 23, 2008, 5 h 38 min) was designated Russian Segment EVA 21 and included the retrieval of experiment samples and deployment of a Langmuir probe to measure electrical and plasma fields close to the docked Soyuz spacecraft. Studies of electromagnetic energy were linked to the ongoing investigations into the problems with the pyrotechnic separation bolts on the Soyuz which had troubled TMA-10 and TMA-11. The EVA crew deployed two experiments on a special platform on the outside of the Zvezda module.

The second EVA (March 10, 2009: 4 h 49 min), designated Russian Segment EVA 21A, was not part of the original mission planning, but when the pair had difficultly installing a ESA experiment on their first EVA, the second excursion was added instead of returning the hardware to Earth. This time, the EXPOSE-R biological exposure samples were installed and the two men took time to clear six straps from the docking area on Pirs to prevent hindrance with future docking operations. Other tasks included closing a loose insulation flap on Zvezda, removing an experiment cassette, and further photo-documentation.

This mission also featured the arrival at station of the latest variant of the venerable unmanned Progress resupply cargo craft, the M-01M (31P), on November 30. Launched on November 26, the longer-than-usual approach allowed ground controllers to fully evaluate the new systems on the vehicle before committing to ISS docking. Outwardly resembling the earlier versions, this upgraded variant included a state-of-the-art digital computer system and more compact avionics, saving 165.375 lb (75 kg) dry mass over previous versions, with 15 fewer components. The upgraded equipment enabled automatic diagnostics between the telemetry and computer systems while also providing digital interfaces for the integration of all systems when docked with the ISS.

With the arrival of the TMA-14 crew (ISS-19) at the end of March, and completion of their science program, it was time once again to exchange the responsibility of command of the station; this was completed on April 2, formally ending the ISS-18 residency after 162 days. Arriving at the station with ISS-19 crew was Space Flight Participant Simonyi, on his historic second 10-day visit to the station. He returned with ISS-18 crew, who undocked in Soyuz TMA-13 on April 8, 2009 after 176 days on board the orbital complex. Their safe landing was achieved later that day near the town of Dzhezkazgan in Kazakhstan.

Milestones

262nd manned space flight
106th Russian manned space flight
 99th manned Soyuz flight
 13th manned Soyuz TMA mission
 17th ISS Soyuz mission (17S)
 15th ISS Soyuz visiting mission
 18th ISS resident crew
  1st flight of a son of a NASA astronaut (Richard Garriott/Owen Garriott)
Chamitoff celebrates his 46th birthday (August 6)

STS-126

International designator	2008-059A
Launched	November 14, 2008
Launch site	Pad 39A, KSC, Florida, U.S.A.
Landed	November 30, 2008
Landing site	Temporary Runway 04, Dryden Flight Research Center, EAFB, California, U.S.A.
Launch vehicle	OV-105 Endeavour/ET-129/SRB BI-136/SSME: #1 2047, #2 2052, #3 2054
Duration	15 da 20 h 29 min 27 s (STS-126 crew) 183 da 00 h 22 min 54 s (Chamitoff)
Call sign	Endeavour
Objectives	ISS assembly flight (ULF2), logistics and outfitting to prepare for a six-person resident crew from 2009, ISS resident crew exchange

Flight crew

FERGUSON, Christopher John, 47, USN, NASA commander, second mission
Previous mission: STS-115 (2006)
BOE, Eric Allen, 44, USAF, NASA pilot
PETTIT, Donald Ray, 53, civilian, NASA mission specialist 1, second mission
Previous mission: ISS EO6/STS-113/TMA-1 (2002–2003)
BOWEN, Stephen George, 44, USN, NASA mission specialist 2
STEFANYSHYN-PIPER, Heidemarie Martha, 45 USN, NASA mission specialist 3, second mission
Previous mission: STS-115 (2006)
KIMBROUGH, Robert Shane, 41, U.S. Army, NASA mission specialist 4

ISS resident expedition crew member transfers
MAGNUS, Sandra Hall, 44, civilian, NASA mission specialist 5 (up only)/ISS-18 flight engineer, second mission
Previous mission: STS-112 (2002)
CHAMITOFF, Gregory Errol, 45, civilian, NASA mission specialist 5 (down only)/ISS-18 flight engineer

Flight log

Dubbed a mission of home improvement and maintenance, this flight was packed with robotic arm operations, repairs, and servicing activities. Designed to deliver construction equipment intended to expand the living conditions in order to accommodate a permanent crew of six, the flight also featured another exchange

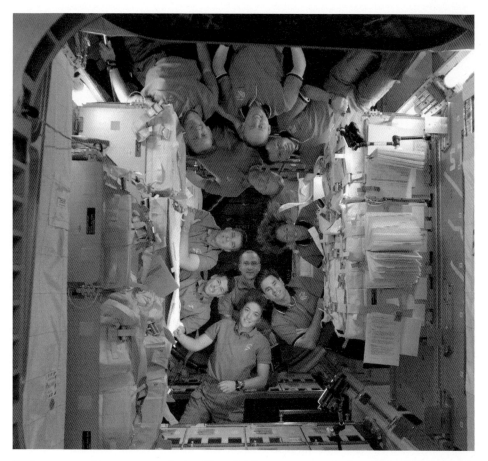

Post press conference crew photo. Don Pettit appears at photo center. Below him is Stefanyshyn-Piper. Clockwise from her position are Kimbrough, Bowen, Boe, Furguson, Fincke, Lonchakov, Magnus, and Chamitoff.

of NASA–ISS flight engineers via the Space Shuttle, as part of the resident crew of station prior to permanent six-person crewing.

In the original planning, Endeavour was intended to support the forthcoming Hubble Service Mission (STS-125) before flying on STS-126. Endeavour was rolled into the OPF on March 27, 2008 for processing. It was then relocated to the VAB on September 11, 2008, for stacking to the ET and SRB. On September 19 the STS-126 stack was relocated to Pad 39B to serve as an emergency rescue vehicle (if needed) for the STS-125 Hubble Service Mission 4, which was preparing to launch from Pad A. This was the first time since 2001 that a shuttle stack had sat on both launchpads at the Cape. When the Hubble mission was postponed into 2009 in late September due to problems with the telescope, all work on Endeavour to support the mission ended and focus shifted back to STS-126. When Discovery was rolled back to the VAB on October 20, Endeavour was moved across to the now vacant Pad 39A on October 23, 2008, the day after its Multi-Purpose Logistics Module (MPLM) payload arrived.

The launch into a night sky was flawlessly performed on November 14, 2008. A 5 h inspection of the vehicle's heat shield was conducted by the crew using the RMS and OBSS, prior to docking with station. Analysis of the imagery and data on the ground revealed that a small piece of thermal blanket was loose on the aft portion of Endeavour. As with previous flights, Endeavour was inverted for documentation and analysis of the underside. Following further analysis of the data, it was determined that there would be no need for a further inspection, as the heat shield looked in good condition for entry.

Endeavour docked with the ISS on November 16 and this was followed a few hours later by Magnus exchanging places with Chamitoff on the resident crew. He had spent 167 days as member ISS-18. The following day, MPLM Leonardo was relocated by the ISS robotic arm to the nadir port on Harmony. On board Leonardo were 6.5 tons of equipment, including two large water-recycling racks, a new "kitchen", a second toilet, two further sleeping stations, extra exercise equipment, and other supplies.

Mission specialists Stefanyshyn-Piper, Bowen, and Kimbrough completed a series of four EVAs during the mission, totaling 26 hours 41 minutes. Bowen logged the most time at 19 hours 56 minutes in three space walks. Piper logged 18 hours 34 minutes in her three space walks, while Kimbrough's two EVAs logged 12 hours 52 minutes.

During the first EVA (November 18; 6 h 52 min), Piper and Bowen spent the majority of their time outside working on the SARJ, removing two of the joint's Trundle Bearing Assemblies (TBAs). They also removed a depleted nitrogen tank from a storage platform on the station, returning it to the payload bay of Endeavour. Other tasks included moving a flex hose rotating coupler from the Shuttle across to the Station Storage Platform and removing insulation blankets from the Cameras Berthing Mechanism (CBM) on the Kibo laboratory.

Approximately halfway into this EVA, one of the grease guns that Piper was preparing to use at the SARJ released some of the Braycote grease into her Crew Lock Bag. This is the bag used by spacewalkers during their activities to retain spare tools and equipment. As she was cleaning the inside of the bag, it drifted away from her towards the aft and starboard of the station. It was soon out of reach, preventing her from retrieving it. Inside this bag were two grease guns, scrapers, several wipes and tethers, as well as several tool caddies. Piper and Bowen spent the remainder of their EVA time sharing a duplicate set of tools from the other Cew Lock Bag (CLB) they had with them.

During the second EVA (November 20; 6 h 45 min) Piper and Kimbrough moved two Crew and Equipment Translation Aid (CETA) carts. They also lubricated the station's RMS latches and an end effector snare (the capture device), then cleaned and replaced four TBAs. The next EVA (November 22, 6 h 57 min) saw Piper and Bowen replace five more TBAs and clean and lubricate race rings on the station's starboard SARJ.

On the final EVA (November 24, 6 h 7 min), Kimbrough and Bowen replaced the final TBA on the station's starboard SARJ, lubricated the race rings on the port SARJ, moved a video camera on the Part 1 Truss, and installed two Global Positioning Satellite Antennas on the Japanese Experiment Module (JEM) Logistics Module. They also retracted the latch on the JEM Exposed Facility Berthing Mechanism and reinstalled the mechanical cover.

In addition to the EVAs conducted outside the station, the crew occupied themselves by working with the resident station crew to convert the orbital facility to support a six-person crew and completed a wide variety of other tasks. Latches on the Exposed Facility Berthing Mechanism on the Japanese Kibo Laboratory were also tested. This mechanism would be used to install the External Science Platform on the Kibo, which would be delivered in 2009.

The astronauts installed a new Water Recovery System, which was designed to treat waste water and then provide recycled water that was clean and safe enough to drink. Supplying any crew with fresh drinking water has always been one of the challenges facing long-duration mission planners, spacecraft designers, and medical staff. Short missions are more easily accommodated, but supporting a crew of six or more 24/7 for 365 days a year is a logistical challenge. Lessons learned from previous space station programs, applied and improved on the ISS, will lay the groundwork for long-duration flights away from Earth, to the Moon, Mars, and the asteroids.

The Urine Processor Assembly (UPA) shut down during initial test operations. The station and Shuttle crews, as well as ground controllers and engineers, investigated possible causes and cures over the next several days. It was determined that the motion of the centrifuge had caused physical interference with the UPA, which resulted in increased power draw and temperatures. The UPA was hand-mounted onto the Water Recovery System (WRS) rack after grommets were removed. Following this remedy, the UPA ran normally. On the 10th day of the mission, the NASA Management Team extended the mission's docked duration by an extra day. This would allow additional time for further WRS troubleshooting if required.

On November 25, the crew were informed that the starboard SARJ had completed a 3 h 20 min test, during which it automatically tracked the Sun for the first time in over a year. In addition, the UPA had completed its second run without shutting down. The combined crew celebrated Thanksgiving aboard station and sent a greeting to American personnel who were serving abroad, away from home and family. The crews thanked members of the armed services for their commitment and dedication and wished them well.

With joint operations nearing completion, MPLM Leonardo was relocated back into the payload bay of Endeavour on November 26 for the return home. Undocking occurred on November 28 after joint operations totaling 11 days 16 hours 46 minutes with Chamitoff logging 179 days on the station. Endeavour's pilot, Eric Boe, completed a standard fly-around of the station for photo-documentation. The next day, the crew conducted an inspection of the thermal protection system with the RMS and OBSS. Following analysis of the images, the heat shield was cleared for entry and landing.

Towards the end of orbital operations on November 29, a small, 15.435 lb (7 kg) USAF satellite was deployed. PICOSat was designed to test and evaluate space environment effects on new solar cell technology. Expected to remain in orbit for several months, it finally de-orbited on February 17, 2010.

Weather concerns at the Florida primary landing site forced two standoffs before finally diverting the landing to California. Endeavour landed on a temporary runway adjacent to the concrete Runway 22 at Edwards Air Force Base. The concrete asphalt runway was 12,000 ft (3657.6 m) long by 20 ft (6.09 m) wide, with a 1,000 ft (348 m) underrun and overrun capability for Shuttle load-bearing support. As this runway was 1,940 ft (3,000 m) shorter than the nominal runways, new braking and rollout techniques had to be employed for this landing, providing new and additional information on landing a Shuttle orbiter.

Milestones

263rd manned space flight
154th U.S. manned space flight
124th Shuttle mission
 26th flight of Endeavour
 27th Shuttle ISS mission
  9th Endeavour ISS mission
First dual-pad Shuttle preparation since 2001
First landing on a temporary runway at Dryden due to maintenance on main runways
Bowen was first USN submarine officer selected for NASA astronaut training, and the second submariner to fly in space (after Mike McCulley on STS-34 in 1989)

STS-119

International designator	2009-012A
Launched	March 15, 2009
Launch site	Pad 39A, KSC, Florida, U.S.A.
Landed	March 28, 2009
Landing site	Runway 15, Shuttle Landing Facility, KSC, Florida, U.S.A.
Launch vehicle	OV-103 Discovery/ET-127/SRB BI-135/SSME: #1 2048, #2 2051, #3 2058
Duration	12 da 19 h 29 min 33 s (STS-119 crew) 133 da 18 h 17 min 38 s (Magnus)
Call sign	Discovery
Objectives	ISS assembly mission 15A, ITS S6, ISS resident crew rotation

Flight crew

ARCHAMBAULT, Lee Joseph, 48, USAF, NASA commander, second mission
Previous mission: STS-117 (2007)
ANTONELLI, Dominic Anthony, 41, USN, NASA pilot
ACABA, Joseph Michael, 41, civilian, NASA mission specialist 1
SWANSON, Steven Roy, 48, civilian, NASA mission specialist 2, second mission
Previous mission: STS-117 (2007)
ARNOLD II, Richard Robert, 45, NASA mission specialist 3
PHILLIPS, John Lynch, 57, USN Reserve (Retd.), NASA mission specialist 4, third mission
Previous missions: STS-100 (2001), ISS-11/TMA-6 (2005)

ISS resident crew members

WAKATA, Koichi, 45, civilian (Japanese), JAXA mission specialist 5 (up only)/ISS flight engineer, third mission
Previous missions: STS-72 (1996), STS-92 (2000)
MAGNUS, Sandra Hall, 44, civilian, NASA mission specialist 5 (down only)/ISS flight engineer, second mission
Previous mission: STS-112 (2002)

Flight log

One of the challenges that a researcher of Shuttle missions has to overcome is the mission numbering system and sequence. For most of the program, the missions

View of completed solar array after delivery and deployment at the ISS.

did not fly in sequence of allocated numbers. This was especially true for the delayed STS-119, which flew after STS-126 but before STS-125!

The inclusion of four crew members whose surnames began with "A" saw the crew being termed the "A" team. Two of these (Acaba and Arnold) were former teachers turned astronauts, who were selected in 2004 to assist NASA to inspire young people to study mathematics and science and hopefully to go on to pursue engineering and aerospace careers.

This mission delivered the fourth and final set of solar array wings, as well as the S6 truss, completing the structural backbone of the station and the main electrical power supply to the facility. With the installation of the final set of arrays, full power capacity could reach 120 kW of electricity, doubling the available power for scientific experiments from 15 kW to 30 kW and allowing the station's permanent resident crew complement to increase from three to six.

Final preparation for the mission began with Discovery being taken into the OPF on June 14, 2008. Rollover to the VAB occurred on January 7, 2009, with transfer to Pad 39A a week later on January 14. The original launch date had been set for February 12, but this was postponed following an issue with the gaseous hydrogen flow control valves. These valves are part of the system that channels gaseous hydrogen from the main engines to the External Tank. The valves had to be replaced and the launch was reset for March 11.

On that date, however, the launch had to be postponed again for at least 24 hours, due to a hydrogen leak in the left-hand vent line between the Shuttle and the ET. Managers and engineers looked at potential repair options and the launch was rescheduled for no earlier than March 15. This meant the flight's docked time at the station would be reduced by two days so that the Shuttle could depart prior to the arrival of the next resident crew on Soyuz TMA-14.

The March 15 launch occurred on time and with no problems during ascent. During FD2 (March 16), the crew completed a close inspection of the orbiter's wing leading edge panels using the RMS and OBSS. The crew installed the orbiter docking system "centerline" camera, tested the rendezvous equipment, and extended the docking ring on the top of the docking assembly. Prior to docking with the station on March 17, the orbiter completed the now familiar backflip maneuver for heat shield damage assessment. Experts in the field and the Damage Assessment Team in Mission Control Houston determined that the heat shield was healthy for reentry.

Discovery docked with the ISS on March 17, with Magnus and Wakata swapping roles and Soyuz seat liners later the same day. Magnus had spent 121 days as a member of the ISS-18 crew by the end of the mission, logging 129 days on board the station and a total of 134 days in space. Her replacement, Koichi Wakata, became the first representative of Japan to join a resident crew. The following day (March 18), the ITS S6 truss structure was relocated across to the station from Discovery's payload bay ready for the series of EVAs to attach it permanently to the station.

There were three EVAs completed during this mission, totaling 19 hours 4 minutes, with three astronauts conducting two EVAs each. Swanson logged 12 hours 37 minutes, Acaba 12 hours 57 minutes, and Arnold 12 hours 34 minutes. On the first EVA (March 19, 6 h 7 min), Swanson and Arnold bolted the S6 truss into place. They then connected the power and data cables that allowed station flight controls to command the segment into operation remotely.

For the second EVA (March 21, 6 h 30 min), Swanson teamed with Acaba. They prepared a work site for new batteries that were scheduled for delivery on STS-127. In addition, they installed a Global Positioning System antenna on the Pressurized Logistics Module attached to the Japanese Kibo laboratory. This would allow the Japanese automatic H-II Transfer Vehicle (HTV) to rendezvous with the station later in 2009. It also set the stage for future assembly tasks by station and Shuttle crews. A misaligned bracket proved too difficult to reposition during the installation of a cargo carrier attach system, so the two astronauts moved to other tasks, including image documentation of the station's radiators.

The third and final EVA (March 23, 6 h 27 min) saw Acaba and Arnold relocate one of the two CETA carts from one side of the Mobile Transporter to the other. Again, difficulty was encountered when they had trouble freeing a stuck mechanism. This would have enabled them to deploy a spare equipment platform, but the task had to be deferred to a future space walk. A similar task on another Payload Attach System was also deleted from the EVA by Mission Control. The astronauts did lubricate the end effecter capture system on the station's RMS. This task had proven effective during the STS-126 mission, preventing the snare from snagging and allowing it to return snugly into its groove inside the latching mechanism.

Inside the station, the crew replaced a failed unit on a system that converted urine to potable water. By March 24, 70 lb (8.38 gallons or 30.09 liters) of urine had been processed in the system, from which 15 lb (1.79 gallons or 0.39 liters) of reclaimed drinking water had been collected. Samples from the Water Recovery System were collected for analysis on Earth to determine if the purified water was suitable for the crew to drink before any was consumed on the station. Two loadmasters (Arnold and Phillips) were assigned the task of keeping track of the transfer of supplies and logistics across to the station and the unwanted gear, experiment results, and samples back into the orbiter.

On March 24, both the Shuttle and station crews (10 astronauts and cosmonauts) gathered in the Harmony Node on station to speak with U.S. President Barack Obama, Members of Congress and schoolchildren from the Washington, D.C. area. Discovery undocked later that day (the 10th day of the mission) after 7 days 22 hours 33 minutes of joint operations. As usual, the mission pilot (in this case, Antonelli) performed the undocking and fly-around maneuver around the station while the rest of the crew photographed the completed truss assembly, now with the final set of solar array wings fully deployed.

On March 26, Antonelli used the Shuttle RMS to hold the OBSS, enabling the lasers and cameras to scan the surface of the orbiter for any signs of damage to the thermal protection system. No such damage was found. The first landing opportunity was waived off due to gusty winds and clouds at the Shuttle Landing Facility at the Cape, but conditions improved enough to allow a successful landing in Florida during the next orbit, 90 minutes later.

Milestones

264th manned spaceflight
155th U.S. manned spaceflight
125th Shuttle mission
 36th flight of Discovery
 28th Shuttle ISS mission
 10th Discovery ISS mission
100th post-Challenger mission
Wakata became the first JAXA/Japanese resident ISS crew member

SOYUZ TMA-14

International designator	2009-015A
Launched	March 26, 2009
Launch site	Pad 1, Site 5, Baikonur Cosmodrome, Republic of Kazakhstan
Landed	October 11, 2009
Landing site	Near the town of Arkalyk, Republic of Kazakhstan
Launch vehicle	Soyuz-FG (serial number IO15000-027), Soyuz TMA (serial number 224)
Duration	198 da 16 h 42 min 22 s (ISS-19/20) 12 da 19 h 25 min 52 s (Simonyi)
Call sign	Altair
Objectives	ISS resident crew transport (18S), ISS-19/20 research program, visiting crew 16 program

Flight Crew

PADALKA, Gennady Ivanovich, 50, Russian Federation Air Force, RSA Soyuz TMA and ISS commander, third mission
Previous missions: Soyuz TM-28/Mir 26 (1998/1999), TMA-4/ISS-9 (2004)
BARRATT, Michael Reed, 49, Civilian, NASA Soyuz TMA and ISS flight engineer 1
SIMONYI, Charles, 60, civilian, U.S.A., space flight participant, second mission
Previous mission: TMA10/TMA9 (2007)

ISS resident crew (Shuttle) exchanges

WAKATA, Koichi, 45, civilian (Japanese), JAXA ISS flight engineer
KOPRA, Timothy Lennart, 46, U.S. Army, NASA ISS flight engineer
(up STS-127, down STS-128)
STOTT, Nicole Maria Passano, 46, civilian, NASA ISS flight engineer
(up STS-128, down STS-129)

Flight log

By the spring of 2009, the station was ready for an increase in the permanent crewing from three to six. It had been decided to overlap main crews between expeditions, to help ease the strain on the station's limited resources while the final Shuttle assembly missions were flown. The plan was to launch a main crew to the station aboard Soyuz ferry craft in two teams of three. One knock-on effect of this would be the serious restriction of the availability of spare Soyuz seats for fare-paying individuals (the space flight participants) for some time. Any available

The first two-flight space flight participant Charles Simonyi floats in the Harmony Node.

seats would now be filled by representatives of the ISS partners (NASA, RSA, CSA, ESA, and JAXA).

Overlapping these crews meant that when the first crew of three returned, there would be a period when only a three-person (skeleton) crew would be in residency on the station until the next trio arrived to restore the crew to six. For brief periods, the ISS could support three Soyuz craft with nine crew members, but not for prolonged periods. In practice, Crew "A" would be joined by Crew "B" to create Residency "X". Once Crew "A" departed, crew "B" would continue alone as Residency "Y" until they were joined by Crew "C". When Crew "B" returned, crew "C" would assume the role of Residency "Z", and so on. The crew of TMA-14 would complete the first move in this new system, along with the next crew on TMA-15, creating the residencies ISS-19 and ISS-20 (or ISS-19/20). The lead crew on any station expedition would be known as ISS "X" Prime.

Pioneering this change and in command of the TMA-14, as well as ISS-19/20, was veteran Mir and ISS cosmonaut Gennady Padalka. Accompanying him was NASA rookie flight engineer Michael Barratt and space flight participant Charles Simonyi on his second visit to the ISS. This was a first for a space flight participant. Simonyi had previously visited the station two years before and this time would complete a 13-day flight (at a reportedly higher price than his first visit) before returning with the ISS-18 cosmonauts in TMA-13. During his second residency, Simonyi conducted a small program of experiments which included photography, a radiation safety experiment, ham radio, and various symbolic activities. Some of these activities were in connection with ESA, the Hungarian Space Office (Simonyi being a Hungarian-born, naturalized U.S. citizen), and the Russian Space Agency. On April 8, 2009 Simonyi and the outgoing ISS-18 crew undocked from ISS and landed safely.

Padalka and Barratt officially joined Japanese astronaut Wakata (already on board the station) as resident crew members on April 2. This trio became ISS-19 and would also assume the lead role for ISS-20 from May until their departure in October. There were still two further Shuttle partial crew exchanges to come after Wakata (who would be followed by Timothy Kopra and then Nicole Stott), so three-person Soyuz crewing would not actually begin until the end of 2009. ISS-19 was an expedition in a period of transition at the station after a decade of assembly. It was also a clear demonstration of the program's increased capabilities in manpower research and the often overlooked level of ground support across the globe.

With all the main laboratory facilities up and running, the research programs could also now step up. The Russian program for ISS-19/20 would see 330 sessions for 42 experiments, including four new research studies. The program encompassed life sciences, geographical research, Earth resources, biotechnology, technical research, cosmic ray research, education and space technology studies, and contract activities. Most of this research would be conducted by Padalka (in preflight planning this amounted to 164 hours), while Barratt focused on U.S. segment experiments. During the ISS-20 phase of his mission, ISS-20/21 flight engineer Romanenko would complete a further planned 160 hours of Russian segment science. In addition, NASA reported 98 experiments in human research, technology development, observations of the Earth, educational activities, and biological and physical sciences. Of these, 39 were new investigations, with 28 others originating from Europe, 16 from Japan, and 5 from Canada. There were 10 ongoing investigations from earlier expeditions.

One of the investigations highlighted by the world's media were the studies in recycling urine into drinking water, clearly not one of the more glamorous aspects of a space explorer's role. After receiving the all clear from ground tests on reclaimed water samples on May 20, a milestone was reached on board the station when ISS-19 crew members drank reclaimed and purified water from the Water Recovery System.

On May 29, 2009, the much anticipated transition from a three to a six-person crew was finally realized with the docking of TMA-15. On board were Belgian Frank De Winne (ESA), Russian Roman Romanenko (son of Salyut 6/Mir cosmonaut Yuri Romanenko), and Canadian Robert Thirsk. For the first time, each of the primary participants (Russia, U.S.A., Europe, Canada, and Japan) were represented in the resident crewing. Long in planning, the symbolic nature of the first six-person crew being comprised of crew members from the major ISS partners was not lost on the international agencies or world's media.

The docking and transfer into the ISS of the new crew of TMA-15 signaled the official end of the ISS-19 residency. The three ISS-19 crew members remained on board, but now ISS-19 officially became ISS-20. It would remain so until shortly before the return of Padalka and Barrett in October, together with the final space flight participant who was scheduled to arrive on Soyuz TMA-16 with the ISS-20/21 crew.

The makeup of the expanded international resident team on station soon changed, however, with the exchange of crew members on Shuttle. In July, Wakata was replaced by Kopra during STS-127, which also delivered the next element of the Kibo laboratory. Kopra remained on station for just over a month until August, when STS-128 delivered his replacement, Nicole Stott. Stott in turn remained on board until November when she returned on STS-129. She was the final Shuttle-transported station resident crew member.

During the ISS-19 phase, no EVAs were accomplished by the resident crew. However, during ISS-20 Padalka and Barratt completed two short excursions totaling 5 hours 6 minutes wearing the new improved Orlan-MK suits. The first of these (June 5, 4 h 54 min) featured additional preparations for the arrival at the Zvezda Service Module of the Mini Research Module-2. During the EVA, Wakata remained inside Zvezda, allowing access to TMA-14 which was docked with the module's rear port. In the event of an emergency, the EVA crew could have proceeded to the Soyuz if they had been unable to reenter Pirs. Romanenko, De Winne, and Thirsk remained in the American segment close to TMA-15 on Zarya. Fortunately, all went well and these well-planned and practiced contingency procedures were not required.

The second activity, on June 10, was a 12 min intravehicular activity (IVA, crew activity while wearing a spacesuit inside an unpressurized spacecraft)—the first on station since 2001. During this IVA, Padalka and Barratt depressurized the Zvezda Node to relocate a conical docking cover over the zenith port so that the MRM-2 could dock there. Later, both Kopra and Stott would assist with Shuttle-based assembly EVAs from the Quest airlock during STS-127 and STS-128. These were not classed as part of the residency EVA program.

TMA-14 was moved on July 2 from the rear port of Zvezda to the recently vacated (by the departing Progress M-02M) port of Pirs. This freed up the Zvezda port for Progress M-67, which would be used to reboost the altitude of the complex. The other three crew members remained in the ISS during the Soyuz relocation operation, which took about 26 minutes, after which the Soyuz crew reentered the station. With half the crew remaining inside, partial shutdown of the station was no longer required, saving both crew time and valuable experiment operating time.

This operation was followed by the arrival of the STS-127 mission (July 15–31) and later STS-128 (August 29–September 12). Another milestone was the arrival of the first Japanese HTV transfer vehicle on September 18, which was grappled by the station's RMS. Its six tons of cargo would be transferred by means of the station's robotic arms later.

In late September, the two ISS-20/21 crew members (Maxim Suraev and Jeff Williams) arrived with Canadian Space Tourist Guy Laliberté aboard Soyuz TMA 16. For a short time, the ISS included eight resident and one visiting crew member. On October 9, the formal change-of-command ceremony took place, with

Padalka handing over the reins to DeWinne. On October 11, Padalka, Barratt, and Laliberté transferred to TMA-14, undocked, and landed a few hours later. This expedition had seen a major milestone achieved in ISS operations and a highly successful period of activity, focused more on science than construction.

ISS-19/20 had logged almost 199 days in flight, of which 197 had been aboard the space station. The ISS-19 formal residency (April 2–May 29) lasted 57 days, while the ISS-20 residency (May 29–October 9) had logged 133 days, totaling 190 days of combined station command time for ISS-19/20.

Milestones

265th manned space flight
107th Russian manned space flight
100th manned Soyuz flight
14th manned Soyuz TMA mission
18th ISS Soyuz mission (18S)
16th ISS Soyuz visiting mission
19/20th SS resident crew
First ISS IVA in Zvezda node since 2001
First flight of Japanese HTV
First ISS six-person residency (ISS-20)
Final planned three-person full resident crew (ISS-19)
First time representative from main ISS partners are represented on resident crew (NASA, RSA, ESA, CSA, JAXA)
Simonyi was the first space tourist to fly twice
Final crew transfers via Shuttle (Wakata, Kopra, Stott)
Padalka celebrates his 51st birthday in space (June 21)

STS-125

International designator	2009-025A
Launched	May 11, 2009
Launch site	Pad 39A, KSC, Florida, U.S.A.
Landed	May 24, 2009
Landing site	Runway 22, EAFB, California, U.S.A.
Launch vehicle	OV-104 Atlantis/ET-130/SRB BI-137/SSME: #1 2059, #2 2044, #3 2057
Duration	12 da 21 h 38 min 09 s
Call sign	Atlantis
Objective	HST service mission 4 (SM4)

Flight crew

ALTMAN, Scott Douglas, 49, Captain USN (Retd.), NASA commander, fourth mission
Previous missions: STS-90 (1998), STS-106 (2000), STS-109 (2002)
JOHNSON, Gregory Carl, 54, Captain USN (Retd.), NASA pilot
GOOD, Michael Timothy, 45, Colonel USAF, NASA mission specialist 1
McARTHUR, Katherine Megan, 37, civilian, NASA mission specialist 2
GRUNSFELD, John Mace, 49, civilian, NASA mission specialist 3, fifth mission
Previous missions: STS-67 (1995), STS-81 (1997), STS-103 (1999), STS-109 (2002)
MASSIMINO, Michael James, 46, civilian, NASA mission specialist 4, second mission
Previous mission: STS-109 (2002)
FEUSTEL, Andrew Jay, 43, civilian, NASA mission specialist 5

Flight log

STS-125 was the fifth and final servicing mission (SM) to the Hubble Space Telescope. It was also the sixth Shuttle mission devoted to the telescope's orbital operations since 1990. The mission, as originally planned, was canceled in the wake of the 2003 Columbia tragedy, but was reinstated following a public and scientific lobby to fly one more Hubble-related mission. For space station Shuttle missions following the loss of Columbia, a rescue vehicle was prepared (normally as part of the next mission's processing) and the station could be used as a safe haven until a rescue flight could be launched. As this was not a space station related mission and would be flying in a different orbit, the ISS was not an option. Instead, a second Shuttle was prepared on an adjoining pad as a potential rescue vehicle. When the Hubble flight was delayed and rescheduled, so too was

John Grunsfeld works on the Hubble Space Telescope on the first of five EVAs during the final servicing mission to the orbiting observatory.

the rescue mission amended accordingly. The Shuttle Endeavour served as the backup rescue vehicle in a mission designated STS-400 (with a four-person crew). It remained on Pad 39B while STS-125 was in space, but once Atlantis was cleared for Earth return, Endeavour reverted to ISS mission preparations.

Atlantis preparations began in the OPF on February 20, 2008. The orbiter was subsequently moved across to the VAB on August 22 that year and the mated stack rolled out to Pad 39A on September 4, 2008. The mission was originally scheduled for October 2008, but was changed to February 2009 when the system that transfers science data from the orbital observatory to Earth malfunctioned. The mission was postponed again when delivery of a second data-handling unit to the Cape was delayed. This resulted in the rescheduled launch date of May 2009.

Prior to the rendezvous with Hubble, a thorough inspection of the Shuttle's heat shield was performed using the RMS. The close-up imagery of the orbiter's surfaces was relayed to MCC for analysis by specialist teams on the ground. During this analysis, the Mission Management Team (MMT) noted an area of damage on the forward part of Atlantis where the wing blends into the fuselage. This was apparently minor damage but additional expert analysis was still conducted to make sure. It was subsequently decided by the MMT that the thermal covering of Atlantis was indeed safe for reentry at the end of the mission.

Atlantis was guided by Altman, assisted by Good and the rest of the crew, to within 50 ft (15.2 m) of Hubble on May 13. The telescope was grappled by the

RMS (controlled by McArthur) without incident and then maneuvered into a Flight Support System (FSS) maintenance platform in the orbiter payload bay. In addition to supporting Hubble, this platform would provide power for thermal control during the service period while aboard the orbiter.

Five EVAs, totaling 36 hours 56 minutes, were conducted, achieving all of the mission's objectives. Two of them went into the record books as the sixth and eighth longest EVAs in history. Two new instruments were installed on the telescope and a further two repaired, restoring them to operational life. The EVA crew also replaced gyroscopes and batteries as well as installing new thermal insulation panels for protection in orbit. Grunsfeld and Feustel logged three space walks each, totaling 20 hours 58 minutes, while Good and Massimino completed two EVAs each totaling 15 hours 58 minutes.

During the first EVA (May 14, 7 h 20 min), Grunsfeld and Feustel replaced the Science Instrument Command and Data Handling Unit (SIC&DHU) and installed the Wide Field Camera 3. Grunsfeld installed the soft capture mechanism for future (though not yet planned) service missions, which would be post-Shuttle retirement. Meanwhile Feustel installed two latches over center kits in order to ease the opening and closing of the large access doors on the telescope. On the second EVA (May 15, 7 h 56 min), Good and Massimino replaced three rate-sensing units which contained two gyroscopes each. Unfortunately, one of the replacement units would not fit, so a spare was used instead. The two astronauts also replaced a battery module from Bay 2 on the telescope.

The Corrective Optical Space Telescope Axial Replacement (COSTAR) was removed by Grunsfeld and Feustel on EVA 3 (May 16, 6 h 36 min). This had been installed during the first Hubble Service Mission in December 1993 to correct and refine the image generated from the telescope's faulty main mirror. As it was no longer required, it was replaced with the Cosmic Origins Spectrograph. This new device would allow Hubble to peer farther into the depths of the universe than ever before, both in the near and far-ultraviolet ranges. Their next task was to repair the advanced camera for surveys, which involved the removal of 32 screws from an access panel, replacing the camera's four circuit boards and installing a new power supply.

EVA-4 by Massimino and Good (May 17, 8 h 20 min) included repairs to the telescope's imaging spectrograph by replacing a power supply board. The astronauts experienced some difficulty with a handrail. This had to be removed before they could fit a fastener capture plate that was designed to retain over 100 screws during the removal of a cover plate. The astronauts found that a stripped bolt was preventing the handrail from coming free. They would receive guidance on how to overcome this problem from engineers at the Goddard Space Flight Center (GSFC). Massimino eventually carefully bent and broke the handrail free to allow installation of the capture plate. This episode raised some concerns in Mission Control over the possibility of the effort involved causing damage to the astronaut's pressure suit. All went well, though, and in fact was much easier than at first thought. However, the astronauts were unable to install a new outer blanket layer on the outside of Hubble Bay 8, so this task was delayed to the fifth and final EVA of the mission. This final EVA (May 18, 7 h 2 min) saw Feustel and Grunsfeld exchange a battery module from Bay 3 for a fresh one, as well as removing and replacing the H2 fine guidance sensor. They were then able to install the new outer blanket layer on three bays outside the telescope.

The result of all this activity was a telescope with six working complementary science instruments. This gave Hubble a capability far beyond what was envisioned when the facility was launched in 1990 and an extended operational life to at least 2014. After Hubble was released on May 19, again using the RMS, a final separation maneuver was made and the berthing mechanism which had held Hubble in the payload bay during the mission was stowed for landing. The crew completed a further RMS-aided examination of the orbiter heat shield to search for any new damage from orbital debris but fortunately no significant damage was discovered.

The following day (May 20), the crew stowed gear and checked the RCS and flight control surfaces for entry and landing. The crew became the first to testify live from space on May 21, during a U.S. Senate Hearing in which they addressed the Senate Operations Committee, Subcommittee on Commerce, Justice, Science and Related Agencies, chaired by Senator Barbara Mikulski (Democrat) of Maryland. She was the political driving force behind getting the STS-125 mission authorized. She and former U.S. Payload Specialist Senator Bill Nelson of Florida also talked to the crew.

Weather concerns and conditions at the Cape during May 22, 23, and 24 forced three consecutive landing waive-offs, requiring Atlantis to land at Edwards AFB, California on May 24, 2009. Just over a week later, the orbiter was ferried back to KSC on the Shuttle Carrier Aircraft, arriving back in Florida on June 2 after a two-day ferry flight. STS-125 had been a highly successful mission. Despite the apprehensions over flying an independent mission which could not visit the space station, the flight passed without major incident, ending a highly successful series of Hubble-related missions.

Milestones

266th manned space flight
156th U.S. manned space flight
126th Shuttle mission
 30th flight of Atlantis
  6th dedicated Shuttle HST mission
  5th HST servicing mission
Heaviest payload carried to Hubble on the Shuttle
Final "solo" Shuttle mission of program
Only post-Columbia-loss non-ISS Shuttle mission

SOYUZ TMA-15

International designator	2009-030A
Launched	May 27, 2009
Launch site	Pad 1, Site 5, Baikonur Cosmodrome, Republic of Kazakhstan
Landed	December 1, 2009
Landing site	57 km from the town of Arkalyk, Republic of Kazakhstan
Launch vehicle	Soyuz-FG (serial number Ю15000-030), Soyuz TMA (serial number 225)
Duration	187 da 20 h 41 min 38 s
Call sign	Parus ("Sail")
Objectives	ISS resident crew (ISS-20/21), transport (19S) to establish six-person crew capability

Flight crew

ROMANENKO, Roman Yuriyevich, 37, Russian Federation Air Force, RSA Soyuz TMA commander, ISS flight engineer
DE WINNE, Frank, 48, Belgian Air Force, ESA (Belgian), Soyuz/ISS flight engineer, second mission
Previous mission: Soyuz TMA-1 (2002)
THIRSK, Robert Brent, 55, civilian (Canadian), CSA Soyuz/ISS flight engineer, second mission
Previous mission: STS-78 (1996)

Flight log

It was with this mission that ISS crewing became a little more complicated to keep track of. On March 29, 2009, Soyuz TM-15 and its three crew members, representing Europe (Belgium), Russia, and Canada, docked to the station after a nominal 2-day flight from Baikonur. Already on board the station was the three ISS-19 crew members, representing Russia, Japan, and the U.S., marking a truly international complement. Unlike previous resident crew arrivals, the ISS-19 crew would not immediately return to Earth, but would remain on board as part of the first six-person resident crew, now designated ISS-20.

The TMA-15 flight was the first complete three-person ISS resident crew complement launched on a Russian launch vehicle and spacecraft since ISS-1 in 2000. The arrival of TMA-15 instigated the change in crew exchange protocol and represented a significant increase in resident crew time for science aboard the station. For the first time, a representative from each of the main ISS partners was part of the main crew.

The six-person Expedition 20 crew in "starburst" formation. Clockwise from right are Padalka, Thirsk, Wakata, Barratt, Romanenko, and De Winne.

At the time of docking with the station, Padalka, Barratt, and Wakata were in residence as ISS-19, and once the hatches were opened, the TMA-15 crew joined them as flight engineers and the six became ISS-20. On July 17, U.S. astronaut Kopra arrived on STS-127 (2J/A) to take over from JAXA astronaut Wakata, who returned at the end of that Shuttle mission. Barely six weeks later on August 31, Nicole Stott replaced Kopra during the STS-128 (17A) activities. Both Kopra and Stott served as flight engineers for the crew of ISS-20 and Stott remained as flight engineer for ISS-21/22 until she came home on STS-129 in November, shortly before the return of the core TMA-15 crew. These were the final Shuttle-delivered crew members. All station resident crew members for the foreseeable future would now be delivered by Russian Soyuz.

The ISS-20 experiment program continued the work conducted during ISS-19. With the Japanese and European laboratories recently installed, the opportunity for broadening the research had increased significantly. By the time this mission flew, most of the station hardware had been delivered by the Shuttle and reflected a change that was taking place. The ISS was evolving from an assembly site to a fully functioning scientific research facility. The Russians noted that during ISS-21/22, there would be 304 sessions on 47 experiments, while NASA reported that over 150 operating experiments would be completed under the station's new role as a U.S. National Laboratory.

Two new crew members for ISS-21/22 arrived aboard TMA-16 on October 2, together with space flight participant Guy Laliberté. He would return with the TMA-14 pair (Padalka and Barratt) on October 11, leaving behind the ISS-21 expedition which would last just six weeks until the TMA-15 crew came home. Officially, ISS-21 began on October 9 and lasted until November 25. During ISS-21 there were no EVAs, but the crew did see the arrival of Progress M-03M and the release of the Japanese HTV-1 with about 725 kg of unwanted equipment and rubbish for burn-up in the atmosphere. On November 12, Mini Research Module-2 (which was delivered by a modified Progress M vehicle) docked with the zenith (upper) port of Zvezda. With all this activity, the crew was certainly not struggling to keep themselves occupied.

STS-129 departed on November 25 carrying Stott home and, just five days later, the TMA-15 trio departed the station for a landing in the early hours of December 1 after a historic and highly successful mission. Following these departures and for the first time since July 2006, the station was down to a skeleton crew of just two until the next Soyuz launched just prior to Christmas 2010.

In a flight of 188 days, the TMA-15 crew had spent all but two of them on the space station. Of these, 133 were as members of the ISS-20 expedition and just 47 days as ISS-21. De Winne became the first European station commander (ISS-20) and Thirsk the first resident Canadian crew member. Future crewing would feature a rota of U.S.–Russian crew members and an allocation of Soyuz seats for Canadian, Japanese, and ESA crew members. NASA reported that the new expedition actually began with the undocking of the outgoing crew on the TMA (though the formal change of command actually took place a few days earlier). In fact, as the crew exchanged command, the outgoing crew became known as the "non-prime crew" for their final few days prior to coming home, while their replacements became the "prime crew", which just seemed to add confusion to those following the formal trail of command and assignment.

Operating a rotating six-person crew and only flying on the Soyuz TMA spacecraft meant that there would be no free seats for fare-paying passengers. Unless separate funds for a complete Soyuz vehicle could be found by those who arranged SFP/tourist flights, it would be very hard for anyone other than a professional astronaut or cosmonaut to fulfill their dream of flying in space for some considerable time.

Milestones

267th manned space flight
108th Russian manned space flight
101st manned Soyuz flight
 15th manned Soyuz TMA mission
 19th ISS Soyuz mission (19S)
20/21st ISS resident crew
Romanenko is the son of cosmonaut Yuri Romanenko (selected 1970), who
flew on STS-26 (1977), Soyuz 38 (1980), and Soyuz TM2 (1987)
Final Shuttle-launched ISS crew members (Kopra and Stott)

STS-127

International designator	2009-038A
Launched	July 15, 2009
Launch site	Pad 39, KSC, Florida, U.S.A.
Landed	July 31, 2009
Landing site	Runway 15, Shuttle Landing Facility, KSC, Florida, U.S.A.
Launch vehicle	OV-105 Endeavour/ET-131/SRB BI-138/SSME: #1 2045, #2 2060, #3 2054
Duration	15 da 16 h 44 min 57 s (STS-127 crew)
	137 da 15 h 04 min 23 s (Wakata)
Call sign	Endeavour
Objectives	ISS-2J/A (JEM EF, ELM-ES, ICC-VLD), ISS resident crew exchange mission

Flight crew

POLANSKY, Mark Lewis, 53, civilian, NASA commander, third mission
Previous missions: STS-98 (2000), STS-116 (2006)
HURLEY, Douglas Gerald, 42, USMC, NASA pilot
WOLF, David Alan, 52, civilian, NASA mission specialist 1, fourth mission
Previous missions: STS-58 (1993), STS-86/89/Mir (1997), STS-112 (2002)
CASSIDY, Christopher John, 39, USN, NASA mission specialist 2
PAYETTE, Julie, 45, CSA, (Canadian) mission specialist 3, second mission
Previous mission: STS-96 (1999)
MARSHBURN, Thomas Henry, 48, civilian, NASA mission specialist 4

ISS resident crew exchange
KOPRA, Timothy Lennart, 45, U.S. Army, NASA mission specialist 5, ISS flight engineer (up only)
WAKATA, Koichi, 45, Japan, JAXA mission specialist 5, ISS flight engineer (down only), third mission
Previous missions: STS-72 (1996), STS-92 (2000)

Flight log

The 29th Shuttle mission to ISS featured a range of robotic operations, most of which focused around the delivery of elements of the Japanese Kibo laboratory module. In addition, U.S. NASA astronaut Tim Kopra was the latest resident crew member to arrive via the Shuttle. He replaced outgoing Japanese (JAXA) astronaut Koichi Wakata, who had been on the station since March.

Japanese Kibo Experiment Module and Exposed Facility.

Final mission preparations began with Endeavour being taken into the OPF on December 13, 2008. The orbiter was then moved across to the VAB on April 10, 2009, where it was mated with the twin SRB and ET before being rolled out to Pad 39B on April 17. This preparation was for its role as a "rescue" vehicle for the forthcoming STS-125 Hubble Service Mission. In the event, this requirement was not called upon after the STS-125 mission proceeded smoothly. On May 31, the STS-127 stack was relocated across to Pad 39A for a planned launch on June 13.

On June 12, a hydrogen leak at the Ground Umbilical Carrier Plate (GUCP) during tanking caused the June 13 attempt to be scrubbed. The seal on the 17 in. (43.18 cm) disconnect valve was replaced and the launch rescheduled to June 17. This attempt was also scrubbed in the early hours of the planned launch day when a similar type of hydrogen gas leak occurred at the GUCP. Troubleshooting of the vent valve took an hour and when it became clear that the problem could not be easily resolved in the current launch window, the launch was eventually rescheduled for July 11. This next attempt was also postponed 24 hours to allow technicians more time to evaluate lightning strikes at the launchpad, which had occurred during a thunderstorm on July 10. It was determined from sensors that 11 lightning strikes had occurred within 0.35 miles (0.56 km) of the pad, which was inside the pad threshold. Launch was rescheduled again for July 12, but once again was scrubbed, this time at $T - 11$ minutes before scheduled launch time due to weather concerns near the Shuttle Landing Facility (SLF) that violated rules for landing. A fifth launch attempt on July 13 resulted in another postponement due to lightning and thunderstorms within the 20 nautical mile circle around the pad. The sixth attempt, rescheduled for July 15, finally occurred without incident.

During the standard two-day flight to the space station, the crew completed the customary survey of Endeavour's thermal protection system using the RMS boom assembly. The pressure suits to be used on the EVAs were also prepared and checked. The now regular backflip maneuver was completed just prior to docking. Shortly after entering the space station, Kopra joined the ISS resident crew and Wakata transferred to the Shuttle crew, ending his formal ISS crew residency after 122 days.

There were five EVAs conducted as part of the STS-127 mission, totaling 30 hours 30 minutes. Four astronauts (including Kopra) participated in the excursions, teaming up in pairs. Marshburn logged 18 hours 59 minutes on three EVAs, Wolf logged 18 hours 24 minutes on his three space walks, and Cassidy amassed 18 hours 5 minutes on his three excursions. Kopra logged 5 hours 32 minutes on his single EVA as part of the STS-127 program.

During the first EVA (July 18, 5 h 32 min), Wolf and Kopra managed to complete all of their primary tasks. These included preparations for the berthing mechanism on Kibo and on the Japanese Exposed Facility, which was transferred to the station. They also deployed an unmanned cargo carrier attachment system on the P3 truss, which had failed to unfurl properly during STS-119 the previous March.

Robotic arm operations continued during and in between the space walks, supporting and supplementing EVA activities. Following the first EVA, the Shuttle RMS and the station robotic arm were used to latch the JEF to Kibo's laboratory. The installation was viewed by the Kibo RMS. Further robotic activities included the installation of the Integrated Cargo Carrier–Vertical Light Deployable (ICC-VLD) cargo pallet. This was located on the port side of the station's Mobile Base System (MBS).

During the second EVA (July 20, 6 h 53 min), Wolf and Marshburn removed a Ku-band space-to-ground antenna, a pump module, and a liner drive unit from the Integrated Cargo Carrier. The two astronauts then attached these items to a storage platform on the P3 truss. Next, Marshburn mounted a grapple bar onto an ammonia tank assembly, which the STS-128 crew would later move with the help of the RMS. He also attached two external power connector insulation sleeves to the Station-to-Shuttle Power Transfer System. The planned installation of a video camera was deferred from this EVA to a later excursion. This EVA had taken place on the 40th anniversary of the famous first lunar surface activity (moonwalk) conducted by Neil Armstrong and Buzz Aldrin during the Apollo 11 mission. Both the station and Shuttle crews honored the legacy of that historic mission and event.

Following the second EVA, Polansky and Payette used the Shuttle RMS to pass the Japanese Logistics Module's Exposed Facility from Endeavour to Canadarm2 on the station. Canadarm 2 was operated by Hurley and Wakata, who then attached it to the Kibo laboratory. Astronauts Hurley and Payette then used the station's robotic arm to move the ICC and to secure the batteries in preparation for the P6 truss battery swap.

The third EVA (July 22, 5 h 59 min) saw Wolf and Cassidy remove insulation covers from the Kibo Laboratory, as well as preparing the JES payloads for transport to the Exposed Facility the following day. Two batteries were also installed but the EVA was curtailed early when CO levels in Cassidy's suit were found to have increased more than expected. The next day, several astronauts took turns to use the Japanese RMS for the first time to move equipment from a Japanese Payload Carrier to the Exposed Facility on the exterior of Kibo. The initial movement of this new arm was faster than expected, so the arm was transitioned to a Slower Manual Mode. The three installed experiments on the facility were the Monitor of All-sky X-ray Image (MAXI), the Inter-orbit Communications System and, the Space Environment Data Acquisition Equipment–Attached Payload.

During the fourth EVA (July 24, 7 h 12 min), Cassidy and Marshburn installed the remaining four batteries on the P6 truss. Four of the older batteries were stored on the ICC for the return to Earth. The completion of robotic work would include the transfer of the ICC back to Endeavour's payload bay by means of the station and Shuttle robotic systems.

The fifth and final EVA (July 27, 4 h 54 min) saw Marshburn and Cassidy install video cameras on the front and back of the Japanese Exposed Facility to assist with the forthcoming rendezvous and berthing of the H-II Transfer Vehicle scheduled to arrive that coming September. The EVA astronauts also secured multilayered insulation around the Special Purpose Dexterous Manipulator (Dextre), split out power channels for the station's CMGs (Control Moment Gyros), tied down cables, and installed handrails and a portable foot restraint to aid future space walks.

Alongside the five EVAs and associated robotic activities, work continued inside the station. Crew members and flight controllers spent their time trouble-shooting the failure of the Waste and Hygiene Compartment (the toilet in the Destiny Laboratory Module). The Carbon Dioxide Removal Assembly on the station also tripped a circuit breaker on July 28 and the ground team switched to manual operation of the backup heater.

Wakata had logged 133 days on the station by the time Endeavour undocked on July 28. After 10 days 23 hours 41 minutes of joint operations, the undocking was followed by the traditional fly-around of the complex, controlled by pilot Hurley, before moving away from the station.

After separation from the station, work was not over for the crew. In addition to a checkout of the Flight Control System and maneuvering engines and stowing of gear, the crew deployed two pairs of small satellites from the Endeavour payload bay prior to the landing at Kennedy Space Center. These were the Dual RF Astrodynamic GPS Orbital Navigation Satellite (DRAGONSat) and the dual Atmospheric Neutral Density Equipment 2 (ANDE-2) satellite. The subsatellites were designed and built by students at the University of Texas in Austin and the Texas A&M University in College stations. The ANDE-2 microsatellites would measure the density and composition of the rarefied atmosphere 200 miles (321.8 km) above the surface of Earth. DRAGONSat de-orbited on March 17, 2010, while the pair of ANDE satellites reentered the atmosphere on March 29 and August 18, 2010, respectively.

Milestones

268th manned space flight
157th U.S. manned space flight
127th Shuttle mission
 23rd Endeavour mission
 10th Endeavour ISS mission
 29th Shuttle ISS mission
  2nd Shuttle mission to feature five space walks

STS-128

International designator	2009-045A
Launched	August 28, 2009
Launch site	Pad 39A, KSC, Florida, U.S.A.
Landed	September 12, 2009
Landing site	Runway 22, Dryden Flight Research Center, EAFB, California, U.S.A.
Launch vehicle	OV-103 Discovery/ET-132/SRB BI-139/SSME: #1 2052, #2 2051, #3 2047
Duration	13 da 20 h 53 min 43 s (STS-128 crew)
	58 da 02 h 50 min 10 s (Kopra)
Call sign	Discovery
Objectives	ISS-17A (MPLM, LMC, ATA), ISS resident crew exchange

Flight crew

STURCKOW, Frederick Wilford, 48, USMC, NASA commander, fourth mission
Previous missions: STS-88 (1998), STS-105 (2001), STS-117 (2007)
FORD, Kevin Anthony, 49, USAF (Retd.), NASA pilot
FORRESTER, Patrick Graham, 52, USAF (Retd.), NASA mission specialist 1, third mission
Previous missions: STS-105 (2001), STS-117 (2007)
HERNANDEZ, José Moreno, 47, civilian, mission specialist 2/flight engineer
OLIVAS, John Daniel, 44, civilian, NASA mission specialist 3, second mission
Previous mission: STS-117 (2007)
FUGLESANG, Arne Christer, 52, civilian (Swedish), ESA mission specialist 4, second mission
Previous mission: STS-116 (2006) ISS resident crew exchange
STOTT, Nicole Maria Passano, 46, civilian, NASA mission specialist 5, ISS-20 flight engineer (up only)
KOPRA, Timothy Lennart, 45, U.S.A., NASA mission specialist 5, ISS-20 flight engineer (down only)

Flight log

According to reports from NASA, this was the mission which marked the beginning of the station's transition from assembly to a continuous scientific research facility. The Leonardo MPLM was packed with supplies, stores, and new equipment to continue outfitting of the station. It was also the final Shuttle

The final Shuttle-delivered ISS resident crew member Nicole Stott in the Quest Airlock.

mission to deliver a new resident station crew member. Nicolle Stott (NASA) would replace Tim Kopra after his 58 days aboard the orbital complex.

Preparations for the mission saw orbiter Discovery return to the OPF on March 28, 2009. It was rolled over to the VAB on July 26 for mating with the SRB and ET, with rollout to the pad occurring on August 4. The original August 25 launch attempt was scrubbed due to adverse weather conditions and the second attempt on August 26 was also postponed (during the refueling of the ET) following indications that a valve on the Main Propulsion System (MPS) had failed to perform as expected. A third attempt (August 27) was also postponed 24 hours to allow engineers more time to resolve an issue with a valve in the vehicle's propulsion system. The fourth launch attempt, on August 28, was accomplished successfully and completed without any major issues.

Prior to docking with the station, the obligatory inspection routines were completed to check the integrity of the heat shield. Following analysis of the imagery on the ground, the thermal protection system was deemed secure for entry and landing.

After the docking and opening of internal hatches, Kopra became part of the Shuttle crew while Stott transferred to the resident crew. Kopra had logged 45 days on board the station as an official resident crew member. The crew also prepared and moved the OBSS across to the station's Canadarm2 to create additional room for maneuvering the Leonardo MPLM from the cargo bay to the Harmony

Node on station the following day. The rest of the day was spent preparing the pressurized cargo module for the transfer of supplies and logistics, an operation which would take the next six days to complete. Olivas, Hernández, and Stott also prepared the tools for the planned EVAs, placing them in the station's Quest airlock.

Three EVAs were completed on this mission, focused upon replacing experiments outside the ESA Columbus Laboratory and installing a new ammonia storage tank. Nicole Stott participated in the first EVA (September 1, 6 h 35 min) with Olivas, during which they removed the depleted ammonia tank on the P1 (Port 1) segment of the truss. The ammonia in these tanks is used to cool the station and expel the heat generated by the resident crew and onboard systems.

The next task was to retrieve two science experiments, EUTEF and MISSE-6, from the ESA Columbus Laboratory. The European Technology Exposure Facility (EuTEF) held a number of different experiments which collected data on the space environment. The sixth Materials International Space Station Experiment (MISSE-6) was housed in two suitcase-sized containers and was used to evaluate the effects of the environment on various materials and coating samples. Olivas reported seeing what he described as MMOD (micrometeoroid and orbital debris) "hits" on a station toolbox and the Quest airlock. He took photos of the area for later evaluation on the ground to determine whether the damage was a cause for concern, although hits for MMOD are not unexpected. For 30 minutes during the EVA, MCC Houston did not have communication with either the station or the Shuttle due to bad weather in Guam that effected the TDRSS reception.

On the second EVA (September 3, 6 h 39 min), Olivas and Fuglesang installed a new ammonia tank on the P1 truss segment and bolted the removed empty assembly into the orbiter's payload bay. Protective lens covers were installed on the station RMS cameras, which would shield them from contamination when the arm was used to dock the Japanese H-11 Transfer Module later that month (September). The two astronauts also installed a portable foot restraint on the station's truss system for use during upcoming missions. The astronauts discovered that the heater cables on the exterior of PMA-3 appeared to be in an incorrect configuration to extend sufficiently for planned relocation and as a result this task had to be deferred.

The final EVA of this mission (September 5, 7 h 1 min) saw Olivas and Fuglesang set up a payload attachment system on the station truss, which would be activated on the next mission. In addition, they replaced a rate gyro assembly and remote power control module, installed two GPS antennas, and removed a slide wire on the Unity Node. The connection of two avionics cables could not be completed on this EVA and had to be rescheduled for a future excursion. These would eventually be connected to Tranquility (Node 3), the final main module that would be delivered to the U.S. segment. The cables were wrapped in insulation, but it was found that one of the connectors on one of the cables would not mate. At the end of the EVA Fuglesang's helmet-mounted video camera and headlight system became detached, so Olivas helped Fuglesang to connect a tether to the system and plans were devised to inspect its latches after they were back inside the spacecraft.

In total, the three space walks accumulated 20 hours 15 minutes. Olivas logged 20 hours 15 minutes in three EVAs, Fuglesang 13 hours 40 minutes on two space walks, and Stott 6 hours 35 minutes on her single venture outside.

On September 2, the Shuttle and station crews transferred the Fluids Integrated Rack, Materials Science Research Rack-1, and Minus Eighty-Degree Laboratory Freezer-2 from Leonardo and installed them in the U.S. Laboratory Destiny. In addition, ISS resident crew member Mike Barrett installed and outfitted the third and fourth planned NASA crew quarters facilities. Inside the station, crew members replaced one of the 16 common berthing mechanism bolts used to secure the Leonardo cargo carrier to the station, as it had not operated correctly earlier in the mission. An old oxygen generation assembly water filter was also opened and inspected. It had been replaced prior to the arrival of the Shuttle. It was found that the filter was 70–80% blocked and the inspection increased confidence that the replacement filter had returned the system to full functionality.

During the docked phase, NASA reported that approximately 7.5 tons of equipment and supplies had been transferred out of Leonardo and 2,400 lb (1,088.64 kg) of discards and experiment results placed back inside the module for return to Earth, all under the supervision of loadmaster Fuglesang. The Shuttle middeck held a further 8,860 lb (4,018.89 kg) of returned items. One of these was the Disney astronaut character Buzz Lightyear from the *Toy Story* films. This was part of the NASA "Toys in Space" project designed to encourage students to pursue studies in science, technology, engineering, and mathematics (STEM). The Disney "Space Ranger" had logged 15 months on board the station. MPLM Leonardo was returned to the payload bay on September 7.

Discovery undocked on September 8 after 8 days 18 hours 32 minutes of joint operations. Following the normal fly-around of the station and pre-landing checks, the orbiter was prepared for landing. Weather concerns prevented a return to the Cape on September 10 and again the next day. The landing was completed on Runway 22 at Edwards AFB in California. On his mission, Kopra had logged a total of 58 days in flight, of which 8 were aboard the space station as a Shuttle crew member and 5 were spent aboard the Shuttle either flying to or coming home from the station. As well as the normal postflight activities for the "human" space crew, the "other" space hero of the mission—Buzz Lightyear—was celebrated with a tickertape parade alongside his space station crew mates and Apollo 11 astronaut Buzz Aldrin at Walt Disney World in Florida on October 2, 2009.

Milestones

269th manned space flight
158th U.S. manned spaceflight
128th Shuttle mission
 37th Discovery mission
 11th Discovery ISS mission
 30th ISS Shuttle assembly mission

SOYUZ TMA-16

International designator	2009-053A
Launched	September 30, 2009
Launch Site	Pad 1, Site 5, Baikonur Cosmodrome, Republic of Kazakhstan
Landed	March 18, 2010
Landing site	60 km north of Arkalyk, Republic of Kazakhstan
Launch vehicle	Soyuz-FG (serial number bl5000-027), Soyuz TMA (serial number 226)
Duration	169 da 04 h 09 min 37 s (Surayev, Williams) 10 da 21 h 16 min 55 s (Laliberté, landed in TMA-14)
Call sign	Tsefay ("Cephus")
Objective	ISS resident crew transport (20S), ISS resident crew 20/21, visiting crew 17 program

Flight crew

SURAYEV, Maxim Viktorovich, 37, Russian Federation Air Force, RSA, TMA commander, ISS flight engineer
WILLIAMS, Jeffrey Nels, 51, U.S.A., NASA TMA flight engineer, ISS-21 flight engineer, ISS-22 commander, third mission
Previous missions: STS-101 (2000), Soyuz TMA-8/ISS-13 (2006)
LALIBERTÉ, Guy, 50, civilian (Canadian), space flight participant

Flight log

With the decision to end Shuttle flights and the increase to a permanent crew of six on the ISS due to the expanded science activity, the availability of Soyuz seats for sale to fare-paying tourists would not extend beyond 2009. With up to six permanent resident crew members on board the station, two Soyuz would always have to be docked at the station in order to serve as Crew Rescue Vehicles. In addition to Russian and American crew members, the remaining crew places would be awarded to European, Japanese, and Canadian astronauts under agreement, rather than for more commercial deals with space flight participants. With three ISS-20/21 and the final Shuttle-delivered crew member already aboard, the two ISS-21/22 crew members were launched as what were expected to be the last space flight participants for some time. The cost of sending a third Soyuz spacecraft to the station, even for a short time, would not be covered by the price of a seat for a single spaceflight participant.

Canadian Guy Laliberté became the lucky passenger on TMA-16, flying as SFP/VC17 on an 11-day mission. Laliberté was the billionaire founder of the Cirque du Soleil Company. His science program included several life science and

The last scheduled space flight participant Canadian Guy Laliberté (center) waves farewell with fellow Soyuz TMA-16 crew members Jeff Williams (top) and Maxim Surayev from the bottom of the launchpad prior to launch from Baikonur. Photo credit: NASA/Bill Ingalls.

public outreach activities during his week aboard the ISS. This included a 2 h long TV session on October 9, designed to highlight both the shortage of clean water across the globe and his work in conservation through his non-profit ONE DROP Foundation. Following a week of activities, Laliberté returned to Earth aboard TMA-14 with the ISS-19/20 cosmonauts Padalka and Barratt on October 11.

The docking of the TMA-16 craft to the aft port of Zvezda on October 2 meant that, for the first time, three Soyuz craft were attached to the facility and nine people were on board as residents, if only for a week! With the return of Padalka and Barratt, Williams and Surayev joined De Winne, Thirsk, Romanenko, and NASA astronaut Stott as the ISS-21 crew members. Both newcomers served as flight engineers on the six-person crew until December 1, when the De Winne trio departed on Soyuz TMA-15. Stott had already left station by then aboard STS-129, the last ISS crew member to launch or land on the Shuttle. For a short time, the station crew became a two-person caretaker crew until the arrival of TMA-17 in late December, increasing the resident crew back up to five.

On November 25, the official change-of-command ceremony took place, with Williams and Surayev becoming the core prime crew of ISS-22 (Williams taking the command position and Surayev as flight engineer). Continuing the science on the Russian segment of the station, the ISS-21 investigation program included 304 sessions for 47 experiments, of which only four were totally new. Over 60 hours were arranged for the crew in the ISS-21 phase and over 148 hours in the ISS-22 phase. Over in the U.S. segment, NASA reported that 150 operating experiments were on board the station, conducting research in human research, biological and physical sciences, development of technology, observations of Earth, and educational activities.

After three weeks with a two-man skeleton crew, the TMA-17 docking on December 23 brought three new crew members to the expedition: Russian cosmonaut Kotov, NASA astronaut Creamer, and JAXA astronaut Noguchi. With the arrival of Noguchi, an expanded Japanese science program was once again possible. Even with a larger crew though, routine maintenance and housekeeping would still take up a lot of crew time.

The first few weeks on station for the TMA-16 crew were quite busy. In late October 2009, HTV-1 was separated from the station using Canadarm2. The unmanned resupply craft, now full of unwanted gear and trash, burned up on reentry on November 1. On November 12, the Progress M-MRM-2 module docked with the zenith port of Zvezda. The following day, the internal hatches were opened and Surayev entered the module for an inspection and to take air samples. This module, called Poisk ("Search"), featured a new Soyuz/Progress docking port and a second EVA airlock for Russian-based EVAs, as well as additional, if limited, storage volume.

On January 14, 2010, Surayev and Kotov conducted a 5 h 44 min EVA to check the exterior of Poisk and the joint docking seals. Soyuz TMA-16 was relocated by Kotov and Williams on January 21, moving the spacecraft from the aft end of Zvezda to Poisk. The flyover took just 19 minutes, with the TMA-17 crew watching and photographing from inside the station as the operation took place. The next Progress (M-04M) arrived at the station on February 5, docking to the aft port of Zvezda. This was the 36th docking of an unmanned supply vehicle from the Russian Progress series and marked the first time on station that four Russian spacecraft (two Soyuz and two Progress vehicles) were docked.

On February 10, after five busy days of cargo transfers by the station crew, Shuttle Endeavour (STS-130) docked with its crew of six astronauts, delivering the ESA-built Cupola and Node 3 Tranquility (see STS-130 entry). While docked, the crews transferred over 1,313 lb (595.57 kg) of supplies across to the station. After the departure of STS-130 on February 20, the TMA-16 crew of Surayev and Williams started to prepare for their return to Earth while assisting the next expedition team to assume command. On March 17, 2010, Jeff Williams formally handed over command of ISS to Kotov. The next day, TMA-16 undocked and followed a normal entry and landing a few hours later. For the record books, the TMA-16 pair had been on board the station for all but two of their 169-day space flight, with their time split between three residencies. They had worked for just seven days as part of the ISS-20 phase, then a further 47 days as flight engineers for ISS-21, before assuming the role of prime ISS crew for ISS-22, which lasted 112 days.

Milestones

270th manned space flight
109th Russian manned space flight
102nd manned Soyuz flight
16th manned Soyuz TMA mission
20th ISS Soyuz mission (20S)
17th ISS Soyuz visiting mission
21/22nd ISS resident crew
First triple docking of Soyuz space craft at ISS: Soyuz TMA-14, TMA 15, TMA-16 featuring nine crew members
First quadruple docking of Russian spacecraft at ISS: Soyuz TMA-15, TMA-16, Progress M-03M, M-04M
7th and last scheduled space flight participant visiting mission
17th and last scheduled Soyuz visiting mission program

STS-129

International designator	2009-62A
Launched	November 16, 2009
Launch site	Pad 39A, KSC, Florida, U.S.A.
Landed	November 27, 2009
Landing site	Runway 33, Shuttle Landing Facility, KSC, Florida, U.S.A.
Launch vehicle	OV-104 Atlantis/ET-133/SRB BI-140/SSME: #1 2048, #2 2044, #3 2058
Duration	10 da 19 h 16 min 13 s (STS-129 crew)
	90 da 10 h 44 min 43 s (Stott)
Call sign	Atlantis
Objectives	ISS-ULF3, Express Logistics Carriers (ELC1 and ELC2)

Flight crew

HOBAUGH, Charles Owen, 48, USMC, NASA commander, NASA, third mission
Previous missions: STS-104 (2001), STS-118 (2007)
WILMORE, Barry Eugene, 46, USN, NASA pilot, NASA
MELVIN, Leland Deems, 45, civilian, NASA mission specialist 1, NASA, second mission
Previous mission: STS-122 (2008)
BRESNIK, Randolph James, 42, USMC, NASA mission specialist 2, NASA
FOREMAN, Michael James, 52, USN (Retd.), NASA mission specialist 3, NASA, second mission
Previous mission: STS-123 (2008)
SATCHER Jr., Robert Lee, 44, civilian, NASA mission specialist 4, NASA

ISS resident crew exchange
STOTT, Nicole Maria Passano, 46, civilian, NASA mission specialist 5, ISS flight engineer (down only)

Flight log

At the time of this mission, there were only six flights of the Space Shuttle left on the manifest. With 86% of station assembly complete, the majority of these remaining missions would focus on the delivery of spares and logistics, as well as the removal of unwanted items of hardware and trash and the return of scientific samples. In addition to addressing the delivery of new supplies, STS-129 also would return the final Shuttle-transported space station resident crew member, Nicole Stott, who already knew she was already assigned to a second flight on the

A crew briefing during FD2 activities on the middeck of Space Shuttle Atlantis.

Shuttle in 2011. Originally, it was planned that Stott would return on a Soyuz, replacing Thirsk who was scheduled to come home on a Shuttle. But delays meant they swapped seats, with Thirsk returning on Soyuz and Stott on the Shuttle instead.

Atlantis was rolled from the OPF to the VAB on October 6, 2009 for mating with the ET and twin SRBs. After a 24 h delay due to an issue with the transfer crane in the VAB, the stack was rolled to the pad on October 14, with only minor issues featuring in a relatively smooth processing schedule and countdown. The payload was moved to the pad on October 29 and installed in the payload bay on November 4.

The launch and approach to station went according to the flight plan, with docking occurring on November 18. Following the hatch opening, the two crews greeted each other in the Harmony Module. With six Shuttle crew and four station crew members, ISS was again quite busy. Shortly after the hatches were opened and celebrations completed, Nicole Stott formally ended her residency on the station by officially joining the STS-129 crew.

Just 90 minutes after the hatches had been opened, the Shuttle RMS was grappling the first of two Express Logistics Carriers (ELC-1), handing it off to Canadarm2 on the station. ELC-1 was then plugged into its new location on the Earth-facing side of the ISS port truss. The ELCs were new platforms designed to support large items of hardware and spares. Subsequent Shuttle flights would add more hardware to the units. This first one had a mass of 6,396 kg (14,100 lb).

Three EVAs were completed, one each by Forman, Satcher and Bresnik, working in pairs with the third acting as IV crew member. This would total 36 hours 15 minutes of experience between the three astronauts. The first EVA (November 19, 6 h 37 min) by Foreman and Satcher featured the installation of a range of hardware and spares and the relocation of a number of items. They worked so efficiently that they found themselves two hours in front of the timeline and were able to perform a number of get-ahead tasks in preparation for EVA 2.

The second excursion outside was performed by Foreman and Bresnik, with Satcher as IV. During this EVA (November 21, 6 h 8 min) all tasks were again completed early, allowing further get-ahead tasks to be performed. This included relocating of the second ELC unit, which was loaded with spare parts and had a mass of 6,136 kg (13,530 lb).

The final EVA (November 23, 5 h 42 min) was performed by Satcher and Bresnik with Foreman taking his turn as IV crew member. The majority of this EVA, as with the earlier two, was taken up with installing new and spare items of hardware and equipment on the exterior of the station. The items relocated by the astronauts across the three EVAs included a spare antenna, an ammonia lines bracket, a bracket on the Columbus laboratory and an additional ham radio antenna. They also installed an antenna on the truss for wireless helmet camera video during future EVAs and relocated the measuring unit which reveals electrical potential around the station. Other tasks included deploying a cargo bracket on the truss, installing a new oxygen tank on the Quest airlock and deploying the next in the series of Material Experiment packing (MISSE 7A&B). Finally, they completed work on the heater cables in advance of the arrival of the Tranquility Node early in 2010.

The three space walks had logged 18 hours 27 minutes. Foreman accumulated 12 hours 45 minutes on his two EVAs, Satcher logged 12 hours 19 minutes in his pair of excursions, while Bresnik logged 11 hours 50 minutes in his two space walks. Shortly after ending his second space walk Bresnik received news of an addition to his family with the birth of his daughter.

While work continued outside, the astronauts transferred over a ton of supplies and logistics into the station and a ton of unwanted equipment, trash, and samples back into Atlantis for return to Earth. With future station operations in mind, the astronauts also armed the Commercial Orbital Transportation Services (COTS) UHF Communication Unit. This was integrated on the station in preparation for forthcoming commercial resupply flights to the ISS by Space Exploration Technologies (SpaceX). This new unit would enable communications between the station and the SpaceX Dragon spacecraft during approaches to the station.

Undocking from the station, with Stott now aboard, occurred on November 25, after 6 days 17 hours and 2 minutes of being docked with the complex. The following day, the crew celebrated the U.S. Thanksgiving holiday aboard Atlantis. With 14 tons of supplies now aboard the station, the mission was deemed a complete success and with Stott aboard the orbiter, the series of Shuttle-delivered crew members that began with Germany's Thomas Reiter (ESA) in July 2006 was now complete. In all, 11 resident crew members—two (ESA) European, one (JAXA) Japanese, and eight (NASA) American astronauts—had been delivered and returned on the Shuttle over a period of three years and five months.

Atlantis landed after an 11-day mission. Stott completed a mission of 91 days in space, 86 of which were on station and 79 as resident crew member during ISS-20/21.

Milestones

271st manned space flight
159th U.S. manned spaceflight
129th Shuttle flight
 31st Atlantis mission
 31st ISS Shuttle mission
First live Tweet up mission from KSC during launch
Second flight of two African Americans on crew (Melvin and Satcher)
Fewest problems reported in processing (54) since STS-125.
Final Shuttle crew rotation mission (Stott landing only)
Bresnik's daughter born November 21, shortly after his second EVA
Stott celebrated her 47th birthday (November 19)

SOYUZ TMA-17

International designator	2009-074A
Launched	December 21, 2009 (Moscow time)
Launch site	Pad 1, Site 5, Baikonur Cosmodrome, Republic of Kazakhstan
Landed	June 2, 2010
Landing site	East of the town of Dzhezkazgan, Republic of Kazakhstan
Launch vehicle	Soyuz-FG (serial number bl5000-031), Soyuz TMA (serial number 227)
Duration	163 da 5 h 32 min 32 s
Call sign	Pulsar
Objectives	ISS resident crew transport (21S), ISS expedition crew 22/23

Flight Crew

KOTOV, Oleg Valeriyevich, 44, Russian Federation Air Force, RSA TMA commander, ISS-22 flight engineer/ISS-23 commander, second mission
Previous mission: Soyuz TMA-10/ISS-15 (2007)
NOGUCHI, Soichi, 44, civilian (Japanese), JAXA TMA/ISS flight engineer, second mission
Previous mission: STS-114 (2005)
CREAMER, Timothy John, 50, U.S.A., NASA TMA/ISS flight engineer

Flight log

Arriving at the ISS on December 23, 2009 (Moscow time), this trio worked as ISS-22 flight engineers for the first part of their mission alongside Jeff Williams and Maxim Surayev. Then, on March 18, 2010, the undocking of the TMA-16 spacecraft signified the end of the ISS-22 phase and the commencement of the ISS-23 phase, although the formal change-of-command ceremony had taken place on March 17. Kotov's crew then served as the three-person ISS-23 residency until they were joined by the TMA-18 crew on April 4, bringing the core crew back up to six persons. The ISS-22 residency continued until June 2, when they undocked from the station after formally handing over the prime role to the ISS-23 crew on May 31. In their 163-day space flight, the TMA-17 crew had resided on the station for 161 days. This was divided into an approximate 85-day tour on the ISS-21 phase and a further 75 days during the ISS-22 phase with just over a day as outgoing crew members.

The formal Russian segment research program for this crew encompassed 363 sessions on 42 experiments. Of these, only two were brand new, with the

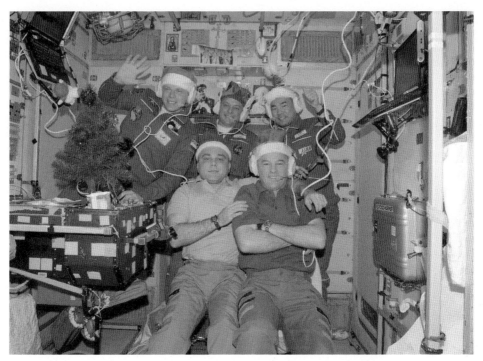

Wearing festive holiday hats Expedition 22 speak to officials from Russia, Japan, and the United States. (Front row) Flight engineer Maxim Surayev and commander Jeff Williams. (Back row) Oleg Kotov, Timothy Creamer, and Soichi Noguchi (all flight engineers).

remainder being continuations of previous experiments, reflecting the longevity of research on the station. In order to achieve this, mission planners had allocated 114 hours of experiment time over the duration of the ISS-23 residency. NASA announced that 45 experiments were being conducted in the U.S. segment. These encompassed 130 investigations from over 400 scientists across the globe. Eight of these experiments were part of the station's role as a U.S. National Laboratory. There were also 55 experiments from ESA, CSA, and JAXA assigned to the expedition.

After only three weeks the trio of TMA-17 cosmonauts were joined by the rest of the ISS-22 resident crew, who arrived on Soyuz TMA-18. The very next day, STS-131 was launched, which docked at the Harmony Module on April 7. This Shuttle mission was a logistics resupply mission, during which the joint crews worked for the next 10 days to unload over 17,000 lb (7,711.20 kg) of cargo for the station and complete four EVAs. With Caldwell-Dyson on the station and three women on the visiting Shuttle crew, a new record was set with four females in space at the same time and on the same vehicle.

When STS-131 departed, the station crew settled to their scientific, maintenance, and housekeeping routines. There was also a flurry of activity relating to the

Progress resupply craft at the station towards the end of April and into May. This included the departure of M-03M (35P) full of discarded items for atmospheric burn-up on April 22. There was also a 20 min 43 s burn of the Progress M-04M (36P) engines to boost the orbital altitude of the complex and, on May 1, the arrival of Progress M-05M (37P) which docked at Pirs. On May 12, the TMA-17 crew relocated their spacecraft in a 27 min flight from the nadir port of Zarya to the aft port of Zvezda, witnessed by the TMA-18 trio from inside the station.

Next for this busy expedition was the arrival of STS-132 on May 16 at the Harmony module, for a week of joint activities and three EVAs designed to support the installation of the Russian MRM-1 module called "Rassvet". This was permanently installed on to the nadir port of Zarya with the aid of the Shuttle and the station's robotic arms on May 18. Two days later, after leak checks, Kotov and Skvortsov entered the Rassvet module for the first time for an initial inspection.

STS-132 departed from the station on May 23 and for the next few days, the crew unpacked the Rassvet module. The TMA-17 crew also prepared their Soyuz for the descent. In order to provide the optimum conditions for the landing, the engines of Progress M-05M were fired for almost 10 minutes on May 26 to lower the orbit of the station by just 1 mile (1.60 km) to 214 miles (344.32 km). NASA called this maneuver a "de-boost". It gave the option of a backup landing site if required.

On May 31, Kotov relinquished command of the station to his Russian colleague Alexander Skvortsov, and the following day the crew prepared their Soyuz TMA-17 for descent. Undocking occurred in the early hours of June 2 and just over three hours later, the Descent Module and crew were back on Earth. All three were in great condition after the mission. Noguchi stated that, compared with his previous brief stay on the station five years earlier during STS-114, it was fun to stay longer in a station which had doubled or tripled in size and habitable volume.

Milestones

272nd manned space flight
110th Russian manned space flight
103rd manned Soyuz flight
17th manned Soyuz TMA mission
21st ISS Soyuz mission (21S)
22/23rd ISS resident crew
Final docking of a Soyuz at the nadir Zarya port
The first time four women are in space at same time (Caldwell-Dyson and three crew from STS-131)
Kornienko celebrates his 50th birthday and Noguchi his 45th birthday (both on April 15)
Skvortsov celebrates his 44th birthday (May 6)

STS-130

International designator	2010-004A
Launched	February 8, 2010
Launch site	LC39A, KSC, Florida, U.S.A.
Landed	February 21, 2010
Landing site	Runway 15, Shuttle Landing Facility, KSC, Florida, U.S.A.
Launch vehicle	OV-105 Endeavour/ET-134/SRB BI-141/SSME: #1 2059, #2 2061, #3 2057
Duration	13 da 18 h 6 min 22 s
Call sign	Endeavour
Objective	ISS-20A (Node 3 Tranquility, Cupola)

Flight crew

ZAMKA, George David, 47, USMC, NASA commander, second mission
Previous mission: STS-120 (2007)
VIRTS Jr., Terry Wayne, 41, USAF, NASA pilot
HIRE, Kathryn Patricia, 50, civilian, NASA, mission specialist 1, second mission
Previous mission: STS-90 (1998)
ROBINSON, Stephen Kern, 54, civilian, NASA, mission specialist 2, fourth mission
Previous missions: STS-85 (1997), STS-95 (1998), STS-114 (2005)
PATRICK, Nicholas James MacDonald, 45, civilian, NASA, mission specialist 3, second mission
Previous mission: STS-116 (2006)
BEHNKEN, Robert Louis, 39, USAF, NASA, mission specialist 4, second mission
Previous mission: STS-123 (2008)

Flight log

Delivering a third connecting node called "Tranquility", fitted out for additional habitation, plus the Cupola robotic control station with its seven-window panoramic view, the STS-130 mission would bring station construction up to 98% complete. Shuttle Endeavour was manifested for the mission, which was originally planned for December 2009, but was postponed until February 2010 following a series of delays.

Final launch preparations began in early October with the twin SRBs stacked on the Mobile Launch Platform (MLP). The ET was attached on November 23. Endeavour was processed in OPF Bay 2 before being rolled over to the VAB on

The Tranquility Node 3 and Cupola in the payload bay of Endeavour prior to docking.

December 11 where it was mated with the twin SRB and ET. This would be the vehicle's penultimate mission. The rollout to Pad 39A was delayed until early January to allow the processing staff to take a break over the Christmas and New Year holiday.

By late November the launch date had already been moved to February 7. This was to allow a few extra days margin in early February to launch the NASA Solar Science Mission on an Expendable Launch Vehicle (ELV). NASA also stated that it was preferable to have the next Russian Progress vehicle (M-4M, planned for a February 3 launch) already docked with the ISS when Endeavour arrived.

On January 6, 2010, Endeavour was rolled out to the pad. During the second week of January there was a small concern when the ammonia jumper hoses ruptured during preflight qualification tests. These pipes were to be used for connecting Node 3 to Node 1, so clearly a solution had to be found quickly. To resolve this issue, the hoses were redesigned and four new ones constructed. Final qualification and acceptance tests were completed just days prior to launch.

In the event, the intended launch on February 7 was delayed for 24 hours due to weather concerns. The following day, Endeavour streaked into the night sky without further incident and proceeded smoothly to orbit. Over the next two days Endeavour chased the space complex, the crew gradually adjusting its orbit to match that of the station. During these two days, the crew checked out the tile protection using the RMS and prepared the EVA equipment. Endeavour safely docked with the station on February 9. Two hours later came the now familiar routine of hatch opening and crew ceremonies.

There were three EVAs during this mission, which totaled 18 hours 14 minutes. All three space walks were performed by Behnken and Patrick, with primary focus on the installation of Tranquility and the Cupola.

Using the Shuttle robotic arm, Tranquility Node 3 was removed from the payload bay of Endeavour and relocated to the left port side of Unity during February 12. The first EVA (February 12, 6 h 32 min) was in support of the transfer operation, with both EVA astronauts working to attach electrical and ammonia cables to the new node. Later that day, members of the crew were able to enter the node from Unity for the first time, conducting preliminary checks of its internal structure, conditions, and systems.

The second EVA (February 13, 5 h 54 min) continued the "plumbing in" of the Tranquility module. Using the newly redesigned ammonia coolant line to link Tranquility to the Destiny science laboratory the new module was hooked up to the station's main coolant system. This EVA also included preparations to relocate the Cupola from its end berthing port location on Tranquility to the nadir (Earth-facing) location on the same node. This transfer was delayed slightly due to minor issues. Once resolved the unit was soon relocated, leaving the end port vacant for a day until PMA-3 was attached to the end of Tranquility. The mission was then extended to 14 days to allow more time for the work being conducted both inside and outside the station.

The third EVA (February 17, 5 h 48 min) was used both to relocate spares for the Dextre robotic arm system and to support the sequenced opening of the seven Cupola window covers. Each cover was tested in turn to confirm its smooth operation before the EVA crew returned inside.

On February 15, U.S. President Obama called the crews to congratulate them on their success. Five days later, Endeavour undocked from the station after 9 days 19 hours 48 minutes attached to the complex. The success of this mission meant that the station was now 98% complete by assembly, and 90% complete by mass. As Endeavour undocked from the station, weather forecasts for the landing day suggested that the return might have to be diverted, but the situation improved to allow the landing back in Florida as planned.

The two new units were the last major habitable modules in the U.S. segment, joining the Destiny laboratory, the Unity and Harmony Nodes, and the Quest airlock. Node 3 was named Tranquility as a result of a number of NASA nominated options for an online public poll. Node 3 was European built and provided much needed additional room for station life support and environmental subsystems. Measuring 23 ft (7.01 m) in length and 14.8 ft (4.51 m) in diameter with a mass of 40,000 lb (18,144 kg), Tranquility replaced the canceled Habitation Module and offered additional living quarters for the expanded station resident crew. The unit provided additional room for air revitalization, oxygen regulation, and waste handling, with waste and hygiene equipment relocated from other parts of the station. The Combined Operational Load Bearing External Resistance Treadmill (COLBERT) exercise device would also be relocated into Tranquility, as would other equipment, freeing up useful volume in other parts of the station.

The Cupola was the other new module to be delivered to the station. This long-planned unit had arrived at the Cape back in 2004. Its seven-window dome provided a 360° panoramic view of the station, the Earth, and space and was called the "Window on the World". The design features one overhead and six side windows allowing the unit to serve as the station's Control Center for robotic and EVA operations and as the focal point for handling the docking, relocation, and undocking of automated cargo craft.

Milestones

273rd manned space flight
160th U.S. manned space flight
130th Shuttle flight
 24th flight of Endeavour
 32nd Shuttle ISS flight
Final nighttime launch of a Shuttle
Mass of ISS grows to 1 million pounds (453,600 kg)

SOYUZ TMA-18

International designator	2010-11A
Launched	April 2, 2010
Launch site	Pad 1, Site 5, Baikonur Cosmodrome, Republic of Kazakhstan
Landed	September 25, 2010
Landing site	Southwest of Arkalyk, Republic of Kazakhstan
Launch vehicle	Soyuz-FG (serial number Ю15000-028), Soyuz TMA-18 (serial number 228)
Duration	176 da 1 h 18 min 38 s
Call sign	Utes ("Cliff")
Objectives	ISS resident crew transport (22S), ISS 23/24 resident crew

Flight Crew

SKVORTSOV, Aleksandr Aleksandrovich, 43, Russian Federation Air Force, RSA, Soyuz TMA commander, ISS-23 flight engineer, ISS-24 commander
KORNIENKO. Mikhail Borisovich, 49, civilian, RSA, Soyuz TMA flight engineer, ISS 23/24 FE
CALDWELL DYSON, Tracy Ellen, 40, civilian, NASA Soyuz TMA flight engineer, ISS 23/24 flight engineer

Flight log

This trio of cosmonauts arrived at the ISS on April 4, 2010. They would serve as flight engineers on ISS-23 under Oleg Kotov as ISS commander until June 2, when the TMA-17 crew departed and their ISS-24 residency began under the command of Skvortsov. On June 18, they were joined on the ISS by the Soyuz TMA-19 crew who became the prime ISS-25 crew after this trio departed. By now, regular rotation of crews had become a feature of station operations and one result of its frequency and seemingly routine nature was that these activities dropped down the news-reporting pecking order outside of the space community.

This of course reflects a safe, regular, and consistent period of flight operations, but does not serve to promote the program to the outside world. It is in this situation that the official websites, new reports, and support information from the partner agencies have to champion the program, after such a long time in development and construction. Up on orbit, the promotion of the program through outreach and educational activities is as important as the baseline science, while the crews were also still hard at work finishing the assembly and completing the transformation of the station into the fully functioning research facility it was intended to be. This work has been aided by the growing phenomena of social

Fresh supplies are always welcome on the ISS. Expedition 23 commander Kotov and flight engineer Tracy Caldwell Dyson enjoy receiving fresh fruit and vegetables during their residency.

media, in part thanks to the regular blogs, tweets, and messages from the crews on board the station.

During this residency, the crew continued the Russian science work begun by the earlier crews, with 363 planned sessions for 42 experiments, of which only two were new investigations. In the ISS-23 phase, over 114 hours of crew experiment time was manifested, with a further 20 hours 15 minutes planned during the ISS-24 phase. The change from assembly to research was becoming more evident with each new expedition, and the subtitle on the ISS-23/24 NASA Press Kit stated that this expedition would include "Science for Six". Therefore, in the U.S. segment there would be 130 investigations from 45 new experiments, as well as those ongoing from earlier expeditions with 8 experiments specific to its role as a U.S. National Laboratory and a further 55 investigations from the international partner agencies.

After the docking at Poisk on April 4, the next couple of months proved to be busy prior to the departure of the ISS-22/23 crew in June and the commencement of the ISS-24 phase. Just three days after the TMA-18 crew had arrived at the station, STS-131 arrived aboard Discovery, which docked at the Harmony Node with more supplies. Then, in May STS-132 delivered the Russian Rassvet module.

With the science work, routine maintenance, and housekeeping, work associated with the Progress resupply craft, and the relocation of accumulated logistics, the new crew had plenty to keep them occupied during the first half of their residency. As a result, light duties were planned for the three crew members until the rest of the ISS-24 crew arrived.

Following the arrival of the TMA-19 crew, the two crews soon completed post-docking safety checks and drills and began an increased science program. On June 28, while the TMA-19 crew relocated their Soyuz from the aft part of Zvezda to the Rassvet module, the TMA-18 crew remained inside the station. On July 1, Progress M-04M was undocked from the station, to be replaced on July 4 by Progress M-06M. The 2-day delay in the docking was caused by a loss of a telemetry lock on M-06M, but its second approach occurred without incident.

Diversity featured in most of the routine operations on the station, with crews working in different modules to cope with the increased science research, maintenance, and housekeeping duties in the Russian and U.S. segments as well as in the Columbia and Kibo laboratories. On July 11, the crew recorded a partial solar eclipse across the world while continuing their preparations for a series of EVAs.

On July 16, Progress M-06M completed a 17 min 45 s reboost to the ISS, increasing its altitude by 2.3 miles (3.07 km). This was necessary to provide the best conditions for docking the next Progress and to ensure the safe return of TMA-18. During July 15–24, the crews observed the 35th anniversary of the joint U.S./U.S.S.R. Apollo–Soyuz Test Project mission.

Three EVAs were planned in July and August, from both U.S. and Russian airlocks. The first EVA of the expedition from the Russian segment, by Yurchikhin and Kornienko, took place on July 27 from Pirs. During the 6 h 42 min excursion, the cosmonauts replaced several items of equipment and visually inspected the exterior of the Russian segment.

The focus now switched to a series of EVAs from the U.S. segment by Wheelock and Caldwell. The first of these took place on August 7 and lasted a record 8 hours 3 minutes—the longest ISS-based EVA and the sixth longest space walk in history. Unfortunately, they failed in their primary goal to remove and replace the ammonia pump module, falling behind the timeline when one of the four coolant lines became stuck. They loosened the stuck valve, but could not totally disconnect the unit as they approached the end of the EVA. An issue with leaking ammonia crystals also required additional cleanup time, leading to the unexpected record EVA duration. Wheelock later admitted that this EVA was "a tough one".

The next EVA (August 11, 7 h 26 min) focused upon removal of the fluid coolant line that had leaked during the first EVA. Using brute force, Wheelock closed and removed the line safely. The pair then disconnected the defunct assembly from the truss and installed it on a payload bracket located on the Mobile Base Assembly. The third EVA (August 16, 7 h 20 min) from Quest featured the installation of a spare ammonia pump module on the S1 truss. The three U.S. segment EVAs totaled 22 hours 49 minutes, and with these excursions completed it was back to the science.

September saw the TMA-18 crew prepare for their return to Earth. A change-of-command ceremony was conducted on September 22, during which Skvortsov handed over command of the ISS to Doug Wheelock. After a short, 24 h delay due to an erroneous signal, Soyuz TMA-18 undocked on September 25. Following a nominal reentry, Soyuz TMA-18 landed some 3 hours 20 minutes after undocking from the station. During a mission of 176 days the crew had resided aboard the station for approximately 174 days. Two days were flown aboard the Soyuz getting to and from the facility. Of the 171 days in residency, 59 days were as part of the ISS-23 expedition and 112 days as the prime ISS-24 expedition. They also spent three days as the outgoing crew prior to undocking from the station.

Milestones

274th manned space flight
111th Russian manned space flight
104th manned Soyuz flight
18th manned Soyuz TMA mission
22nd ISS Soyuz mission (22S)
23/24th ISS resident crew
Record longest ISS-based EVA (August 7, 8 h 3 min)
Caldwell-Dyson celebrates her 41st birthday (August 14); this was her second birthday spent in space having marked her 38th birthday during STS-118 in 2007

STS-131

International designator	2010-012A
Launched	April 5, 2010
Launch site	LC39A, KSC, Florida, U.S.A.
Landed	April 20, 2010
Landing site	Runway 33, Shuttle Landing Facility, KSC, Florida, U.S.A.
Launch vehicle	OV-103 Discovery/ET-135/SRBs BI-142/SSME: #1 2045, #2 2060, #3 2054
Duration	15 da 2 h 47 min 10 s
Call sign	Discovery
Objectives	ISS-19A (MPLM, LMC), ISS logistics resupply mission

Flight crew

POINDEXTER, Alan Goodwin, 48, USN, NASA commander, second mission
Previous mission: STS-122 (2008)
DUTTON Jr., James Patrick, 41, USAF, NASA pilot
MASTRACCHIO, Richard Alan, 50, civilian, NASA mission specialist 1, third mission
Previous missions: STS-106 (2000), STS-118 (2007)
METCALF-LINDENBURGER, Dorothy Marie, 34, civilian, NASA mission specialist 2
WILSON, Stephanie Diana, 43, civilian, NASA mission specialist 3, third mission
Previous missions: STS-121 (2006), STS-120 (2007)
YAMAZAKI, Naoko, 39, civilian (Japanese), JAXA mission specialist 4
ANDERSON, Clayton Conrad, 51, civilian, NASA mission specialist 5, second mission
Previous missions: STS-117/ISS-15/16/STS-120 (2007)

Flight log

With only four or five manifested Shuttle flights to the ISS before their retirement in 2011, the chances of carrying large items to and from the station on the orbiter were diminishing rapidly. Though the majority of the main hardware had been delivered (certainly on the U.S. segment), there still remained a few bulky items to be launched. Time seemed to have flown by since the start of construction just under a dozen years previously and now the countdown to assembly completion was ticking away. One of the main objectives for the payload capacity in these few remaining missions was to stock up the station with supplies and spares. Another was to remove as much unwanted equipment, waste, discarded items, and experi-

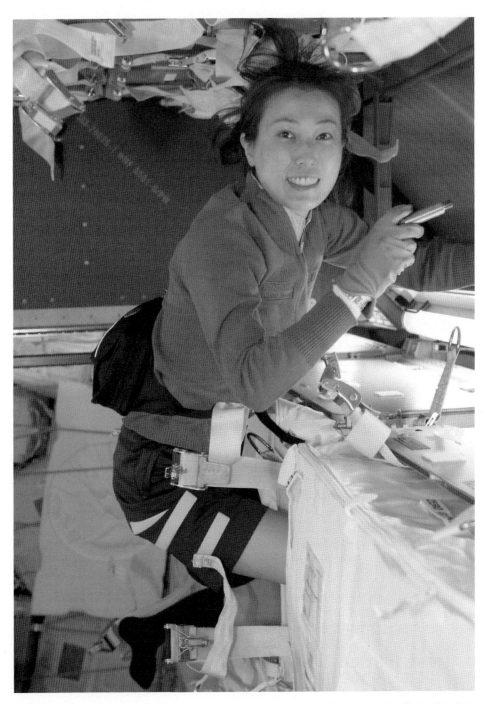

Loadmaster Naoko Yamazaki works in the Leonardo Multi-Purpose Logistics Module (MPLM) linked to the ISS during the Discovery mission.

ment results as possible to free up the internal volume of the station while the Shuttle's large load capacity was still available. On this mission, therefore, Discovery was carrying the Leonardo Multipurpose Logistics Module (MPLM), which was filled with about 8 tons of supplies and hardware. It would return to Earth with valuable experiment results and samples, unwanted equipment, and as much trash as possible.

As with most previous flights, final preparations for the mission began with the arrival of Discovery back at KSC following its last mission. Two weeks after landing in California at the end of the STS-128 mission in September 2009, Discovery was returned to the Cape. Initial inspections conducted inside the OPF revealed relatively few issues that needed to be addressed in processing for the next mission. Having the MPLM as the primary payload made the preflight processing somewhat easier as well, as the logistics carrier would be installed in the payload bay when Discovery was on the pad.

The stacking of the twin SRBs began in early October and the ET had been mated with the boosters by late November. Everything was ready for the move of Discovery across to the VAB but the weather refused to play ball, with exceptionally cold temperatures being recorded. As a result, the move was delayed until February 22. The mated stack was then moved out to Pad 39A on March 3. The delay shifted the planned launch from March 18 to April 4 but this happened to be the Easter weekend. This was impractical for launch teams, so April 5 was chosen instead. This also gave the new residents on the station, who were scheduled to arrive via Soyuz TMA-18 on April 4, additional time to acclimatize to their new home before the Shuttle arrived.

Launchpad preparations proceeded smoothly, with the MPLM placed on board Discovery on March 19. After an on-time launch on April 5, 2010, Discovery was back in orbit within 8 minutes to begin a 2-day chase to station. Docking occurred on April 7. When the hatches were opened and the familiar ceremonies observed, the mission was already adding new milestones to the history books. For the first time, four women were in space at the same time and now they were all aboard the same spacecraft. Two Japanese astronauts were also flying together for the first time as well. The orbiter crew also included the final rookies that would fly on a Shuttle mission—Metcalf-Lindenburger, Yamazaki, and Dutton.

Nine days of joint activities were planned following the docking. The MPLM was moved to the Earth-facing port on Harmony on April 7 for unloading. The loadmaster on the crew, in charge of moving the 17,000 lb (7711.20 kg) of cargo between the spacecraft and the station, was Yamazaki. With cargo floating both ways, she would be kept very busy during her stay on board the station.

The major elements of cargo transferred were a Muscle Atrophy Research and Exercise System Rack, a Window Operational Research Facility, an ExPRESS Rack and Zero-G Storage Racks, Resupply Storage Racks, the final four resident crew sleeping quarters (intended for installation in Harmony), the third Minus Eighty Degree Laboratory Freezer, and equipment for a new water production system. Other, smaller items of equipment, supplies, and stores were also transferred. With Leonardo emptied, the cargo intended for return to Earth was loaded back into the MPLM.

While work continued inside the station, the crew was also occupied outside, with Anderson and Mastracchio completing three EVAs totaling 20 hours 17 minutes. The first of these (April 9, 6 h 27 min) began the work of exchanging an old Ammonia Tank Assembly (with a mass of 1,800 lb or 816.48 kg) with a new unit. This took up most of the EVA timeline, but the two men worked efficiently and were able to also repair a Rate Gyro Assembly and retrieve a Material Experiment Exposure Device from the exterior of the Japanese module. The following day was a planned rest day, during which the crew were informed that their mission would be extended by 24 hours to facilitate the RMS inspection of the heat shield while docked with the station instead of after undocking. This was due to a failed Ku-band communication antenna on the orbiter.

The second EVA (April 11, 7 h 26 min) continued the work on the Ammonia Tank Assembly. Despite some difficulty with the installation of the hold-down bolts, the pair were able to complete most of their tasks, with just a few delayed to their third space walk. Electrical cables were connected but the ammonia and nitrogen lines were not. Two micromaterial debris shields were retrieved for analysis back on Earth.

The crew rest day of April 12 was also the 49th anniversary of Gagarin's flight and the 29th anniversary of the first Shuttle flight. These events were noted in communication sessions with ground control centers, one of which featured a call from Russian President Dmitry Medvedev. The final EVA (April 13, 6 h 24 min) began with the tasks carried over from EVA 2, plus the return of the old Ammonia Tank into the Shuttle's payload bay. The crew then completed several smaller tasks before winding up the exterior activities for the mission.

In the closing four days of the docked phase, the joint crews completed the relocation of cargo, returning the refilled MPLM back into the payload bay on April 16. They also held press conferences and enjoyed a day off. The undocking on April 17, after 10 days 5 hours 8 minutes of joint operations, was followed shortly afterwards by the traditional fly-around maneuver before the orbiter departed from the vicinity of the orbital complex.

Discovery flew a descending node reentry on April 20 and, in the daylight hours, took the orbiter over most of the continental U.S.A. This profile had been flown only once before (on STS-120 in 2007) since the loss of Columbia in 2003, but it was a journey that afforded the flight deck crew a spectacular panorama as they approached the landing site in Florida.

Milestones

275th world manned space flight
161st U.S. manned space flight
 33rd Shuttle ISS mission
131st Shuttle flight
 38th Discovery flight
 12th Discovery ISS flight
 10th and final round trip MPLM flight
  7th Leonardo MPLM flight
First time three females fly on same Shuttle mission
First time four females in space at same time (with ISS resident crew member Caldwell-Dyson)
First time four females on the ISS at same time
First time two JAXA astronauts in space at same time
First time two JAXA astronauts on the ISS same time
Dutton, Metcalf-Lindburger, and Yamazaki become the final rookies to fly on a Shuttle

STS-132

International designator	2010-019A
Launched	May 14, 2010
Launch site	LC39A, KSC, Florida, U.S.A.
Landed	May 26, 2010
Landing site	Runway 33, Shuttle Landing Facility, KSC, Florida, U.S.A.
Launch vehicle	OV-104 Atlantis/ET-136/SRB BI-143/SSME: #1 2052, #2 2051, #3 2047
Duration	11 da 18 h 29 min 9 s
Call sign	Atlantis
Objective	ISS-ULF4 (Russian Mini Research Module 1 (MRM1), ICC-VLD)

Flight crew

HAM, Kenneth Todd, 45, USN, NASA commander, second mission
Previous mission: STS-124 (2008)
ANTONELLI, Dominic Anthony, 42, USN, NASA pilot, second mission
Previous mission: STS-119 (2009)
REISMAN, Garrett Erin, 42, civilian, NASA mission specialist 1, second mission
Previous mission: STS-123/ISS-16/17/STS-124 (2008)
GOOD, Michael Timothy, 47, USAF, NASA mission specialist 2, second mission
Previous mission: STS-125 (2010)
BOWEN, Stephen George, 46, USN, NASA mission specialist 3, second mission
Previous mission: STS-126 (2008)
SELLERS, Piers John, 55, civilian, NASA mission specialist 4, third mission
Previous missions: STS-112 (2002), STS-121 (2006)

Flight log

The STS-132 mission was significant in that the primary payload was not American, but the Russian-built Mini Research Module-1 (MRM-1), also known as Rassvet ("Dawn"). This module was to be installed on to the lower (nadir, Earth-facing) port of Zarya. The secondary payload was the second Integrated Cargo Carrier (ICC), packed with further spare supplies and equipment.

Inside the VAB, the External Tank was attached to the twin SRBs on March 29. The rollover of Atlantis to the assembly building on April 13 recorded only 22 problems being tracked since the orbiter's return from STS-129. The payload arrived at the pad inside the payload canister on April 15. Rollout to the pad had

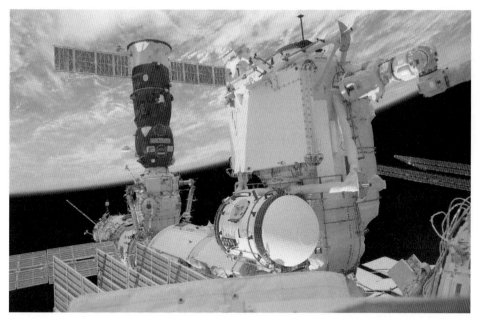

Rassvet ("Dawn"), the Russian-built Mini Research Module-1 (MRM-1), is seen (at right) attached to Zarya.

been scheduled for April 19, but bad weather delayed transfer until late on April 21, with the stack arriving after a 6.5 h journey in the early hours of April 22. The payload was installed in the cargo bay of the orbiter three days later.

Atlantis blasted off from KSC on time with an all-veteran crew aboard. Just over eight minutes later, the flight entered orbit to begin the chase to station. The following day was taken up with an RMS inspection of the heat shield and preparing the EVA suits and equipment for the planned space walks. Prior to docking, the now traditional backflip maneuver was completed for visual checking and imagery by the station crew. Atlantis docked at the PMA-2 port of Harmony on May 16, less than a month after Discovery had departed at the end of mission STS-131. Two hours later, both crews were inside the station preparing to embark on a week of joint activities.

The Integrated Cargo Carrier was transferred to the station by Canadarm2 and placed on the Mobile Transporter. This unit was packed with spares and equipment for installation during the three EVAs. The unit also held spares designed to support the life of the station towards (and hopefully beyond) 2020. These included a spare Ku-band antenna and truss, six NiH batteries, and spare hardware components for the Dextre manipulator system.

The three EVAs logged 21 hours 20 minutes, with three astronauts (Reisman, Bowen, and Good) completing two space walks each. The first EVA was by Bowen and Reisman (May 17, 7 h 25 min) and featured a number of hardware installations, including a space-to-ground Ku-band antenna on the station truss and a new tool platform for Dextre. There was time at the end of the EVA for a get-ahead task, with the crew loosening several bolts holding the batteries that would be exchanged over the next two space walks.

On May 18, the Rassvet module was grappled by the RMS, handed over to the space station RMS, and then attached permanently to the nadir port on Zarya. The Rassvet module features eight workstations inside its pressurized compartment. It was designed for a variety of scientific experiment operations and research. Taking advantage of the payload and launch capacity of the Shuttle, the Rassvet had 1.5 tons of cargo, supplies, and scientific gear for relocation to the U.S. segment packed inside. The Russians reported that the scientific research to be conducted in the new module included developing technologies, biological sciences, fluid physics, and educational research.

The second EVA (May 19, 7 h 9 min) was by Bowen and Good, who began by releasing a snagged cable on the Orbiter Boom Sensor System (OBSS). The pair then began the exchange of five of the six batteries, a process known as "shepherding", with the old batteries intended for return to Earth. The team then completed a couple of small chores before wrapping up their excursion. The next day, cosmonauts Kotov and Skvortsov opened the inner hatches and entered Rassvet for the first time.

The final EVA (May 21, 6 h 46 min) by Good and Reisman was primarily devoted to completing the exchange of batteries. The original units had a design life of six and a half years but had been in operation for nine years. Prior to closing out the space walk, the astronauts left a Power Data Grapple Fixture in the Quest airlock and prepared the ICC for return to the payload bay of Atlantis, which occurred on May 22. In total, Bowen accumulated 14 hours 34 minutes in two space walks, Reisman logged 14 hours 11 minutes on his two EVAs, and Good completed his two excursions in 13 hours 55 minutes.

Following a couple of rest days, completion of the transfer of cargo signaled the end of joint work with the station crew. During their week of joint activities, the crews had moved over 2,879 lb (1305.91 kg) of cargo into the station and some 8,229 lb (3732.67 kg) back into Atlantis. The orbiter was undocked on May 23 after 7 days 0 hours 54 minutes. Following the normal fly-around to photograph the station and Shuttle, the two vehicles separated, allowing the Atlantis crew to prepare for the return home and the station crew to resume their science program.

On May 26, Atlantis swooped to a spectacular landing on Runway 33 at the Cape. Following the visit of Atlantis, the station had grown to a mass of 815,000 lb (369,684 kg) and was now 94% complete by volume and over 98% complete by mass.

Although this was originally to be the final flight of Atlantis, there were plans to prepare the orbiter to be a launch-on-need rescue vehicle (designated STS-335) for STS-134, then scheduled as the final Shuttle mission of the program. However, discussions were ongoing over using the additional hardware for one more flight (STS-135). NASA had already bought an extra ET and SRB and needed only Congressional agreement and funding to mount the extra mission.

Milestones

276th world manned space flight
162nd U.S. manned space flight
 34th Shuttle ISS mission
132nd Shuttle flight
 32nd Atlantis flight
 11th Atlantis ISS flight
Only Russian ISS segment component launched by U.S. Shuttle

SOYUZ TMA-19

International designator	2010-029A
Launched	June 16, 2010 (Moscow time)
Launch site	Pad 1, Site 5, Baikonur Cosmodrome, Republic of Kazakhstan
Landed	November 26, 2010
Landing site	52 miles northeast of Arkalyk, Republic of Kazakhstan
Launch vehicle	Soyuz-FG (serial number bl5000-032), Soyuz TMA-19 (serial number 229)
Duration	163 da 7 h 10 min 47 s
Call sign	Olympus
Objective	ISS-24/25 resident crew transport (23S)

Flight Crew

YURCHIKHIN, Fyodor Nikolayevich, 51, civilian, RSA, Soyuz commander, ISS flight engineer, third mission
Previous missions: STS-112 (2002), TMA-10 (2007)
WALKER, Shannon, 45, civilian, NASA, Soyuz/ISS flight engineer
WHEELOCK, Douglas Harry, 50, NASA, Soyuz/ISS flight engineer, second mission
Previous mission: STS-120 (2007)

Flight log

On arrival at the station on TMA-19 on June 17, this crew served as flight engineers on ISS-24 before taking over as the prime core crew of ISS-25 on September 22, when Douglas Wheelock assumed ISS command from the outgoing Skvortsov. Under the ISS-25 residency, the crew continued the extensive scientific program as a three-person crew until early October, when the TMA-01M trio arrived to complete the ISS-25 complement. During their 163-day space odyssey, the TMA-19 crew would spend approximately 160 days aboard the station, 97 of them as members of the ISS-24 crew and then a further 63 in prime command of ISS-25.

The TMA-19 crew relocated their spacecraft at the station very early in the residency. The docking at the aft port of Zvezda on June 17 was followed just nine days later by the relocation of their Soyuz to the Rassvet module, allowing future arrivals to use the aft Service Module port. The 25 min operation was delayed by 75 minutes due to difficulties feathering the P4 truss solar wings to allow the smooth passage of the Soyuz. Following the docking, the crew inspected the docking cone of Rassvet to document any scuff marks as a result of the linkup. This was the first time a Soyuz had docked with the Rassvet module.

Soyuz TMA-19 docks with the Rassvet MRM-1.

When the second half of the ISS-25 crew arrived in the first TMA-M vehicle, the science program returned to its full potential. As with all new crews arriving on the station, formalities and zero-g adaptation took a few days, but the science work had to continue, as did preparations for receiving the next Shuttle mission (STS-133). On October 18, the Russian members of the international crew took part in an all-Russian census, confirming they were Russian nationals. Yurchikhin, who had participated, during 2002, in a previous census from orbit, revealed that he also had Greek roots.

On October 20, the Progress M-07M engine fired for a 3 min 49 s burn to raise the orbital altitude of the complex by just 890 m (2920 ft), a small but essential alteration to assist with the upcoming docking of Progress M-08M and STS-133. Five days later, on October 25, Progress M-05M was undocked from the Pirs port and placed in a parking orbit until it reentered on November 15. On October 30, a new resupply craft, Progress M-08M, docked at Pirs. Aboard the new craft were 6,320 lb (1,293.07 kg) of supplies and a few treats for the upcoming Halloween holiday.

On October 31, the 10th anniversary of the launch of the first resident crew to the station (ISS-1 aboard Soyuz TM-31) was observed, followed on November 2 by the anniversary of the docking and transfer of the first expedition into the station to start continuous occupation. In 10 years of successive crew exchanges, 24 resident crews comprising 196 crew members had logged 1.5 billion miles (2.415 billion km) or 57,361 manned orbits of Earth. NASA Administrator and former

Shuttle astronaut Charles Bolden likened the achievement to a modern day *Star Trek*.

With the news that Shuttle mission STS-133 had been delayed to the end of November at the earliest, the crew focused on preparing for a Russian section EVA, as well as maintaining the routine-but-necessary housekeeping and maintenance program that had kept the station operating successfully for 10 years.

On November 15, Yurchikhin and Skripochka conducted a 6 h 27 min EVA from the Pirs module wearing Orlan suits. A small workstation was installed on the starboard side of Zvezda and samples were taken from underneath the insulation covering on both Pirs and Zvezda for later analysis on Earth. A new materials experiment was deployed on Pirs and a robotic experiment was cleaned and removed for return inside the station. The cosmonauts found it difficult to remove some insulation on Rassvet that was blocking the installation of a TV camera, so the camera was returned to the station while the problem was evaluated. The day after the EVA saw the cosmonauts performing post-EVA maintenance on the suits, including drying them, performing systems checks, and discharging the suit batteries.

The return of the ISS-25 crew was scheduled four days earlier than planned due to an Organization for Security and Cooperation in Europe (OSCE) summit in Astana, Kazakhstan during the first two days in December. This would require clear air space in the vicinity, even from descending spacecraft! Anticipating their homecoming, Wheelock was looking forward to a shower not having had one since June. Walker was told, not very encouragingly, that a Soyuz landing was very similar to "a series of explosions followed by a car crash!" After conducting a "considerable amount of science" on their expedition, the TMA-19 crew's stay on the station was coming to an end. Their Soyuz was checked over and Kelly officially took over command of the station on November 24, beginning the 26th expedition.

Late on November 25, the three returning crew members entered their Soyuz and closed the hatches. Undocking occurred on November 26 and they landed 3 hours 23 minutes later in Kazakhstan after a 163-day mission. In just over 10 years, a total of 25 expeditions had been completed successfully. Now, the first crew of the second decade of operations to occupy the station was on board, with several other crews in various stages of training across the globe.

Milestones

277th manned space flight
112th Russian manned space flight
105th manned Soyuz flight
 19th manned Soyuz TMA mission
 23rd ISS Soyuz mission (23S)
24/25th ISS resident crew
 100th launch dedicated to ISS operations since November 1998
Walker becomes first Houston, Texas, U.S.A. citizen in space
First Soyuz docking with Rassvet module
First time two women were on main ISS resident crew (Walker and Caldwell Dyson)
Ten years of constant resident crew operations completed (November 2)

SOYUZ TMA-M

International designator	2010-052A
Launched	October 8, 2010 (Moscow time)
Launch site	Pad 1, Site 5, Baikonur Cosmodrome, Republic of Kazakhstan
Landed	March 16, 2011
Landing site	50 miles (80.45 km) north of Arkalyk, Republic of Kazakhstan
Launch vehicle	Soyuz-FG (serial number Ы5000-035), Soyuz TMA-M (serial number 701)
Duration	159 da 8 h 43 min 5 s
Call sign	Ingul
Objective	ISS 25/26 resident crew transport (24S)

Flight crew

KALERI, Aleksandr Yuriyevich, 54, civilian, RSA Soyuz TMA-M commander, ISS flight engineer, fifth mission
Previous missions: Soyuz TM-14 (1992), Soyuz TM-24 (1996), Soyuz TM-30/Mir-28 (2000), Soyuz TMA-3 (2003)
SKRIPOCHKA, Oleg Ivanovich, 40, civilian, RSA Soyuz TMA-M flight engineer, ISS flight engineer
KELLY, Scott Joseph, 46, USN, NASA ISS-25 flight engineer; ISS-26 commander, Soyuz TM-M flight engineer, third mission
Previous missions: STS-103 (1999), STS-118 (2007)

Flight log

On October 8, 2010 (Moscow time), a new, modified Soyuz TMA-M was launched on its first mission with a three-man crew. It docked with ISS at Poisk on October 10. Such was the confidence in the system and the internal system upgrades of Soyuz, an unmanned TMA-M mission was deemed unnecessary, although several systems had been test-flown on earlier unmanned Progress missions.

In command of the new vehicle was veteran Russian civilian engineer cosmonaut Alexander Kaleri, who had already logged 610 days in space on his three flights to Mir and an earlier mission to the ISS. He had worked on the development of the TMA-M upgrades which enabled him to take the coveted command of the inaugural mission. With him were rookie cosmonaut Oleg Skripochka and veteran American Space Shuttle commander Scott Kelly, who became the first NASA pilot–astronaut to serve on an ISS residency crew since Ken Bowersox in 2003.

NASA astronaut Scott Kelly is pictured inside the Soyuz TMA-M Descent Module on docking day.

The mission of TMA-M was, of course, to transport the next resident crew to and from the station and serve as a rescue vehicle should it be required, but this was also an important test flight of a new vehicle which would be the mainstay of Russian and ISS manned operations for some years to come. It was imperative that all went well in this maiden flight.

The external appearance of TMA-M was similar to earlier versions of the craft; the upgrades were mainly within the avionics of the spacecraft. A new, lighter, digital command and control system freed up mass to allow an increase in payload capacity by 154 lb (69.85 kg). The old Argon analog computer system, used since 1974, was finally upgraded to the new TSVM-101 system, which meant that just one qualified pilot could now fly Soyuz rather than having two fully trained crew members, saving on training time. One-person Soyuz rescue (return) capability had been available for some years, of course, though it had never been called upon in flight.

After 45 days working as part of the ISS-25 crew, Kelly assumed command of the station on November 24, beginning the ISS-26 residency. For the next three weeks, they continued the science and maintenance programs as a three-person crew while awaiting the arrival of their colleagues on TMA-20. The new crew, who would take over as ISS-27 in the spring, arrived on December 17 and docked with the Rassvet module.

The station was reboosted on December 22, using the eight thrusters of the Progress M-07M for 21 minutes 11 seconds to raise the orbit of the complex by 2.6 miles (4.18 km) to 219 miles (352.37 km) in preparation for the arrival of the second Japanese unmanned resupply craft, HTV-2. On Christmas Eve, Skripochka celebrated his 41st birthday and the crew had a day off on Christmas Day. The closing days of the year were spent on a significant amount of preparation work for 2011 docking and joint flight operations, which would also see the retirement of the Space Shuttle.

The New Year arrived with shocking news. On January 8, U.S. Congresswoman Gabrielle Gifford was shot at a political rally in Tucson, Arizona. Six other people, including a 9-year-old girl, were fatally wounded. Gifford is the sister-in-law of Scott Kelly and her husband, Kelly's twin brother Mark, was due to command STS-134. That mission was originally scheduled to fly to the station during March while Scott was aboard the station, making for a historic meeting in orbit. In memory of the victims one minute's silence was held aboard the station and across the United States on January 10. It was also announced that processing and payload delays would result in the STS-134 mission being postponed until April. The opportunity for the two brothers to meet aboard the space station was lost.

On January 21, Kondratyev and Skripochka conducted a Russian segment EVA (5 h 23 min) from Pirs, installing an antenna on Zvezda as part of the Russian Radio Telescope System for Information Transfer which would allow radio technicians to send large files at 100 megacycles per second from computers on the station to Earth. They also removed a failed generator on the Expre-R camera from Zvezda and finally installed the docking camera on the outside of Rassvet.

Following the EVA, things became busier at the station, with operations to restock the station accelerating in lieu of the retirement of the Shuttle later in 2011. On January 24, Progress M-08M was undocked and this was soon followed by the arrival of the second Japanese HTV unmanned resupply craft, Kounotori 2 ("White Stork"). The HTV was grappled by Canadarm2 on January 27 and was initially attached to the nadir port on Harmony. The crew entered the module for the first time, wearing masks as a safety precaution for a new vehicle, on January 28. On board the HTV were 6,455 lb (2,927.98 kg) of cargo. On January 30, Progress M-09M docked with the Pirs module bringing a further three tons of supplies. On February 4, another anniversary was marked in the program as Zarya, the original ISS element, completed 70,000 orbits of Earth since its launch on November 20, 1998.

A second EVA (February 16, 4 h 5 min) by Kondratyev and Skripochka continued the installation of exterior experiments outside Zvezda and the retrieval of panels of exposed materials. The Japanese HTV was relocated from the nadir port of Harmony to the module's zenith port on February 18 to make room for the forthcoming docking of Discovery (STS-133). Less than a week later, on

February 24, the second European unmanned resupply craft Johannes Kepler (ATV-2) docked with Zvezda's aft port with a further 3,500 lb (1,587.60 kg) of cargo aboard. While docked with the station for the next three months, it was planned to use the ATV for station reboost. This would also give the crew time to unload the supplies and utilize the extra volume, before filling it with unwanted material prior to undocking for destructive burn-up in the atmosphere.

Shuttle Discovery, flying STS-133, was the next arrival at the station, docking with the PMA-2 of Harmony on February 26 to deliver more cargo. The mission would also transfer the former MPLM Leonardo—now designated the Permanent Multipurpose Module (PMM)—and the ExPRESS Logistic Carrier-4 (ELC-4) across to the station. The docking of Discovery created a unique moment in ISS history, as for the first time all the current resupply craft were docked with the space station—Soyuz, Progress, ATV, HTV, and the Space Shuttle. The ISS was at this point the biggest it had ever been. Unfortunately, a fly-around of one of the Soyuz spacecraft to photograph the historic linkup was not possible. The Russians were rightly cautious in that the next planned departure vehicle—Soyuz TMA-M—was the inaugural flight of the vehicle and it was deemed too risky to violate safety protocols. It was hoped that an opportunity would arise for such a unique photograph before the Shuttle retired later in the summer. Discovery undocked on March 7.

On March 11, Kounotori was relocated again from the zenith side of Harmony back to the nadir side of the module. That same day, an 8.9 magnitude earthquake and tsunami struck northern Japan. The Tsukuba Flight Control Center, some 30 miles (48.27 km) northeast of Tokyo, was shut down for 3 days. The center suffered a little damage but fortunately no casualties. Communication links with Houston were also disrupted, reducing regular operations on Kibo. Three JAXA controllers flew to the United States to establish temporary Kibo control in Houston. Until the communications links could be fully restored, the HTV could not be unberthed, delaying its departure from the station. In the interim, the Japanese vehicle continued to be packed with additional unwanted material and trash. The HTV departed from ISS for its destructive atmospheric reentry on March 28.

On March 14, Kondratyev assumed command of the station for the ISS-27 phase from Kelly, effectively ending the ISS-26 prime residency after 110 days. Two days later, Soyuz TMA-M undocked from the ISS, landing later the same day on the snowy steppes of Kazakhstan. The Descent Module landed on its side and was dragged about 75 feet (22.86 m) by its recovery parachute. Despite this, it was a highly successful initial flight for the new vehicle.

Milestones

278th　manned space flight
113th　Russian manned space flight
106th　manned Soyuz flight
1st　Soyuz TM-M mission
24th　ISS Soyuz mission (24S)
25/26th　ISS resident crew
Skripochka celebrate his 41st birthday (December 24)
First time all main station resupply craft are docked with the ISS at same time—Soyuz, Progress, Shuttle, ATV, and HTV

SOYUZ TMA-20

International designator	2010-067A
Launched	December 15, 2010
Launch site	Pad 1, Site 5, Baikonur Cosmodrome, Republic of Kazakhstan
Landed	May 24, 2011
Landing site	Near town of Dzhezkazgan, Republic of Kazakhstan
Launch vehicle	Soyuz-FG (serial number Ы5000-034), Soyuz TMA (serial number 230)
Duration	159 da 8 h 17 min 15 s
Call sign	Varyag
Objective	ISS resident crew transport ISS-26/27 (25S)

Flight crew

KONDRATYEV, Dmitri Yuriyevich, 41, Russian Federation Air Force, RSA Soyuz TMA commander, ISS-26 flight engineer, ISS-27 commander
COLEMAN, Catherine Grace, 50, USAF (Retd.), NASA–Soyuz TMA and ISS-26/27 flight engineer, third mission
Previous missions: STS-73 (1995), STS-93 (1999)
NESPOLI, Paolo, 53, civilian (Italian), ESA–Soyuz TMA and ISS-26/27 flight engineer, second mission
Previous mission: STS-120 (2007)

Flight log

The next resident crew to fly to the ISS launched to the station on one of the last TMA versions of the venerable Soyuz spacecraft. The trio was another truly international crew. Commander of the Soyuz was rookie cosmonaut Kondratyev, who would serve as commander of ISS-27 after he and his two Shuttle veteran colleagues served as flight engineers on ISS-26. Docking occurred on December 17 at the Rassvet module with the hatches opened three hours after docking for the crew to join their ISS-26 colleagues.

The Descent Module in which they had flown to the station was not the one they had planned to fly. The original Descent Module of TMA-20 was damaged in October 2009 during transportation to the Baikonur Cosmodrome from the Energiya factory where it had been fabricated. Fortunately, Soyuz is comprised of three separate but integrated elements and, as several other components were in various stages of preparation, the Descent Module planned for TMA-21 was available as a replacement. The planned launch date only slipped by two days. This demonstrated the flexibility and versatility of both the Soyuz design and the Russian spacecraft processing system.

Cosmonaut Dmitry Kondratyev conducts an EVA at the Russian segment.

The damage was apparently due to "sloppiness" on the part of the transport team, which resulted in serious damage to the transport container and a 1.5 mm displacement in the base of the Descent Module. This was sufficient to create a micro-fracture in the pressure compartment, which would need detailed examination back at Energiya. It was not clear if this would result in taking the affected Descent Module out of the flight manifest permanently. Energiya reported that about 30 different elements of the TMA vehicle were in various stages of production at the time of the incident. Once the new element had been incorporated into the processing flow, preparations for the mission continued without further incident.

Once safely aboard the station, the new crew received their required safety and update briefings. They were given a light-duty weekend before joining their three colleagues in their six-person science program. There were now three cosmonauts working the Russian segment experiments, two Americans handling the U.S. segment, and Nespoli in Columbus (assisted by the Americans where necessary). The joint program for ISS-26/27 was stated to include 504 sessions of 41 experiments in the Russian segment, of which 7 were brand new investigations. There would be over 366 hours of work conducted during the ISS-26 phase. Over in the U.S. segment, the expedition would work on 111 experiments, of which 73 were from NASA. Of these, 22 came under the auspices of the National Laboratory status and a further 38 from other partner agencies. This entailed over 540 hours of planned crew time.

Following Christmas, New Year, and the Russian Orthodox Christmas on January 7, the crew prepared equipment for a Russian EVA on January 21. The 5 h 23 min EVA by Kondratyev and Skripochka saw them install and repair equipment. A second EVA was completed on February 16 lasting 4 hours 51 minutes during which the two cosmonauts installed Earth monitoring experiments to the exterior of Zvezda and removed two exposure panels from the same module and discarded a foot restraint. The two space walks logged 10 hours and 14 minutes of EVA time for the pair of cosmonauts.

The first weeks in the New Year were a busy time for the crew with the arrival of HTV-2, ATV-2, and STS-133, as well as departure and arrival of Progress craft. On March 14, Kondratyev assumed command of the ISS from Kelly. When the TMA-M crew departed on March 16, the Soyuz TMA-20 crew became the ISS-27 expedition, initially as a three-person residency. They would be joined by their three new colleagues on April 6, 2011 with the arrival of Soyuz TMA-21.

April saw much to celebrate on board the station. Nespoli celebrated his 54th birthday on April 6 and this was followed on April 12 by two important anniversary celebrations. The first was the 50th anniversary of Yuri Gagarin's historic first manned space flight and the second was the 30th anniversary of the first Shuttle mission. On April 17, new arrival Andrei Borisenko celebrated his 47th birthday on orbit. Yet another anniversary was celebrated on April 19 as the crew observed the 10th anniversary of the launch of the station's robotic arm systems. This was also the 40th anniversary of the launch of Salyut 1, the world's first space station, something that was overlooked somewhat by the world's media. The TMA-20 mission was full of celebrations, and actually missed two as well. Coleman had turned 50 the day *before* launch (had the mission launched as planned she would have celebrated her birthday in orbit), and Kondratyev celebrated his 42nd birthday the day *after* landing.

On April 29, the STS-134 mission was scrubbed for about a month due to technical issues, which meant that it would arrive at the station towards the end of this residency. On May 3 came the sad news of the death, aged 78, of Nespoli's mother Maria Motta, in Verano Brianza, northern Italy. The astronaut had been aware that his mother was ill and, as a mark of respect, the combined crew of six gathered the next day in the Cupola for a minute's silence in her memory as they gazed out over the Earth below them. The STS-134 mission arrived at the station on May 18 and remained docked until May 30, delivering the ExPRESS Logistics Carrier-3 and Alpha Magnetic Spectrometer-02. What was different on this mission was that the TMA-20 departed the station *before* the Shuttle, thus offering the opportunity for the Soyuz crew to photograph from a distance the almost complete complex with a Shuttle orbiter docked with it for the first time.

On May 22, Kondratyev passed the command of station to fellow cosmonaut Andrei Borisenko, formally ending the ISS-27 program which officially ceased with the undocking of TMA-20 two days later. During the fly-around, Nespoli took a series of stunning and unique photos of the ISS complex with the Soyuz TMA, Progress, ATV, and Endeavour docked to it. Never again would such a photo be possible. Only one mission remained on the Shuttle manifest and no Soyuz departures were planned during that flight.

It had been a busy expedition, reflecting the changes in the program as the final Shuttle missions arrived and new resupply craft were being introduced. The TMA-20 crew had spent over 157 days of their mission duration on board the station, with 87 days as part of the ISS-26 crew and about 71 days as lead ISS-27 crew.

Milestones

279th manned space flight
114th Russian manned space flight
107th manned Soyuz flight
 20th manned Soyuz TMA mission
 25th ISS Soyuz mission (25S)
26/27th ISS resident crew
Nespoli celebrates his 54th birthday (April 6)
Borisenko celebrates his 47th birthday (April 17)
Distant photography conducted of ISS with Shuttle and other current transport vehicles docked to it for the first and only time

4

Commencing the sixth decade: 2011–2012

STS-133

International designator	2011-008A
Launched	February 24, 2011
Launch site	LC39A, KSC, Florida, U.S.A.
Landed	March 9, 2011
Landing site	Runway 15, Shuttle Landing Facility, KSC, Florida, U.S.A.
Launch vehicle	OV-103 Discovery/ET-137/SRBs BI-144/SSME: #1 2044, #2 2048, #3 2058
Duration	12 da 19 h 3 min 51 s
Call sign	Discovery
Objective	ISS flight ULF-5

Flight crew

LINDSEY, Steven Wayne, 50, USAF, NASA commander, fifth mission
Previous missions: STS-87 (1997), STS-95 (1998), STS-104 (2001), STS 121 (2008) BOE, Eric Allen, 46, USAF, NASA pilot
DREW Jr., Benjamin Alvin, 48, civilian, NASA mission specialist 1, second mission
Previous mission: STS-118 (2007)
BOWEN, Steven George, USN, NASA mission specialist 2, third mission
Previous missions: STS-126 (2008), STS-132 (2010)
BARRATT, Michael Reed, civilian, NASA mission specialist 3, second mission
Previous mission: Soyuz TMA-14/ISS-19/20 (2009)
STOTT, Nicole Maria Passano, 48, civilian, NASA mission specialist 4, second mission
Previous mission: STS-128/129/ISS-20/21 (2009)

Flight log

When this crew was named, they were also announced as the final Shuttle crew. At the time, this was indeed planned as the final Shuttle mission, manifested to fly after STS-134. However, as had been the way of the Shuttle program since its inception, the manifest changed and the flight sequence altered. The main payload for STS-134 was delayed and the mission slipped in the launch schedule to fly after STS-133. Then STS-135 was added to the manifest as the new final Shuttle mission. The change in flight sequence was not the only one, as there was also a milestone alteration to the crew. In January 2011, mission specialist Tim Kopra was injured in an off-duty bicycle accident and his lengthy recovery saw Steve Bowen take his place on the mission. Bowen thus became the first (and only)

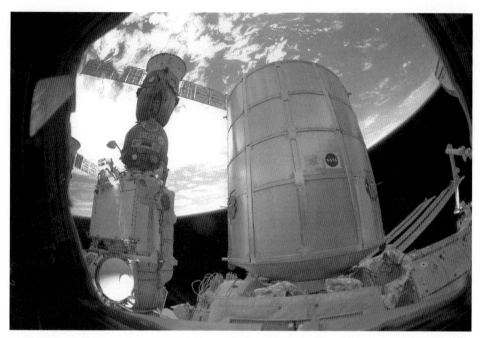

The newly attached Permanent Multipurpose Module (PMM) and a docked Soyuz are featured in this image.

NASA astronaut to fly back-to-back Shuttle missions, having just completed a flight as mission specialist on STS-132.

Aboard Discovery for its final voyage was the Leonardo Multi-Purpose Logistics Module (MPLM), which had been converted into the Permanent Multi-purpose Module (PMM) that would be attached to the station as an additional storage facility. Previously, MPLMs were returned back to Earth in the Shuttle payload bay full of unwanted equipment and trash; but, with volume at a premium on station, it had been decided to convert one of the three available MPLMs for permanent attachment. The Shuttle's cargo also included the ExPRESS Logistics Carrier 4, which was filled with equipment and spares. Among the delivered cargo was the Robonaut R2 humanoid robot, which was to be evaluated inside the station for its potential as a support for future EVAs or for activities outside the station that were potentially risky or inaccessible for an astronaut in a pressure suit. Reports suggested that later variants of the Robonaut could be used to support future operations on the Moon, at Mars, or the asteroids.

Discovery was rolled over to the VAB on September 9, 2010 and mated with the ET two days later. Discovery's final rollout to the launchpad occurred on September 20, with a planned launch for the end of October. However, problems with a leak in the Orbital Maneuvering System (OMS), followed by a main engine controller problem and a leak from a ground umbilical plate pushed the mission into 2011.

Prelaunch preparations were blighted with niggling problems, especially with the ET, where inner stringers had to be strengthened. Things did not bode well when further leaks were found in the tank's insulation and a seal had to be replaced. A loose screw in an inspection tool caused it to fall on to the ET and it was thought that another delay would ensue. Fortunately no serious damage was found and processing continued without further problems. Another issue, however, was the upcoming launch and docking of ATV-2 with the station and the launch of an ELV (Delta IV) from the Cape. To prevent these conflicts, a 24 h launch slip was proposed for STS-133, to allow time to dock the ATV with the station and still allow for the flight rule of 72 hours between station dockings. However, a slip on the ATV launch moved the Shuttle docking closer again, so NASA decided to return to the original schedule for the Shuttle—launching just 6 hours after the ATV docked with the space station. The scheduled date to launch the Delta, March 11, would require the Shuttle to land by March 10. This still allowed the mission plan, with a landing at the Cape planned for either March 8 or 9 and a 2-day contingency for safety. Launching vehicles into space, bringing them together in orbit, and returning them home again is never straightforward.

The ascent to orbit occurred without incident on February 24, and over the next two days the crew checked the orbiter's heat shield and EVA equipment. Following the backflip for further heat shield inspection by the station crew, Discovery docked on February 26 at the Harmony module. Within 2 hours, the internal hatches were open and the combined crew of 12 astronauts and cosmonauts completed the ceremonial greetings before getting straight down to the joint work program.

Another space first for this mission was the combined docking of all available resupply craft at the station at the same time—Shuttle, Progress, Soyuz, ATV, and the recently arrived Japanese HTV—something that would not be achieved again. A planned fly-around of the new Soyuz TMA-M was canceled by the Russians as an unnecessary risk for the new spacecraft on its maiden flight, a safety issue agreed to by both the American and Russian partners.

From inside the docked vehicles, the astronauts used the Shuttle RMS and station robotic arm to move the ELC-4 across to the truss structure on February 24 for unloading at a later date. There were two EVAs (totaling 12 h 48 min) completed during this mission, by Drew and Bowen.

The first EVA (February 28, 6 h 34 min) featured the installation of a backup power cable between the Unity and Tranquility nodes. The two astronauts also moved the now redundant failed 800 lb (362.88 kg) ammonia pump to the External Stowage Platform-2 for return to Earth (possibly during the STS-135 flight at this point) for postflight analysis and determination of its unexpected and unexplained July 2010 failure. The astronauts also installed a Japanese education exposure experiment that would be retrieved on the very next EVA.

Between EVAs, on March 1, the PMM was moved to its permanent position on the Earth-facing (nadir) port on Unity. Protection shields had been fitted to its exterior to ensure it would endure at least 10 years in orbit as part of the ISS. The second EVA (March 2, 6 h 14 min) featured a range of maintenance tasks and the retrieval of the Japanese education exposure experiment.

During the docked phase, logistics transfers continued and the crew assisted in outfitting the station to expand its scientific operations. The Robonaut unit, which was still boxed up in foam packaging, raised a few smiles during the crew's conversation with U.S. President Barack Obama when Lindsey joked that the crew was sure that every now and again they could hear scratching from inside the crate! The crew also tested a SpaceX DragonEye sensor, essentially a Light Detection and Ranging (LIDAR) system, designed to evaluate alternative technologies for use in future automated and manned spacecraft docking with the station.

The crew enjoyed a couple of days rest prior to undocking on March 6 after 7 days 23 hours 55 minutes of joint activities. The landing occurred during the night of March 9 and with it Discovery completed its final mission into space.

In a 27 yr career which began with the STS-41D mission during August and September 1984, the orbiter had logged 39 missions, completed 15,830 orbits, and flown 148,221,675 miles. A few hours after landing, Discovery was towed to the OPF for the final time, where it would be de-processed, decontaminated, and finally decommissioned before relocation to a museum for public display. Sadly, the final acts of the operational Shuttle era were being played out.

Milestones

280th world manned space flight
163rd U.S. manned space flight
 35th Shuttle ISS mission
133rd Shuttle flight
 13th Discovery ISS flight
 39th and last Discovery flight
First back-to-back Shuttle flight by an astronaut (Bowen)
First time public helped to choose crew wake-up songs

SOYUZ TMA-21

International designator	2011-012A
Launched	April 4, 2011
Launch site	Pad 31, Site 5, Baikonur Cosmodrome, Republic of Kazakhstan
Landed	September 16, 2011
Landing site	92 miles southeast of Dzhezkazgan, Republic of Kazakhstan
Launch vehicle	Soyuz-FG (serial number И15000-036), Soyuz TMA (serial number 231)
Duration	164 da 5 h 41 min 19 s
Call sign	Gagarin (in honor of the 50th anniversary of the first manned space flight)
Objective	ISS resident crew transport ISS-27/28 (S26)

Flight crew

SAMOKUTYAEV, Alexander Mikhailovich 41, Russian Federation Air Force, RSA Soyuz TMA commander/ISS-27/28 flight engineer
BORISENKO, Andrei Ivanovich, 46, civilian, RSA Soyuz TMA flight engineer/ISS-27 flight engineer/ISS-28 commander
GARAN Jr., Ronald John, 49, USAF (Retd.), NASA Soyuz TMA and ISS-27/28 flight engineer, second mission
Previous mission: STS-124 (2008)

Flight log

It was poignant that the next manned launch was Russian, as it occurred just eight days before the 50th anniversary of the world's first manned space flight, by Yuri Gagarin in Vostok on April 12, 1961. The radio call sign for Soyuz TMA-21 became "Gagarin" in celebration of that historic event. It was also fitting that the crew emblem of this Russian/American crew featured the name of (Alan) Shepard, the first American in space (suborbital), just three weeks after Gagarin's flight. It was an excellent way of linking the early pioneers to the modern day space explorers. The contrast between these two eras is readily apparent in the flight durations. The combined flights of Gagarin and Shepard logged just over 123 *minutes* in flight, whereas the TMA-21 crew were embarking on a 165-*day* mission, joining the crew that was already on the station when the "Gagarin" Soyuz left the pad.

The TMA-21 mission continued the uninterrupted science work on board the station, and would also be noted for receiving the final Shuttle missions on the manifest. In the Russian segment, the ISS-27/28 increment was planned to

Celebrating the 50th anniversary of Gagarin's historic journey. Garan, Samokutyaev, and Borisenko pose outside their Soyuz TMA-21 spacecraft, which bears the likeness of the first cosmonaut, and was given the call sign Gagarin in honor of the celebration. Photo credit: NASA/Victor Zelentsov.

conduct a program of 725 sessions covering 50 experiments, of which only three were new. To accomplish this, the Russian crew members were assigned 174 hours 25 minutes of experiment time during the ISS-27 phase and 359 hours 20 minutes under the ISS-28 program. Across in the American segment, there were a further 111 experiments, supported by a network of over 200 researchers around the world. NASA was sponsoring 73 of these experiments, with 22 under the auspices of the U.S. National Laboratory program and another 38 experiments sponsored by other partner agencies. These would take up over 540 hours of crew time.

This increment continued the rotational 3/6/3/6 crewing sequence of earlier expeditions. A three-person crew operated ISS-27 from March 16 to April 6, with the crew of six operating between May 7 and 24. The ISS-28 crew continued as a three-person occupation from May 24 to June 9, returning to a six-person crew between June 10 and September 16. This was of course very positive for the operational side of the ISS program and the overall long-term expansion of human space exploration at large. However, for those who record the assignments and activities for each space explorer or expedition, it was becoming more difficult to keep track of individual records as one expedition blended into the next.

The April 7 docking with Poisk was followed three hours later by crew transfer into the main station compartments. The combined crew completed the usual welcome and safety briefings procedures, highlighted by speaking to their families at Mission Control in Korolev (Moscow). One amusing incident occurred when Garan's wife reminded him that she was safely holding his credit card while he was "out of town". The Soyuz was mothballed on April 7 and the crews got down to their well-orchestrated blend of science, maintenance, housekeeping, safety, and some time out for sleep and personal hygiene.

The first scheduled visitors, on STS-134, were delayed by technical problems preventing the launch, so the resident crew continued with their own program, demonstrating the flexibility of the timeline of long space flights. The Shuttle mission finally arrived on May 18 and this was followed by the departure of the TMA-20 crew, signaling the end of the ISS-27 expedition with their undocking on May 23. Handover of command between Kondratyev and Borisenko had taken place the day before.

The crew continued as a three-person residency until the arrival of the Soyuz TMA-02M on June 9. During the residency, the complex was reboosted several times using the engines on the ATV-2, to maintain its operational altitude while the science program continued on board the station. In July, the final Shuttle mission (STS-135) visited the station, marking the end of Shuttle operations. From this point, the Soyuz and Progress vehicles of Russia, Europe's ATV, and Japan's HTV would be the only operational resupply systems for the station. There were plans under way to launch new commercial vehicles to the complex, in order to test the feasibility of such systems for future use. It also emerged that detailed studies were under way to evaluate the use of the station up to 2028, if the partners could verify that the various components could work effectively and safely for that long.

On August 1, Samokutyaev and Volkov completed a 6 h 23 min EVA, deploying scientific experiments and an experimental high-speed laser communication system outside the Russian segment. The two cosmonauts also removed a rendezvous antenna which was no longer needed and deployed by hand a small, 57 lb (25.85 kg) ham radio satellite, as part of the 50th anniversary celebrations of Gagarin's flight in Vostok. A planned relocation of the Strela-1 ("Arrow-1") crane was postponed until 2012. The final task was to have photographs taken, with the two men holding pictures of Gagarin, spacecraft designer Sergei Korolev, and space theorist Konstantin Tsiolkovsky and using the Earth as a backdrop. The three images had been displayed inside Zvezda for years and were returned there after their trip into open space.

Later that month, on August 24, station operations were dealt a blow with the loss of Progress M-12M some 5 minutes 25 seconds into the launch phase. The Soyuz-U (R-7) launch vehicle's third stage ignited for 25 seconds but, following a loss of pressure, promptly shut down resulting in a loss of velocity and subsequent crash. This was the first ever loss of a Progress craft during launch since the series began back in 1978. On board the resupply craft was 5,863 lb (2,659 kg) of supplies, propellant, and oxygen. Repercussions from this loss included delaying the landing of the TMA-21 crew by a week, but the TMA-02M crew would return as planned in mid-November. The next manned flight would have to be delayed from September 22 to late October or early November.

The Soyuz TMA had an on-orbit operational life (mothballed and docked with the ISS) of 210 days and there were more than enough supplies on board the station (thanks to the final Shuttle flights) to keep a crew sustained for over a year, so there was no immediate risk to the crew on board. Nevertheless, it was still a difficult time for the Russians, with talk of lack of confidence in their space hardware and manufacturing/processing systems and the potential prospect of abandoning the station by all crew members if the R-7 could not be recertified for operations. After the next two unmanned launches, one commercial and another Progress, this requalification would come during the ISS-29 crew duty shift; the ISS-28 crew was due to come home.

After the delay due to the loss of the Progress resupply craft, the handover of station command from Borisenko to Fossum took place on September 14. The TMA-21 crew returned to their Soyuz craft and closed the hatches late on the following day, with undocking occurring in the early hours of September 16. There was some anxiety in Russian Mission Control after the planned 3 min black-out period, when communications with the cosmonauts in the Descent Module could not be established. Fortunately, contact was soon restored and the trio landed safely, apparently unaware of any communication problems.

During the 165-day flight, the crew logged approximately 162 days on board the station, with 160 days as part of expeditions. This included 45 days as members of the ISS-27 phase and 115 days as the prime ISS-28 crew.

Milestones

   281st manned space flight
   115th Russian manned space flight
   108th manned Soyuz
    21st manned Soyuz TMA
    26th ISS Soyuz mission (26S)
27/28th ISS resident crew
Became the last crew to host a visiting Shuttle mission (STS-135)
Borisenko is accredited with being the 200th person to enter the ISS facility

STS-134

International designator	2011-020A
Launched	May 16, 2011
Launch site	LC39A, KSC, Florida, U.S.A.
Landed	June 1, 2011
Landing site	Runway 15, KSC, Florida, U.S.A.
Launch vehicle	OV-105 Endeavour/ET-122/SRB BI-145/SSME: #1 2059, #2 2061, #3 2057
Duration	15 da 17 h 38 min 22 s
Call sign	Endeavour
Objective	ISS ULF-6

Flight crew

KELLY, Mark Ehward, USN, NASA commander, fourth mission
Previous missions: STS-108 (2001), STS-121 (2006), STS-124 (2008)
JOHNSON, Gregory Harold, USAF (Retd.), NASA pilot, second mission
Previous mission: STS-123 (2008)
FINCKE, Edward Michael, USAF, NASA mission specialist 1, third mission
Previous missions: Soyuz TMA-4/ISS-9 (2004), Soyuz TMA-14/ISS-18 (2008)
VITTORI, Roberto, Italian Air Force, ESA mission specialist 2, third mission
Previous missions: Soyuz TM-34 (2002), Soyuz TMA-6 (2005)
FEUSTEL, Andrew Jay, civilian, NASA mission specialist 3, second mission
Previous mission: STS-125 (2009)
CHAMITOFF, Gregory Errol, civilian, NASA mission specialist 4, second mission
Previous mission: STS-124/126/ISS-17/18 (2008)

Flight log

The 25th mission of Endeavour, the last vehicle to join the fleet in 1992, was also to be its final space flight. Designated STS-134, this was a utilization and logistics mission that had originally been manifested as the final flight in the Shuttle program, until STS-135 was added. Endeavour's swansong delivered the Alpha Magnetic Spectrometer-2 (AMS-02) experiment to the space station. This is a particle physics detector designed to search for a range of unusual matter by measuring cosmic rays. It was planned that the gathered data would be used in research into the study of the formation of the universe, in the search for evidence of dark matter, strange matter, and antimatter. In addition to the AMS, Endeavour carried the ExPRESS Logistics Carrier-3 (ELC-3), a platform full of spares. The mission would also see the final scheduled EVAs by Shuttle crew members, ending an impressive 28 yr series of space walks by orbiter crews.

One of the first pictures of a Shuttle docked with the ISS from the perspective of a Soyuz spacecraft (TMA-20).

The original launch date in March was delayed due to technical issues. This was also compounded by the tragic shooting in Arizona of Congresswoman Gabrielle Gifford, the wife of mission commander Mark Kelly. The event also placed NASA in a difficult situation, if Kelly required more time with his wife as she recovered. Replacing a commander of a mission so close to launch had never occurred in previous missions, so veteran Shuttle commander Richard Sturckow was assigned as backup commander as a precaution. Gabrielle Gifford's recovery was remarkable, to the point that she was able to attend the planned April 29 launch attempt. However, as a result of technical problems with an Auxiliary Power Unit on the Shuttle, the launch was canceled four hours prior to liftoff and postponed until May. Fortunately, Gifford was also able to make it to the Cape to witness the May 16 launch, unlike U.S. President Barrack Obama and his family, who had witnessed the April 29 abort, toured the center, and met with the Gifford's but were unable to reschedule a visit for the May launch. The series of delays also meant that Mark Kelly would not join his twin brother Scott in orbit. Scott had been on the space station since the previous October and was the serving space station commander, but returned home before STS-134 launched.

Endeavour and its experienced crew of six left the pad at KSC for the final time on May 16, 2011, just over 50 years after Alan Shepard became the first

American in space on May 5, 1961, and nine days prior to the 50th anniversary of President John F. Kennedy's commitment to place Americans on the Moon by the end of 1969. Though the entire Endeavour crew consisted of space flight veterans, Fincke and Vittori were making their first flights on the Shuttle, having previously flown to the ISS atop Russian rockets aboard Soyuz spacecraft. Vittori became the final non-NASA astronaut to fly a Shuttle mission.

During the standard 2-day flight to the ISS, the Shuttle crew checked out the RMS and used it to examine the Thermal Protection System. Meanwhile, the EVA crew prepared the EVA suits and equipment. The crew was also scheduled to evaluate the Sensor Test of Orion Relative Navigation Risk Mitigation (STORRM) during their mission. This equipment evaluated sensor techniques for routine spacecraft docking with the ISS. Evaluations would be taken during rendezvous and docking and a later re-rendezvous with the station following the undocking towards the end of the mission.

Endeavour completed its 12th and final docking with the ISS on May 18 at PMA-2, with internal hatches opened just a few hours later. Following the usual welcoming ceremony and safety briefings by the resident station crew, the joint program of activities began. The ELC-3 pallet was transferred from the RMS to Canadarm2 some 5 hours after docking and was then installed on the P3 truss. Loaded on the pallet were two communication antennas, a high-pressure gas tank, and spare parts for the Dextre robotic device. The next day, the AMS-02 unit was transferred to the top of the S3 truss, where it is scheduled to remain to at least 2020. To avoid interference with other systems and storage platforms, the unit was installed at a 12° angle. The AMS science program is a global program involving 600 scientists and technicians from 56 institutions across 16 nations. The simpler AMS-01 flew on the Shuttle in June 1998, as part of the STS-91 payload.

During the mission, Italian astronaut Roberto Vittori carried out a program of six ASI-sponsored experiments under the DAMA Mission (named for the AMS search for dark matter). Vittori served as a test subject of two ESA experiments which studied possible changes to his body after the flight. This was part of a program of investigations in the fields of technology, nuclear power, biology, and materials. The Italian astronaut also assisted in the transfer of cargo into the station and unwanted material back into the orbiter.

The mission included four EVAs, totaling 28 h 33 min, shared between Feustel, Chamitoff, and Fincke. The first EVA (May 20, 6 h 19 min) was conducted by Feustel and Chamitoff. They installed an ammonia jump cable that would connect the coolant loops of the station's P3 and P4 segments, installed a cover on an SARJ and removed the MISSE 7A and 7B experiment packages from the Express Logistics Carrier-2, replacing them with the MISSE-8 experimental package. An external communication antenna was also installed on the Destiny Laboratory, to provide a link between the various ExPRESS Logistics Carriers mounted on the outside of the station. An issue with a carbon dioxide level sensor on Chamitoff's suit caused concern in the latter stages of the EVA, with some tasks delayed to a subsequent space walk to maintain safety. Prior to this there had been no indication of a CO_2 problem, but the EVA ended a little earlier than planned and the issue did not reoccur.

The second EVA (May 22, 8 h 7 min) by Feustel and Fincke featured a program of maintenance work. They refueled one of the station's port side cooling loops with 5 lb (1.26 kg) of ammonia and also lubricated the port side SARJ and one of the "hands" on Dextre. Storage beams were fixed on the S1 truss segment and a camera cover was installed on Dextre to end the space walk.

Between the second and third EVAs, the resident station crew was reduced from six to three with the departure of the ISS-27 crew in Soyuz TMA-20 on May 23. The delay in launching Endeavour, coupled with a delay in undocking the Soyuz, had created the unique situation of a Shuttle being docked with the station during the departure of a Soyuz crew. The remaining three TMA-21 crew members on the station now became the ISS-28 resident crew. As the TMA-20 retreated to about 600 ft (182.88 m) away from the Rassvet Module, Nespoli took a series of stunning digital images and video of Endeavour docked with the station from the viewing port in the Soyuz Orbital Module. At the same time, the whole station was rotated about 130°, which was a rare maneuver in itself, allowing the Italian astronaut some of the best views possible during a 30 min period.

Looking towards the future, the third EVA made use of a new EVA pre-breathe protocol, known as the In-Suit Light Exercise (ISLE). Instead of the normal Campout Pre-breathe Protocol System, where the astronauts breathe pure oxygen for 60 minutes in the airlock, this new technique saw the air pressure of the airlock lowered to 10.2 psi (703 hPa). The astronauts then put on their space suits and performed light exercise, before resting for an additional 50 minutes, breathing pure oxygen all the while prior to exiting the airlock to conduct the EVA.

Activities during EVA 3 (May 25, 6 h 54 min) completed by Feustel and Fincke included an upgrade to Canadarm2, installing a power and data grapple feature on the Russian Zarya module that would enable the station arm to "walk" across to the Russian segment and conduct robotic operations there. This had not been possible before and would now extend the range of the station's robotic arm system. The astronauts also installed additional cables between the American Unity module and the Russian Zarya module, which would provide backup power to the Russian segment. A series of photos were taken of the effects of the Zarya thrusters on the skin of the module, along with verbal information on the condition at various work sites which was relayed to the ground. The astronauts also completed the work postponed from the first EVA due to the suit malfunction.

The fourth and final EVA (May 27, 7 h 24 min) by Fincke and Chamitoff was the last scheduled EVA by a Shuttle crew, ending a program which had started back in 1983 during STS-6. It had included 162 excursions, many of these essential for station assembly since 1998. Ironically, the final "Shuttle" EVA was not actually from a Shuttle Orbiter, but from the Quest airlock on ISS. The last EVAs directly from a Shuttle Orbiter had occurred in 2009 during the STS-125 Hubble Servicing Mission. Since 2001, most Shuttle crew EVAs had actually been from the Quest airlock on the station rather than through the middeck airlock/hatch system on the orbiter. For this final EVA, the astronauts assisted with the transfer of the 50 ft (15.24 m) Orbiter Boom Sensor System (OBSS) from the orbiter across to the starboard truss and installed a new grappling system, extending the reach of the station's robotic arm even farther. This EVA also saw station EVA operations pass the 1,000 h mark since assembly began in December 1998.

In three space walks, Fincke logged 22 hours 25 minutes, while Feustel accumulated 21 hours 10 minutes. Chamitoff recorded 13 hours 43 minutes in his two trips outside. The EVAs were followed by the completion of cargo transfers and logistics to and from the ISS.

The astronauts conducted a number of media and public outreach activities before undocking from the station on May 29, after 11 days 17 hours and 41 minutes attached to the complex. A fly-around and photo-documentation of the station's exterior was followed by a final test of the STORRM rendezvous system from 1,044 ft (318.21 m) below and 300 ft (91.44 m) behind the station. When Endeavour landed at KSC, the youngest orbiter had completed the 25th mission of its career, having traveled 122,883,151 miles (19,771,898 km) in 4,677 orbits of Earth and 299 days in space. Endeavour would remain at the Cape for decommissioning before being dispatched to its new museum home.

With the installation of the AMS-2 payload, the official assembly complete point of the U.S. segment was met. In fact, most of the station was now complete, with only a couple of Russian modules and some desired large experiments to be launched. The station was no longer a construction site but a research facility. This was made possible by the series of Shuttle assembly missions since 1998 ... and now there was only one left on the manifest.

Milestones

282nd world manned space flight
164th U.S. manned space flight
 36th Shuttle ISS mission
134th Shuttle flight
 25th and final flight of Endeavour
 12th Endeavour ISS flight
Vittori first Italian ESA astronaut to fly on both Soyuz and Shuttle missions
Last non-NASA astronaut to fly on a Shuttle mission (Vittori)

SOYUZ TMA-02M

International designator	2011-023A
Launched	June 7, 2011
Launch site	Pad 1, Site 5, Baikonur Cosmodrome, Republic of Kazakhstan
Landed	November 22, 2011
Landing site	Near the town of Arkalyk, Republic of Kazakhstan
Launch vehicle	Soyuz-FG (R-7) (serial number И15000-037), Soyuz TMA-M (serial number 702)
Duration	167 da 6 h 12 min 5 s
Call sign	Eridanus
Objective	ISS resident crew transport ISS-28/29 (7S)

Flight crew

VOLKOV, Sergei Alexandrovich, 38, Russian Federation Air Force, Russian, RSA, Soyuz TMA-M commander, ISS-28/29 flight engineer, second mission
Previous mission: Soyuz TMA-12/ISS-17 (2008)
FURUKAWA, Satoshi, 47, civilian (Japanese), JAXA, Soyuz TMA-M and ISS-28/29 flight engineer
FOSSUM, Michael Edward, 53, NASA, Soyuz TMA-M flight engineer, ISS-28 flight engineer, ISS-29 commander,third mission
Previous missions: STS-121 (2006), STS-124 (2008)

Flight log

Flying the second upgraded TMA-M spacecraft into space were another truly international trio, who docked their spacecraft with the Rassvet module on June 9 (June 10, Moscow time). They entered the space station in the early hours of the following day to complete the required safety and update briefings. During the 2-day flight to the space station, the crew had completed the flight development tests begun on Soyuz TMA-M, further qualifying the new vehicle for operational missions.

In the early part of their residency, the three new crew members participated in preparations for the departure of the ATV-2 resupply vehicle from the aft Zvezda port on June 20 and the arrival of Progress M-11M at that same docking location three days later. On June 28, the six crew members had to shelter in their respective Soyuz craft (TMA-21/TMA-02M) as an unidentified piece of orbital debris (designated Object 82618, unknown) passed within just 820 ft (249.93 m) of the station, possibly the closest near miss in the station's history.

The crew's experiment program included, in the Russian segment, 725 sessions on 50 experiments, of which 47 were continuations from the previous increments.

Mike Fossum and Satoshi Furukawa make preparations to power up Robonaut 2.

To accomplish this target, there were 359 hours 20 minutes planned for the crew, as well as supportive work during the exchange of crews. There was no dedicated ISS-29 press kit released (instead the kits went from ISS-27/28 to 30/31) to identify research work in the American segment, but it is clear that the work continued without interruption in all sections of the station during this period.

The final Shuttle mission (STS-135) arrived at the station on July 10 and remained docked with the station until July 18. On July 12, resident crew members Fossum and Garan donned American EMUs and exited the Quest airlock for a 6 h 30 min space walk (see STS-135 entry) which was the final space walk of the Shuttle era. On August 1, Volkov accompanied Samokutyaev on a 6 h 23 min EVA from the Russian segment (see Soyuz TMA-21 entry).

On August 14, the Zarya module, the first element of the station launched in November 1998, completed 73,000 orbits of Earth.

On September 14, Fossum assumed command of the outpost from Borisenko, ending the ISS-28 phase and commencing the ISS-29 phase. The official end of the outgoing expedition occurred with the undocking of TMA-21 two days later. With the departure of the ISS-28 prime crew, the ISS-29 crew continued as a three-person residence pending the arrival of TMA-22. For the next few weeks, science and maintenance occupied the crew's time. This included work with Robonaut 2, putting the unit through some at times difficult mobility tests of its hand and neck joints.

On September 29, after a decade of being the only space station in orbit, the ISS was joined by a new neighbor—the Chinese Tiangong-1 ("Heavenly Palace-1") mini space module. This was similar to, but slightly smaller than, the early Soviet Salyut space stations launched in the 1970s. A month later, on October 31, the unmanned Shenzhou 8 was launched into orbit on a mission to evaluate the new space station's docking mechanism. Clearly, a new era of space station operations had begun.

Back on the ISS the science work continued. This trio would be joined by their colleagues when Soyuz TMA-22 docked at the Poisk module on November 16. This was to be a short six-person residency, due to the delays caused by the loss of Progress M-12M the previous August. The first few days of the full crew were a busy time, as the TMA-02M trio prepared to return home. The formal transfer of command from Fossum to Dan Burbank occurred on November 20 and the ISS-29 trio undocked their Soyuz in the early hours of November 22.

The atmospheric burn-up of the discarded modules was captured on video by the station residents as the Descent Module containing the three crew members continued its descent towards Earth. The crew landed safely, although in subzero temperatures. Shortly afterwards, Volkov was flown back to the Cosmonaut Training Center near Moscow and Fossum and Furukawa flew on a NASA jet back to Houston, Texas for postflight readaptation and debriefings.

The crew had spent 166 days on board the station out of the 168 days logged in space. Of these, 97 days were spent as part of the ISS-28 expedition and 67 as prime ISS-29 residents.

Milestones

283rd manned space flight
116th Russian manned space flight
109th manned Soyuz
2nd manned Soyuz TMA-M and 2nd test flight of the new variant
27th ISS Soyuz mission (7S)
28/29th ISS resident crew

STS-135

International designator	2011-031A
Launched	July 8, 2011
Launch site	LC39A, KSC, Florida, U.S.A.
Landed	July 21, 2011
Landing site	Runway 15, Shuttle Landing Facility, KSC, Florida, U.S.A.
Launch vehicle	OV-104 Atlantis/ET-138/SRB BI-146/SSME: #1 2047, #2 2060, #3 2045
Duration	12 da 18 h 27 min 52 s
Call sign	Atlantis
Objective	ULF-7

Flight crew

FERGUSON, Christopher John, 49, USN Retired, NASA commander, third mission
Previous missions: STS-115 (2006), STS-126 (2008)
HURLEY, Douglas Gerald, 44, USMC, NASA pilot, second mission
Previous mission: STS-127 (2009)
MAGNUS, Sandra Hall, 46, civilian, NASA mission specialist 1, third mission
Previous missions: STS-112 (2002), STS-126/ISS-18/119 (2008/2009)
WALHEIM, Rex Joseph, 48, USAF (Retd.), NASA mission specialist 2, third mission
Previous missions: STS-110 (2002), STS-122 (2008)

Flight log

The finale of the 30 yr Space Shuttle program came in July 2011 with the flight of STS-135. This was a mission added to the manifest to utilize the remaining available hardware and was a final opportunity to stock up the station and remove a quantity of unwanted material and trash before finally retiring the fleet.

When Ken Ham and the rest of the STS-132 crew returned in May 2010, discussions were already under way over the possibility of flying one more Atlantis mission. In light of this, Ham called his recent mission "the first last flight of Atlantis" and so it proved. The orbiter had one processing cycle to go through, as rescue orbiter for STS-134—the "first" last Shuttle flight—but once the new flight had been authorized, this became the final processing cycle as, following the STS-134 mission, the vehicle was processed for STS-135.

Towards the end of 2010, it had become more likely that the mission would indeed fly once the funding had been organized, and by January 2011 the mission was included in the internal flight roster for planning purposes. In February,

The final Space Shuttle launch, July 8, 2011.

NASA management was told that the mission would fly even if adequate funds were not found, but the budget for the mission was authorized in April after saving funds in other areas. By that time, preparations for the mission were well on the way towards completion anyway.

For the final time, the Shuttle ground processing team geared up for a launch. The stacking of the SRBs began towards the end of March 2011, with the ET being attached on April 25. Atlantis was rolled from the OPF across to the VAB on May 16, and by May 18 the stack was completed. The rollout to the pad occurred on the night of May 31/June 1, with the Rafaello MPLM being installed in the payload bay of Atlantis on June 20. The manifest also included a significant number of commemorative items and the U.S. flag that had flown on STS-1, along with a special 9/11 flag. With everything ready for a planned July 8 launch, the July 4 Independence Day weekend was kept free to allow for some extra processing margin. All eyes looked at the weather, which appeared to be the only concern (as it had been so many times) but when NASA affirmed the July 8 launch date, even nature cooperated to see the Shuttle program off in fine style.

Designated ULF-7, the payload included the Raffaello MPLM packed with over 9,000 lb (4082.4 kg) of supplies and the Robotic Refueling Mission (RRM) experiment. This was an experiment to demonstrate and test the tools, techniques, and technologies required to develop a robotic satellite-refueling capability, even though the target vehicle might not be designed to be refueled. The astronauts were also to return a failed ammonia pump for evaluation by engineers prior to refurbishment for relaunch at some future date.

After an incident-free launch and ascent to orbit, the crew prepared for the ISS docking by checking the vehicle, inspecting the TPS, and setting up the rendezvous tools and EMUs. Throughout the mission, the crew received well wishes from family, friends, fellow workers, and the general public as the final Shuttle flight continued. Docking with the station occurred on July 10, with the prime ISS resident crew tolling the ship's bell for an incoming spacecraft one final time for a Shuttle orbiter. One hour and 40 minutes after docking, the hatches were opened and the two crews welcomed each other, followed by the mandatory safety briefings and status updates. The only EVA during the mission would be conducted by the station crew from Quest.

Work began almost immediately, with the RMS used to relocate the 50 ft (80.45 m) external boom to Canadarm2 for the inspection of the Shuttle Thermal Protection System. As the boom was now permanently part of the station, this inspection could not be completed by the crew the day after launch. On July 11, the Raffaello module was transferred to the Harmony Node in a 30 min operation. The supplies included 2,677 lb (1,214.28 kg) of food, enough for a full year for the station crews. The crew relocated some of the cargo from Raffaello into PMA-3, with the supplies packed in 17 different racks inside the pressurized logistics module. These included eight Resupply Stowage Platforms (RSP), two Inter-mediate Stowage Platforms (ISP), six Resupply Stowage Racks (RSR), and one Zero Stowage Rack. An additional 2,228 lb (1,010.62 kg) of cargo was stowed on the middeck of Atlantis which also had to be transferred to the station. Sandra Magnus was loadmaster over a planned 130 hours of unloading time during the docked phase. Once empty, the module would be refilled with 5,666 lb (2,570.09 kg) of equipment no longer needed on the station, plus discontinued logistics and trash. After an evaluation of available consumables, an extra day was added to the mission to give the crew additional time to relocate all the cargo and supplies between the vehicles.

The EVA from the Quest airlock was performed by ISS astronauts Fossum and Garan on July 12. It lasted 6 hours 3 minutes. The reason for the station crew performing the EVA was essentially one of time and experience. Confirma-tion that STS-135 would actually launch came late in the cycle, so the training of the crew focused mainly on getting to and from the station and handling the massive cargo transfer. Contingency EVA training was included, but as the astro-nauts were all Shuttle veterans this made the compressed training cycle much easier to accomplish. The two station resident astronauts, Garan and Fossum, had logged nine previous EVAs between them, three of which were performed together during STS-124 in 2008, so they were used to working together as a team. It was also possible that with so many supplies being delivered, the weight saved from flying no more than four crew members could be reallocated to the logistics manifest.

The objective of this EVA was to retrieve a faulty 1,400 lb (635.04 kg) ammonia pump module which had failed in 2010 and had been stowed in the

External Storage Module 2 during STS-133 earlier in the year. The two ISS astronauts relocated it to the cargo bay of Atlantis to be returned and refurbished as a spare unit. They also set up the RRM experiment on an external pallet and released a stuck latch on the Data Grapple Fixture at the front of Zarya. This would extend the operating envelope of Canadarm2 across to the Russian segment to support robotics work. A further material experiment was also deployed from a carrier located on the station's truss. This was the eighth such experiment, with this one focused upon optical reflection materials. Originally attached during STS-134, it had not been deployed due to concerns from outgassing from the AMS unit. Finally, to close out the EVA, insulation was installed on the end of the Tranquility PMA in an area that was exposed to the effects of sunlight.

Throughout the mission, the Shuttle crew received a number of special wake-up calls in celebration of the end of the program. On July 11, much was made in reports of an "all American meal" which featured grilled chicken, corn, baked beans, cheese, and the traditional apple pie. This was also reported on the NASA website. The meal was originally planned for July 4, but the launch delays postponed it for a week.

On July 15, almost at the end of the 30 yr Shuttle program, U.S. President Barack Obama called Atlantis, wishing them well on their mission. The crew later solved a problem with the fourth general purpose computer on Atlantis, which required it to be rebooted to get it up and running again. Later, the EVA suits were reconfigured in order to leave them behind on the station. As the checklist of tasks remaining shortened, so the four Shuttle astronauts supported the station crew in relocating some of the cargo they had delivered to ease the post-docking workload for the resident crew as much as they could. The Shuttle crew also repaired a broken access door to the Shuttle air revitalization system, where the lithium hydroxide canisters that purified the air inside the orbiter were exchanged. On July 16, the 42nd anniversary of the launch of Apollo 11, Ferguson formally presented the station crew with the historic U.S. flag that had flown aboard Columbia on STS-1 30 years before. This flag will remain on board the space station until the next crew launched from the soil of the United States arrives at the orbiting facility to return it to Earth once again.

Hatches between the two spacecraft were closed for the final time on July 18 after 7 days 21 hours and 41 minutes. The next day, after a few hours' sleep on board Atlantis, the crew undocked from the station after 8 days 15 hours and 21 minutes of being attached to the orbital facility, ending a period of Space Shuttle station dockings that had begun 16 years before, with the flight of STS-71 to Mir in 1995. Safely undocked, the crew backed the orbiter away for the formal fly-around maneuver to photo-document the exterior of the station. The station was yawed 90 degrees for an optimum view during the 27 min photo opportunity, which captured never-before-seen views of the longitudinal axis of the station from the Shuttle. With this completed, it was time to fire the separation engines and depart from the vicinity of the station to begin final preparations for the flight home.

On the day before landing (July 20), the crew performed the traditional pre-landing checks of the Thermal Protection System, Flight Control Systems, and RCS engines for the final time on a Shuttle mission. The last science objectives of the program were completed with the deployment of the PicoSat technology demonstration satellite from a small canister in the payload bay and an onboard experiment on osmosis was also conducted by the crew. On July 21, to the wake-up call of *God Bless America* by Kate Smith, there was a tribute to all those who had been involved in the program since its inception over 40 years ago. Even the weather was cooperating, helping to celebrate the end of an era of American manned space flight in fine style as Atlantis swooped to a perfect pre-dawn landing at the Cape.

By the end of its last mission, Atlantis had traveled 125,935,769 million miles (20,263,063 km) over 33 missions, logging over 307 days in space, and completing 4,848 orbits of Earth. When the crew disembarked, there remained only the period of decommissioning after the mission and then a program of preparations for shipping the Atlantis to its new museum home. But before the vehicle had cooled down from its fiery reentry, the celebrations and emotional recollections had begun. For commander Ferguson, the realization that the Shuttle program was over came when the wheels stopped on the runway and the vehicle was powered down. In the pre-dawn darkness the displays went blank and the vehicle fell silent, creating a "rush of emotion" for the commander.

The Shuttle program had created many milestones and memories over 30 years, but never again would an orbiter of that design venture into space. Its work was done and it was time to move aside for new generations of human spacecraft to write the next pages in space history. The Shuttle era was finally over.

Milestones

284th world manned space flight
165th U.S. manned space flight
 37th and final Shuttle ISS mission
135th Shuttle flight
 33rd and final Atlantis flight
 12th Atlantis ISS flight
Final Shuttle flight of program
First four-person Shuttle crew since STS-6 in April 1983

SOYUZ TMA-22

International designator	2011-067A
Launched	November 14, 2011
Launch site	Pad 1, Site 5, Baikonur Cosmodrome, Republic of Kazakhstan
Landed	April 27, 2012
Landing site	Approximately 56 miles (90 km) northeast of the town of Arkalyk, Republic of Kazakhstan
Launch vehicle	Soyuz-FG (R-7) (serial number И15000-038), Soyuz TMA (serial number 232)
Duration	165 da 7 h 31 min 34 s
Call sign	Astra
Objective	ISS resident crew transport craft ISS 29/30 (28S)

Flight crew

BURBANK, Daniel Christopher, 50, U.S. Coast Guard (Retd.), NASA Soyuz TMA/ISS-29 flight engineer, ISS-30 commander, third mission
Previous missions: STS-106 (2000), STS-115 (2006)
SHKAPLEROV, Anton Nikolaevich, 39, Russian Federation Air Force, RSA Soyuz TMA commander, ISS-29/30 flight engineer
IVANISHIN, Anatoly Alekseevich, 42, Russian Federation Air Force, RSA Soyuz TMA and ISS-29/30 flight engineer

Flight log

The delayed launch of TMA-22 finally took place on November 14 through a raging snowstorm, but reached orbit without too much difficulty. When NASA tried to launch Apollo 12 through a thunderstorm exactly 42 years earlier, on November 14, 1969, things were very different. That vehicle was hit by lightning and almost suffered a major systems failure seconds after launch. It was a tense few minutes that convinced the American agency never to launch in the rain again. The Russians do not seem to share the same concerns, so the Soyuz was launched exactly on time for its rendezvous with the Poisk module of ISS, with which it docked on November 16.

Following the leak checks, the hatches were opened to begin a very full four and a half days of briefings and handover operations before the outgoing resident crew came home. In those four days, Burbank took over command of the station from Fossum, who returned to Earth with his TMA-02M colleagues on November 22, ending the ISS-29 expedition and starting the ISS-30 phase. The delays caused by the loss of the Progress in August had shortened the overlap of the two crews.

An unpiloted Progress resupply vehicle docks with the ISS, providing regular deliveries to each resident crew and a method of disposing of unwanted trash.

The first month for the new team was spent settling in to their new home and continuing the science program. The two Russians on this crew were planned to conduct just 28 hours of work in support of the abbreviated ISS-29 phase of science, but 204 hours under the ISS-30 science program in the Russian segment. This amounted to 356 sessions over 46 experiments, including two new investigations. Most of the U.S. segment work in the Destiny, Columbus, and Kibo laboratories was continuations of earlier investigations.

Apart from the science, work continued on Robonaut 2 (or R2). During December 15 and 16 things did not go well, with fault messages regularly appearing in the android's systems. Further work was delayed until January. For night passes over two days from December 22, Burbank observed and photographed the Comet Lovejoy, which he described as the "most amazing thing he had witnessed," echoing the sentiments of the Skylab 4 astronauts when they observed Comet Kohoutek 38 years earlier.

The next resident crew arrived in the middle of these observations on December 23, docking at the Rassvet module on the nadir side of Zarya. The hatches opened and the day was spent in briefings, ceremonies, and bringing the newcomers up to speed before the 3-day Christmas holiday. The New Year celebrations in orbit actually extended for 24 hours, as the station ventured over the International Date Line 16 times each Earth day. The celebrations were soon followed by Ivanishin's 43rd birthday on January 15.

The mission progressed with more science, a new arrival in the form of Progress M-14M, more work with R2, and preparations for a planned February EVA from the Russian segment. On January 27 it was announced that the six-person presence on the station would be extended for a while. The next launch of a crew had been delayed until mid-May, as the planned Soyuz vehicle had to be exchanged with the next one in the sequence. This meant that the Burbank crew would not return home until late April and the Kononenko crew would also have to stay a little longer than planned, as their replacements were delayed from May to mid-July.

On February 15, Burbank's hard work with R2 finally paid off, as the human astronaut shook hands with the robotic one for the first time inside the Destiny module, while NASA proudly announced: "Man meets machine aboard the ISS." Further tests were planned over the next few weeks, but it was a great start for a machine that was hoped would assist in delicate operations on future spacecraft. The R2 device later used sign language to say: "Hello, world."

The Russian EVA took place from Pirs on February 16. Kononenko and Shkaplerov relocated the Strela-1 ("Arrow-1") crane from Pirs to Poisk in preparation for the replacement of the Pirs module with the new Russian Multipurpose Laboratory Module Nauka ("Science") component. This was planned for later in 2012 but subsequently delayed once again. The two cosmonauts also installed a material science experiment on Poisk, collected organic test samples from Zvezda, and installed five debris shields on the Service Module. The EVA ended after 6 hours 15 minutes.

More celebrations occurred on February 20. Not only was it Shkaplerov's 40th birthday, it was also the 50th anniversary of the first American orbital flight of John Glenn aboard Friendship-7. Burbank, Pettit, and Kuipers spoke to Glenn from orbit as part of the celebrations of his flight. On March 7, the highly anticipated Robotic Refueling Mission (RRM) began, with several days of external operations using Dextre and Canadarm2 coordinated by the Canadian Space Agency and the NASA Goddard Space Flight Center (GSFC). This was a demonstration of the potential for robotic complexes to refuel satellites and included opening and closing valves, cutting through wires with millimeters of clearance, removal of insulation, and fuel transfer. The hardware had been delivered to a pallet during STS-135. Canadarm2 and Dextre returned to the Mobile Base System on March 12 at the end of the RRM exercise after about 43 hours of activity. Early results were deemed a success, which bodes well for future developments in this field.

After a 2-week delay in launch due to incorrectly stowed cargo bags, the third Automated Transfer Vehicle (named "Edoardo Amaldi") was launched on March 23 by Ariane 5 from the Kourou Launch Center in French Guiana, South America. The docking occurred on March 28, with the ATV delivering 7.2 tons of supplies to the station. In the closing phase of the ISS-30 residency, most of the activities focused upon unpacking both the ATV and Progress M-14M. The latter was undocked from the Pirs module on April 19 and was replaced by Progress M-15M three days later.

As their science program wound down, Burbank formally handed over command of the station to Kononenko on April 25. Two days later they undocked from the Poisk module in the final Soyuz TMA spacecraft, followed just over three hours later by what was reported as a "bulls-eye landing". It was a good way to end an impressive record of TMA missions begun a decade earlier with TMA-1.

Statistically, this expedition was a little unbalanced, although the overall mission logged as much as many others. The delays in launching the mission, caused by the loss of the Progress in August and qualification of the R-7/Soyuz-U and FG vehicles, meant that this crew spent only 6 days as formal members of the ISS-29 expedition, but 155 days as prime ISS-30 crew. In their 165-day flight, 163 days were spent aboard the station.

Milestones

285th manned space flight
117th Russian manned space flight
110th manned Soyuz
 28th ISS Soyuz mission (28S)
 22nd and final Soyuz TMA flight
29/30th ISS resident crew
First post-Shuttle era ISS mission
Shortest ISS expedition residency (6 days on the ISS-29 phase by this trio)
Ivanishin celebrates his 43rd birthday (January 15)
Shkaplerov celebrates his 40th birthday (February 20)

SOYUZ TMA-03M

International designator	2011-078A
Launched	December 21, 2011
Launch site	Pad 1, Site 5, Baikonur Cosmodrome, Republic of Kazakhstan
Landed	July 1, 2012
Landing site	Near the town of Dzhezkazgan, Republic of Kazakhstan.
Launch vehicle	Soyuz-FG (R7) (serial number Л15000-39), Soyuz TMA-03M (serial number 703)
Duration	192 da 18 h 58 min 21 s
Call sign	Antares
Objective	ISS resident crew (ISS-30/31) transport 29S

Flight crew

KONONENKO, Oleg Dmitryevich, 47, civilian, RSA, TMA commander, ISS-30 flight engineer, ISS-31 commander, second flight
Previous mission: Soyuz TMA-12/ISS-17 (2008)
PETTIT, Donald Roy, 56, civilian, NASA Soyuz TMA flight engineer, ISS-30/31 flight engineer, third flight
Previous missions: STS-113/ISS-6/TMA-1 (2002/3), STS-126 (2008)
KUIPERS, André, 53, civilian (The Netherlands), ESA Soyuz TMA flight engineer, ISS-30/31 flight engineer, second flight
Previous mission: Soyuz TMA-4/ISS-VC6/TMA-3 (2004)

Flight log

This was the third flight test of the new Soyuz TMA-M and the only qualification test flight of the vehicle. Following this mission, the TMA-M would be confirmed in its operational roles as both the primary crew transport to and from the ISS and as the Crew Rescue Vehicle for resident crew expeditions.

The three-man crew docked their spacecraft with the Rassvet module on December 23. The upcoming three-day Christmas break allowed the crew time to adjust to the station's environment and to catch up with the work being conducted by the prime ISS-30 crew. The assignment of experienced NASA and ESA astronauts to the crew would help increase the science in the U.S. segment (and in particular the European Columbus lab) once again. Dr. Pettit, a chemical engineer, had spent over 158 days on the station as a member of the ISS-6 crew almost a decade before and, while much had changed on the station since then, his experience soon began to show in his regular, informative blogs from space as he delved into the expanding American segment science program with Burbank.

ESA astronaut André Kuipers is inside the European Columbus laboratory during the PromISSe mission.

Kuipers was also working on the station for a second time, but this time as a resident not a visitor. He also blogged his experiences to followers on Earth. Kuipers' science program was called PromISSe, a name that was a reflection of the efforts and expectations placed on human space missions. It also continued the trend of the four previous European missions of including the acronym "ISS" as part of the science program name. This package included 30 investigations covering a range of disciplines in human research, fluid physics, materials science, radiation and solar research, and biology and technology demonstrations. In addition, Kuipers participated in over 20 experiments for NASA and JAXA using over 30 facilities spread across the station.

In the Russian segment, work continued on the experiments that were running during ISS-29, with Kononenko planned to assist for 56 hours 25 minutes in science operations during the ISS-30 phase. ISS-31 Russian segment science plans totaled 146 hours 20 minutes of science for Kononenko, Padalka, and Revin. The latter pair would arrive on Soyuz TMA-04M in May.

Following the Christmas holidays and New Year celebrations it was down to work for the TMA-03M crew as flight engineers for ISS-30. Kononenko assisted Shkaplerov on the 6h 15min EVA from Pirs on February 16. As well as the science, maintenance, and housekeeping programs, the crew practiced required safety drills and supported the now familiar exchange of Progress resupply craft, as well as the arrival of the third ATV vehicle in March. Burbank passed command to Kononenko on April 25, effectively ending the ISS-30 expedition and starting the ISS-31 phase. Formal closure of ISS-30 occurred when the last Soyuz TMA (No. 22) was undocked two days later on April 27. The Kononenko crew had spent 122 days as flight engineers for ISS-30 and now took the prime role for ISS-31 for three weeks, until they were joined by the TMA-04 crew on May 17, returning the station to six-person operation.

On May 25, a new milestone was reached in ISS history with the arrival of the first "commercial" mission (though still largely funded by the U.S. government). The Dragon unmanned resupply vehicle was grabbed by Canadarm2 and attached to the nadir port on the Harmony Node. This vehicle was the first step on the road to replacing the Shuttle as a U.S. resupply vehicle. Operated by SpaceX from their Mission Control in Hawthorne, California, the vehicle had been launched on the Falcon 9 rocket from LC-40 at the Cape Canaveral Air Force Station, Florida, three days before. A trial rendezvous with the station was completed on May 24. The crew "entered the Dragon" as the media put it, on May 26 to begin the unloading process.

The spacecraft delivered 1,104 lb (500.77 kg) of cargo and was subsequently loaded with 1,367 lbs. 620.07 kg) of hardware for the return to Earth. This was an important difference with Dragon. Unlike Progress, ATV, and HTV, this new vehicle could reenter the atmosphere and be recovered, greatly increasing the cargo return capacity over Soyuz (22.67 lb or 50 kg) and to a small degree compensating for the loss of the Shuttle's significantly larger cargo return capability. The Dragon was grappled by Canadarm2 on May 31 and then released to begin its journey back to Earth, completing a successful splashdown under three parachutes in the Pacific Ocean some 563 miles (905.86 km) west of Baja, California, in the Pacific Ocean. The successful flight had logged 9 days 7 hours 57 minutes and was the first operational splashdown associated with a returning American spacecraft (admittedly unmanned) since Apollo 18 returned at the end of the Apollo Soyuz Test Project on July 24, 1975.

Aboard the ISS, the science work continued, along with preparations for the return of the TMA-03M crew at the end of their residence. On June 16, the international crew was joined in orbit (but not on board) by three Chinese Shenzhou 9 crew on a mission to the Tiangong-1 space laboratory, that nation's first space station mission. Meanwhile, on board the ISS station on June 25 Pettit logged his 365th cumulative day in space across his three missions, and in so doing became the 28th person and only the fourth American to achieve this feat.

On June 29, Kononenko handed command of the station over to Padalka, who became the first person to command three separate expeditions to the ISS. For his final hours on the station, Kononenko became flight engineer 4. The prime ISS-31 expedition had been in command for 65 days which, added to their ISS-30 phase, meant that the trio had logged 187 days across the two expeditions. The formal end of the ISS-31 residency was achieved two days later, with TMA-03M making a safe and nominal return to Earth.

Milestones

286th manned space flight
118th Russian manned space flight
111th manned Soyuz
    3rd Soyuz TMA-M flight; completed the TMA-M test program
30/31st ISS resident crew
First commercial flight of SpaceX Dragon spacecraft
Kononenko celebrates his 48th birthday in space (June 21)

SOYUZ TMA-04M

International designator	2012-022A
Launched	May 15, 2012
Launch site	Pad, 1 Site 5, Baikonur Cosmodrome, Republic of Kazakhstan
Landed	September 17, 2012
Landing site	85 km north of Arkalyk, Republic of Kazakhstan
Launch vehicle	Soyuz-FG (serial number Л15000-041), Soyuz TMA-04M (serial number 705) 30S
Duration	124 da 23 h 51 min 30 s
Call sign	Altair
Objective	ISS resident crew transport (ISS-31/32)

Flight crew

PADALKA, Gennady Ivanovich, 54, Russian Federation Air Force (Retd.), RSA Soyuz TMA-M commander, ISS-31 flight engineer, ISS-32 commander, fourth flight
Previous missions: Soyuz TM-28 (1998), Soyuz TMA-4 (2008), Soyuz TMA-14 (2009)
REVIN, Sergey Nikolayevich, 46, civilian, RSA Soyuz TMA-M flight engineer, ISS-31/32 flight engineer
ACABA, Joseph Michael, 45, civilian, NASA Soyuz TMA-M flight engineer, ISS-31/32 flight engineer, second flight
Previous mission: STS-119 (2009)

Flight log

Arriving at the station on May 17, which also happened to be Acaba's 45th birthday, the Soyuz TMA-04M brought three new residents to supplement the three-member ISS-31 crew already on board the complex. Less than four hours after docking, the six astronauts and cosmonauts of the ISS-31 phase were together inside the station, progressing through the welcoming routines and ceremonies. They soon began concentrating on the more formal work schedule, which included receiving the first SpaceX Dragon unmanned supply vehicle on May 25.

By the time this mission flew, the Shuttle had been retired for about a year and media coverage of major launches and events had become sparse at best, coupled with the shift in emphasis of the program away from the "drama" of assembly to the more "mundane" scientific operations and resupply. True, there remained some further Russian components to be delivered to the station, but science and research now moved to the forefront. Even the promotional science

The SpaceX Dragon commercial cargo craft is grappled by Canadarm2.

material for the mission emphasized a "beehive of activity" for the crew, with delivery of new research facilities and testing for a new microsatellite deployment system.

There were over 240 experiments planned (over 80 of which were brand new), supported by over 400 investigators across the globe. In the Russian segment, there were to be 303 sessions covering 38 experiments, with cosmonauts planned to work on the experiment packages for over 350 hours across the ISS-31 and 32 phases. The experiments included human research, biological and physical sciences, technology development, Earth observation, and education. The NASA press packs included explanations of overlapping science studies beyond ISS-32 into the ISS-33 and 34 expeditions. For ISS-31/32, there were 201 separate investigations planned, of which 123 were brand new and 82 were NASA led. A further 118 were internationally supported research investigations.

On May 31, after 9 days 23 minutes docked with the station, the Dragon spacecraft was unberthed using Canadarm2 to begin its return to Earth. While the new spacecraft was docked with the station, the crew had unloaded over 1,100 lb (499 kg) of fresh supplies and then refilled the vessel with over 1,300 lb (590 kg) of hardware. This time, however, it would be returned to Earth rather than burned up in the atmosphere as with the other types of resupply craft.

On June 21, there was a double birthday celebration on board the ISS, as Padalka celebrated his 55th birthday in space and Kononenko his 48th. For

Padalka this was a very special celebration, as it was the third time he had marked his birthday in space (previously celebrated in 2004 and 2009). It was also the second such occasion for Kononenko, having previously celebrated his birthday in space in 2008.

During the final two weeks of June 2012, the ISS crew was accompanied in space, if in different orbits, by the first Chinese space station crew aboard Tiangong-1. This milestone was noted in the press, though there would be no direct communications between the two crews.

Padalka took over formal command of the station from Kononenko during June 29. A couple of days later, during the early hours of July 1, Soyuz TMA-03 undocked from the station, at which point Expedition 32 officially began. Following a light-duty weekend, the remaining station crew of three resumed their schedule, although there were further light-duty shifts for the July 4 U.S. Independence Day and the weekend of July 7/8.

On July 17, Soyuz TMA-05M docked with the Rassvet module bringing the three Expedition 33 crew members. For the first two months of their mission, they would serve as Expedition 32 flight engineers under the command of Padalka. With the crew readjusting to six-person operations again, the Japanese HTV-3 was launched on July 20, carrying almost 4 tons of supplies for the station. The latest cargo craft was grappled on July 27 by Canadarm2 and attached to the nadir port of Harmony.

While the crew brought the Japanese resupply craft in to the station successfully, they experienced difficulties in redocking the Progress M-15M space-craft on July 24. M-15M had been undocked two days before and placed in a parking orbit and was due to attempt a redocking to test the new Kurs-NA system. The system failed at 9.3 miles from the station, so the unmanned resupply craft was "parked" a safe distance below the station while the failure was investigated. A second attempt was completed successfully, docking with the station in the early hours of July 29. The cause of the original failure was determined to be a fail-safe test which aborted the docking. Once the vehicle had been reattached to the station, Padalka dismantled the Kurs-NA avionics box and then stowed it aboard the Zarya module for later return to Earth for analysis. The Progress was undocked for a final time on July 30.

A new Progress, M-16M, docked with the Pirs module on August 2, delivering over 5,800 lb (2600 kg) of cargo and propellant to the station. This was the 48th docking of a Progress to the station since August 2000. The difference with this flight was that the Progress took only four orbits (6 hours) to reach the station instead of the normal two days. This involved four very precise rendezvous maneuvers in the first 2 hours 40 minutes of flight, completed as a test for a proposed shortening of the journey to the ISS by Soyuz TMA-M flights in the hope of reducing the discomfort for the crew of two days in the cramped capsule. This new profile was not expected to be operational on manned flights for over a year.

For most of the month of August, the ISS crew kept busy with their science and preparations for the expedition's first EVA. Performed by Padalka and Malenchenko, this took place on August 20 (for 5 h 51 min) and featured the relocation of the Strela-2 cargo crane from the Pirs to the Zarya module. This was in readiness for the eventual undocking of Pirs to make room for the new Russian Nauka Multi-Purpose Laboratory. The cosmonauts also deployed a small TEKh-44 Sfera ("Sphere") satellite by hand, which would be used for ground tracking tests over a two-to-five-month period to evaluate orbital debris and decay. The two cosmonauts also installed five micrometeoroid shields outside Zvezda, retrieved an exposure package, and installed support struts on the Pirs EVA ladder ready for relocation across to the Nauka module when it arrived. A second exposure experiment could not be retrieved, as the cosmonauts were unable to close the package enough for it to fit through the Pirs airlock hatch. It was left for a later crew to retrieve.

The Expedition 32 phase drew to a close in September, so while the three Soyuz TMA-04M crew members wound up their research and increased their conditioning routine for the return home, the other half of the crew prepared to take over prime command. They conducted two EVAs from the Quest airlock and supported the unberthing of Kounotori-3 from the station on September 12. The unmanned Japanese resupply craft, filled with 16.5 tons of unwanted material, performed a destructive reentry on September 14.

Five days after HTV-3 departed it was time to bid farewell to the TMA-04M crew, who undocked in the early hours of September 17 (Moscow time). They landed less than four hours later, after a 125-day mission. They had resided in the station for 123 days, of which they spent 43 days as part of the Expedition 31 crew and 78 days as the prime Expedition 32 crew. Formal handover of station command occurred on September 15 between Padalka and Sunita Williams.

By the end of the flight, Padalka had accumulated over 710 days in space, on three missions to the ISS and his visit to Mir in 1998/1999. This made him the fourth most experienced space explorer; he also had nine EVAs to his credit. During the post-landing press conference, Padalka (who is unlikely to fly a fifth mission) reportedly spoke openly about the condition of the Russian segment. He described the living conditions as sparse, noisy, cold, and overcrowded, with only one-seventh of the room afforded to the U.S. astronauts. It would not, he felt, be suitable for the proposed 1 yr missions that were being discussed for future expeditions. Some of these differences between the American and Russian segments, especially the noise levels, have been clearly revealed in recent video tours of the station. The noise levels differ noticeably as the guided tour passes from the American to the Russian segment and back again.

Clearly there remains much to do in creating a universal comfortable environment for a crew on long international expeditions. This needs to be addressed before we attempt to venture into deep space.

Milestones

287th manned space flight
119th Russian manned space flight
112th manned Soyuz
 30th ISS Soyuz mission (30S)
 26th ISS Soyuz visiting mission
  4th Soyuz TMA-M flight
31/32nd ISS resident crew
Acaba celebrated his 45th birthday (May 17—the day Soyuz TMA-4M docked with the ISS)
Padalka celebrates his 55th birthday (June 21)
Kononenko celebrates his 48th birthday (also June 21)
Padalka first three-time ISS commander

SHENZHOU 9

International designator	2012-032A
Launched	June 16, 2012
Launch site	Pad 921, South Launch Site 1, Jinquan Satellite Launch Center, China
Landed	June 29, 2012
Landing site	Siziwang Banner, Inner Mongolia Autonomous Region, China
Launch vehicle	Long March 2F (CZ-2F)/Y9
Duration	12 da 15 h 24 min 0 s
Call sign	Unknown
Objectives	First manned docking and occupation of Tiangong-1 space laboratory, first flight of Chinese female taikonaut

Flight crew

JING Haipeng, 45, Chinese PLA Air Force, commander, second flight
Previous mission: Shenzhou 7 (2008)
LIU Wang, 43, Chinese PLA Air Force, flight engineer
LIU Yang, 34, Chinese PLA Air Force, flight engineer

Flight log

This mission came three years after Shenzhou 7 and provided China with a number of space "firsts" and a significant leap in manned space flight experience and operations. The primary objective was to place the first crew on board the inaugural space laboratory. There was also a female taikonaut in the crew, who became the first Chinese female in space. Launch of Shenzhou 9 occurred on the 49th anniversary of the launch of Valentina Tereshkova's Vostok 6, the first to carry a female cosmonaut into space. Liu Yang's entry into the record books also came two days before the 29th anniversary of Sally Ride becoming the first American woman in space, aboard STS-7. Liu Yang had been selected as a member of the second (2010) group of taikonauts.

Forecasts of the flight had been circulated for some time before the hardware was brought together to fly the mission. The Chinese had indicated as early as 2003 their desire to create a space laboratory, supplied by Shenzhou spacecraft. In the West, this seemed very reminiscent of the Soviet Soyuz–Salyut missions of 1971–1985. The Shenzhou 9 mission was part of a four-spacecraft program designed to provide the Chinese with experience in space station operations. First, the pre-fitted space laboratory, called Tiangong ("Heavenly Palace"), would be launched unmanned into Earth orbit. This would be followed by Shenzhou 8, also unmanned, which would test the docking system and docking port. Shenzhou 9

Liu Yang, the first Chinese female to fly in space, pictured at the 2012 IAF Congress in Rome, Italy, October 2012. Photo copyright: Brian Harvey, used with permission

would then take the first crew to occupy Tiangong and, if successful, a second manned mission, Shenzhou 10, would complete the program.

The launch of Tiangong-1 (2011-053A) by the upgraded Long March 2F (T1) occurred on September 29, 2011. Over the following month, the systems of the station were activated, evaluated, and tested prior to the launch of Shenzhou 8 (2011-063A), also by a Long March 2F (G), on October 31. The Shenzhou performed an automated docking on November 3 and remained docked with the station for structural integrity tests between the two docked vehicles over the next two weeks.

On November 14, Shenzhou 8 undocked, backed away, re-rendezvoused, and docked a second time with the station as a further test of the automated systems. Shenzhou 8 was undocked a second time on November 16 to complete a short solo flight and landing the next day after a flight of 18 days. These successful steps paved the way for the manned attempt at docking with Tiangong but, as the months slipped into 2012, little information was forthcoming other than that the crew may include a female. The delays caused some in the West to suggest that there were problems either with Shenzhou 8, the station, or in the preparations for Shenzhou 9. But this overlooked the cautious nature of the Chinese program and the absence of the "race" situation that was a prominent part of the early Soviet and American years.

Behind the scenes, preparations for Shenzhou 9 were well under way. The crew assignments were made in March 2012 but remained unannounced until just prior to launch, although many Western space sleuths were able to deduce the likely candidates ahead of the official announcements. The spacecraft assigned to the flight arrived at the Jinquan launch center for processing on April 9, and then the launch vehicle was delivered to the launch site a month later on May 9.

With typical Chinese efficiency, the combined spacecraft and launcher was rolled 1.5 km to the launchpad on June 9, in an operation that took one hour to complete. The sequence of previous missions suggested that a launch could occur sometime between June 14 and 16. The Chinese authorities confirmed this and indicated a planned mission of about 13 days, including an automated docking with Tiangong-1 and 10 days of joint operations, during which the three-person crew (still unidentified officially) would work inside the station. Towards the end of the mission, it was stated, the crew would conduct a manual docking test before final separation and a short solo flight, with reentry and landing the following day.

The three-day countdown began on June 13 and the names of the crew were formally announced in the days prior to launch, making headlines around the world. The launch on June 16 went flawlessly and it took only 9 minutes 45 seconds to place Shenzhou 9 in orbit to begin its 2-day chase towards Tiangong-1. On June 18, the spacecraft made its final approach to the station 140 minutes prior to the planned docking time. There had been five maneuvering burns to adjust the spacecraft's orbit prior to start of the automated rendezvous. The approach and docking was fully automated, although Liu Wang was ready to take over manual control if necessary. The automated system worked perfectly, however, with preplanned holds at 5 km, 140 m, and 30 m. The docking system was very similar to the Androgynous Peripheral Attach System with two rings first used on Apollo–Soyuz in 1975. The docking between Shenzhou 9 and Tiangong-1 occurred on the second day after launch and was followed a short time later by the crew transferring across to the space laboratory.

During their stay on board the station, the trio rotated their sleeping cycle so that at least one crew member was awake at all times to monitor onboard systems. Most of their time was taken up with evaluations and tests of the new space station, including several small maneuvering engine burns. The science program of 10 experiments included five medical studies of the taikonauts' own physical condition during China's first extended duration space flight. A series of air samples were taken to evaluate the status and condition of the station and the crew also completed a series of questionnaires on their health and operational tasks. They were also able to communicate with the ground via email. Much was made of the Chinese food available, of their enjoying weightlessness, and of Liu Yang performing tai chi for the cameras. At one point, Liu Wang played a harmonica and all three seemed to be adapting well to their new environment.

The medical experiments focused upon physical exercise, physiology, cell biology, and sleep studies. The air purification system and other onboard systems were also tested and evaluated. Tiangong-1 is the first of a scheduled three stations in the series, leading up to the launch of a larger station (about the size of the U.S. Skylab) due in 2020. These studies in Tiangong-1 will go a long way towards determining which procedures or equipment will be best suited for inclusion on those larger, longer duration stations.

Several celebrations were marked during the mission. On June 26, the crew held a conversation with President Hu Jintao. They also celebrated the Dragon Boat Festival and talked with the oceanauts on the Jiaolong submersible (named after a mythological sea dragon) 7,020 meters beneath the sea in the Mariana Trench in the Pacific Ocean, part of China's Deep Dive program. The three taikonauts also had regular contact with family members, who visited Mission Control.

On June 24, the crew mothballed the station and reentered the Shenzhou to undock after 5 days 21 hours and 1 minute. They backed the Shenzhou away some 400 meters before bringing the vehicle back in under manual control. They halted again at 140 m and then 30 m before completing the first Chinese manual docking. The two craft were separate for about 1 hour 30 minutes. Once the docking connections and seals had been checked for integrity, the hatches were opened and the crew reentered the space lab for a few more days of work before returning to Earth. The Shenzhou was undocked a second time on June 28 after 4 days 21 hours 13 minutes, giving a total docked time across the two periods of approximately 10 days 8 hours 14 minutes.

Shenzhou 9 completed its expected short solo flight following the undocking, allowing the crew time to prepare for entry and landing the next day. The recovery of the spacecraft was completed on June 29, with the spacecraft landing safely but heavily and apparently bouncing and rolling before coming to a halt.

Subsequent reports indicated that the Descent Module had actually missed its intended landing target by 9.94 miles (16 km), though this was still within the planned 22.37 miles (36 km) by 22.37 miles (36 km) landing footprint. The DM landed near a small river, hitting a slope on one of the riverbanks before coming to a rest. Rescue crews were soon on the scene and the three occupants seemed none the worse for their ordeal. They departed the landing zone a few hours after landing and then completed a 2-week postflight recuperation and debriefing period. The mission was a huge success for the program and for China on the world stage, with talk of the next stage—Shenzhou 10 visiting the station—being likely as early as 2013, reflecting a renewed confidence in the Chinese program.

As a new pioneer was feted, another was mourned. Less than a month after the landing of Shenzhou 9 and the flight of the first Chinese woman in space, the American lady with that honor, Sally Ride (STS-7, STS-41G), sadly died on July 23, 2012 after a long battle against pancreatic cancer. She was just 61.

Milestones

288th manned space flight
  4th Chinese manned space flight
  4th manned Shenzhou mission
  1st manned Chinese automated docking mission (June 18)
  1st Chinese manual docking (June 24)
  1st resident crew on Tiangong-1
  1st Chinese taikonaut to make two missions (Jing Haipeng)
  1st Chinese female in space (Liu Yang)

SOYUZ TMA-05M

International designator	2012-037A
Launched	July 15, 2012
Launch site	Pad 1, Site 5, Baikonur Cosmodrome, Republic of Kazakhstan
Landed	November 19, 2012
Landing site	Northern Kazakhstan landing zone (near to the town of Arkalyk)
Launch vehicle	Soyuz-FG (R-7) (serial number Л15000-042), Soyuz TMA-05M (serial number 706)
Duration	126 da 23 h 13 min 27 s
Call sign	Agat
Objectives	ISS resident crew transport (ISS-32/33), Soyuz 31S

Flight crew

MALENCHENKO, Yuri Ivanovich, 50, Russian Federation Air Force, RSA ISS-32/33 flight engineer, Soyuz TMA-M commander; fifth mission
Previous missions: Soyuz TM-19 (1994), STS-101 (2000), Soyuz TMA-2 (2003), Soyuz TMA-11 (2007)
WILLIAMS, Sunita Lyn, 46, NASA, U.S.A., ISS-32 flight engineer, ISS-33 commander, Soyuz TMA-M flight engineer, second mission
Previous mission: STS-116/ISS/STS-117 (2006/2007)
HOSHIDE, Akihiko, 43, JAXA, (Japanese) ISS-32/33 flight engineer, Soyuz TMA-M flight engineer, second mission
Previous mission: STS-124 (2008)

Flight log

In the Expedition 33 preflight Mission Summary, the flight was described as "action-packed", including the arrival of the first commercial resupply mission and research across a variety of experiments including muscle atrophy. Expedition 33 would continue to expand the research program, looking into the radiation levels aboard the outpost and the effects of microgravity on the human spinal cord. The Agricultural Camera would investigate dynamic processes on Earth (such as melting glaciers), seasonal changes, and how the ecosystem is affected by human intervention. The crew experiment program would encompass further experiments in human research, biological and physical sciences, development of new technologies, Earth observations and education.

Calling the expedition "action-packed" may have been stretching the description a little at the start, but the crew was certainly never at a loss for

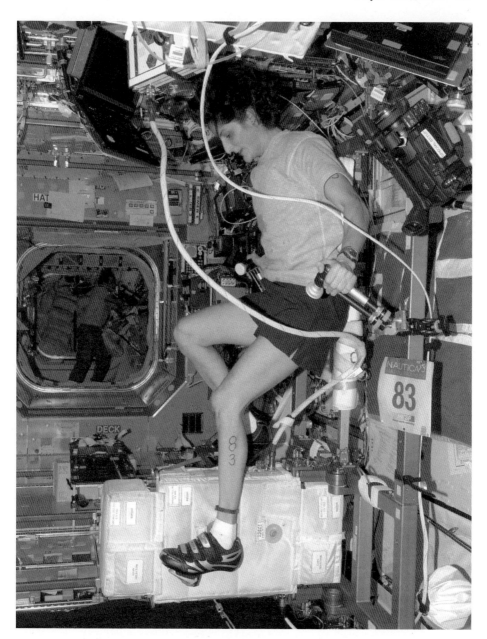

Expedition 33 commander competes in the first triathlon in space.

things to occupy their time. With difficulties encountered on their first EVA, there was soon plenty of unplanned "action" for them to deal with. Their mission emblem description explained that the work on the ISS was "heading into the future". Now that the space station was almost complete and the Shuttle retired, Expedition 33 was part of the push for new goals in space, even though it was not exactly clear where those goals were heading.

The launch occurred on the 37th anniversary of the launch of Soyuz 19 and Apollo 18 under the Apollo–Soyuz Test Project, the first joint U.S.S.R./U.S. manned space flight program. This was not lost on the crew or officials recalling the event in their pre and postlaunch speeches. The arrival at the space station on July 17 was also exactly 37 years after the docking of Soyuz and Apollo and gave rise to further celebrations and comments on how far the joint programs had progressed since that time. The Soyuz TMA-05M spacecraft was docked success-fully with the Rassvet MRM1 module and, after the hatch opening, normal safety briefings, and welcoming ceremonies, the three new crew members were soon unloading equipment from the Soyuz, powering down their spacecraft, and getting up to speed on the various science and research activities across the station.

For most of their first two months on board the station, the trio were designated flight engineers as part of the ISS-32 Expedition. They were involved with activities associated with the Progress, HTV, and ATV resupply craft, as well as various science activities and general housekeeping duties. As August pro-gressed, Malenchenko assisted Padalka on a Russian segment EVA from Pirs (August 20), while Williams and Hoshide prepared for their own space walk from the Quest airlock using U.S. EMU suits.

That EVA (August 30, 8 h 17 min) became the third longest space walk in history. The main objective of the EVA was to install a new Main Bus Switching Unit (MBSU) to the S0 truss segment. This unit was one of four which routed electricity from the solar arrays on the truss to the station. After removing the failed unit, the astronauts found it difficult to install its replacement, chiefly because securing the bolts proved to be much harder than anticipated. Indeed, they could not secure one particularly stubborn bolt, so they used a long-duration tie down tether to secure the unit temporarily until a second EVA could be undertaken to finish the task. Unfortunately, with the MBSU out of service and two arrays out of action, the power available on the ISS was reduced by 25%. The EVA crew was able to connect one of two power cables in preparation for the arrival of the new Russian module, but the replacement of a camera on Canadarm2 also had to be postponed. Ironically, an unconnected drop-off of the station's power system on September 1 meant that a third panel went off-line, reducing the station to five out of eight power channels for the first time in several years. The second EVA, on September 5 (6 h 28 min) was more successful, with the crew able to secure the MBSU and install the Canadarm2 camera.

On September 15, the Altair ISS-32 crew handed over command of the station to the Agat ISS-33 crew shortly before departing the station and ending their 125-day mission. Once again, the resident crew compliment was down to just three. Commander Sunita Williams became only the second female station expedition commander in 12 years and over 30 expeditions. The formal start of ISS-33 operations occurred when Soyuz TMA-04M undocked from the station to begin their return to Earth.

The handover occurred on the same weekend that Williams became the first person to complete a triathlon in space. After "participating" in the Boston Marathon during her first stay on the station in April 2008, Williams "participated" in the Nautica Malibu Triathlon, held in Southern California on September 16. Orbiting some 240 miles (386 km) above the other competitors, she used special exercise equipment designed to keep astronauts fit during their mission and specially formulated to simulate the triathlon experience in space. Using a treadmill and stationary bike, she ran for 4 miles and cycled for 18 miles. To simulate "swimming", Williams used the Advanced Resistive Exerciser Device (ARED), which allowed her to complete weightlifting and resistance exercises that approximated swimming in microgravity for "half a mile". Her total time taken for the three disciplines was 1 hour 48 minutes 33 seconds.

The science work gathered pace for the crew through the rest of the month. Other tasks included preparing ATV-3 for undocking from the station. This was accomplished on September 28, with the vessel completing its destructive descent in the atmosphere on October 4. On October 10, the SpaceX Dragon CRS-1 cargo ship (which had launched on October 7) was grasped by the station's RMS and attached to the Harmony Node, making it the first operational commercial resupply mission to arrive at the space station. On board were 882 lb (400 kg) of cargo to replenish supplies at the station. The crew loaded about 1,600 lb (726 kg) of cargo for return to Earth when the Dragon spacecraft detached from the station on October 28. It splashed down in the Pacific about six hours after undocking.

The next event was the arrival, on October 25, of the other three Expedition 33 crew members on board Soyuz TMA-6M. They were to take over from Williams and her colleagues in November and continue as the Expedition 34 trio for the remainder of the year. With the new crew safely docked and integrated into the main residency program the emphasis shifted to preparations for the next EVA planned for November 1. On this EVA Sunita Williams and Akihiko Hoshide were allocated 6 hours and 30 minutes to repair an ammonia leak on one of the station's port side radiators. The ammonia, which is circulated through the external thermal control system of the orbital facility, is used to cool the electronics and other systems.

The November 1 EVA (designated U.S. EVA-20) performed by Williams and Hoshide was accomplished in 6 h 38 min accomplishing all the assigned and one get-ahead tasks. The pair completed both parts of the EAS (Early Ammonia System) jumper reconfiguration; demated the PVR 2B FQDC (Photovoltaic Radiator Flight Quick Disconnect Coupling); removed the cover from the spare TTCR (Trailing Thermal Control Radiator), then released and deployed the device. They also took documentary photography of the IEA (Integrated Equipment Assembly) and the PVR, as well as conducting the get-ahead task of inspecting the port SARJ (Solar Array Joint).

With the EVA completed the "Agat" trio prepared to hand over command of the station to the "Kazbek" crew and end their residency. Formal handover of the command of the ISS from Williams to Kevin Ford took place on November 17. The official ending of the ISS-33 phase and start of the ISS-34 phase took place on

November 19 with the undocking of Soyuz TMA-05M. The residency had accumulated 127 day in space with approximately 60 days spent as part of the ISS-32 expedition and then 63 days as the ISS-33 expedition.

Milestones

289th manned space flight
120th Russian manned space flight
112th manned Soyuz
  31st ISS Soyuz mission (31S)
  5th Soyuz TMA-M flight
32/33rd ISS resident crew
Williams celebrated her 47th birthday in space (September 19)
Williams becomes only the second female ISS expedition commander
Williams also surpasses Whitson's EVA record for a female astronaut setting a new cumulative EVA record of 50 h 40 min (seven EVAs)
Williams becomes the first person to complete a "triathlon" in space' on September 16, adding the achievement to her space marathon run completed in April 2008

SOYUZ TMA-06M

International designator	2012-058A
Launched	October 23, 2012
Launch site	Pad 31, Site 6, Baikonur Cosmodrome, Republic of Kazakhstan
Landed	March 15, 2013 (planned)
Landing site	North Kazakhstan landing zone (planned near to the town of Arkalyk)
Launch vehicle	Soyuz-FG (R-7) (serial number Л15000-044), Soyuz TMA-06M (serial number 707)
Duration	144 da (planned)
Call sign	Kazbek
Objectives	ISS resident crew transport (ISS 33/34), ISS Soyuz 32S

Flight crew

NOVITSKIY, Oleg Victorovich, 42, Russian Federation Air Force, RSA Soyuz TMA-M commander, Soyuz 33/34 flight engineer

TARELKIN, Evgeny Igorevich, 37, Russian Federation Air Force (Retd.), RSA Soyuz TMA-M flight engineer, ISS 33/34 flight engineer

FORD, Kevin Antony, 52, USAF (Retd.), NASA Soyuz TMA flight engineer, ISS-33 flight rngineer, ISS-34 commander, second mission
Previous mission: STS-128 (2009)

Flight log

The next crew to launch to the space station created a small footnote to history by using a launchpad not utilized for manned missions since July 1984 (for the Soyuz T12 mission). The majority of manned launches from Baikonur have occurred from Pad 1 on Site 5. This is known as "Gagarin's Start" in recognition of the historic 1961 mission and is steeped in cosmonautics history and tradition, but was in need of significant overhaul since its previous upgrade in 1983. It was even more important to improve the facilities at Pad 1 once the Shuttle had retired, as this became the *only* operational launchpad for manned missions to the International Space Station. The upgrades had already been completed at Pad 31, Site 6 and used for an unmanned launch earlier in the summer. When the work is completed at Pad 1 it will once again be used to support future manned launches.

Docking with the station occurred on October 25 at the Poisk module two days after launch to join the Expedition 33 crew as flight engineers. Following the return of the Agat trio in November the Kazbek team assumed command of the station as Expedition 34 until their return to Earth in March 2013. In December

Not exactly the usual way to fit-check a Sokol suit. From left to right Oleg Novitsky, Yevgeni Tarelkin and Kevin Ford, hoist their suits above their heads during preparations for the launch day dress rehearsal two weeks prior to lift off. Photo credit: NASA/Victor Zelentsov

they were joined by the Soyuz TMA-7 trio who will return the residency to six and take over as Expedition 35 continuing the occupancy of station into 2013.

For Expedition 33/34 their five months on the station will include a number of visits by unmanned resupply craft. They will assist in the very first arrival at the ISS of the Orbital Sciences Corporation's commercial cargo vehicle Cygnus. In addition this expedition is expected to host two further commercial SpaceX Dragon spacecraft, as well as four Russian Progress resupply vehicles. There are no space walks planned by this crew but they will be kept busy (in a "hive of activity" according to the press kit) as they continue the program of experiments conducted on previous expeditions and, during the Expedition 34 phase, initiate several new experiments on the station.

Soyuz TMA-06M became the 129th mission to the space station and the 84th Russian (RSA) mission including the failed Progress launch in August 2011. Since the launch of Zarya, the first element of the ISS in November 1998 there had been

37 U.S. (NASA) missions, three European (ESA) ATV missions, three Japanese HTV missions, and two U.S. commercial (SpaceX missions), an impressive total over a 14 yr period.

As the new crew members settled on board, the NASA *ISS On-Orbit Status Report* for October 25, 2012 stated that effective as of 04:00 EDT that day, the ISS was orbiting at 264.0 miles (424.9 km) × 249.8 miles (402.3 km) inclined at 51.65 degrees, with a period of 92.84 minutes, meaning that it was completing 15.51 orbits every 24 h Earth day.

Since its launch in November 1998 the Functional Cargo Block Zarya had logged 79,823 revolutions of Earth and accumulated 5,088 days in space, or 13 years 11 month 5 days of orbital operations. The total cumulative resident crew time, from docking with the ISS-1 aboard Soyuz TM-31 on November 2, 2000, was 4,375 days or 11 years 11 months 23 days.

With new missions on the manifest the story continues ...

Milestones

290th manned space flight
121st Russian manned space flight
115th manned Soyuz
32nd ISS Soyuz mission (32S)
6th Soyuz TMA-M flight
33/34th ISS resident crew
First manned launch from Pad 31/Site 6 since Soyuz T12 in July 1984

5

The immediate future: 2012–2020

In the earlier edition of this log, the closing chapter outlined "The Next Steps", which were quite clear at the time of writing in 2006. They mainly featured the completion of the International Space Station and the retirement of the American Space Shuttle. Originally planned for 2010, the final flights of the Shuttle stretched into 2011 and included not only the final station assembly missions, but also the reinstated fourth Hubble Telescope servicing mission. The book also forecast the first Chinese EVA and the launch of a small Salyut-class space station module.

As we write the closing lines of this edition of the *Manned Spaceflight Log*, the Shuttle has retired after 135 missions, the majority of the ISS assembly is complete, and the station has now become an orbital research facility. The Hubble Space Telescope has been visited and upgraded once again, and the Chinese also delivered on their announced plans for EVA capability and a small space station.

Now that these goals have been met, what are the most likely plans for manned space flight for the rest of this sixth decade of operations and what will those missions establish to move forward in the coming decades?

SOYUZ TMA-07M

International designator	2012-074A
Launched	December 19, 2012
Launch site	Pad 31, Site 6, Baikonur Cosmodrome, Republic of Kazakhstan
Landed	May 14, 2013 (planned)
Landing site	North Kazakhstan landing zone (near the town of Arkalyk)
Launch vehicle	Soyuz-FG (R-7) (serial number Л15000-040)
	Soyuz TMA-07M (serial number 704A)
Duration	Approximately 146 da (planned)
Call sign	Parus ("Sail")
Objective	ISS resident crew transport (ISS 34/35) Soyuz (33S)

Flight crew

ROMANENKO, Roman Yuriyevich, 41, Russian Federation Air Force, RSA, TMA-M commander; ISS-34/35 flight engineer, second mission
Previous missions: Soyuz TMA-15/ISS-20/21 (2009)
HADFIELD, Christopher Austin, 53, Canadian Air Force (Retd.), CSA TMA-M flight engineer; ISS-34 flight engineer/ISS-35 commander, third mission
Previous missions: STS-74 (1995), STS-100 (2001)
MARSHBURN, Thomas Henry, 52, civilian, NASA Soyuz TMA flight engineer, ISS 34/35 flight engineer, second mission
Previous missions: STS-127 (2009)

Flight log

Docking occurred successfully at the Rassvet module on December 21, 2012, restoring the station residency to six.

Chris Hadfield (left), Tom Marshburn (center), and Roman Romanenko participate in a cake-cutting ceremony (featuring their mission emblems) at NASA JSC, marking the end of their U.S. training program.

Milestones

291st manned space flight
122nd Russian manned space flight
116th manned Soyuz
   7th Soyuz TMA-M mission
  32nd ISS Soyuz mission
Hadfield is scheduled to become the first Canadian ISS expedition commander
All three crew members have birthdays during August: Romanenko (August 9) whilst Hadfield and Marshburn share a birthday (August 29) but 12 months apart

SPACE STATION OPERATIONS

The main focus for humans in space until at least 2020 will be in low Earth orbit and in operations associated with space stations; primarily the International Space Station but also the Chinese series of stations called Tiangong ("Heavenly Palace"). At both facilities, successive expeditions are expected to perform a wide range of experiments and research programs focusing attention towards our own planet Earth. Some will investigate the mysteries of the space environment, while others will develop the technologies to support operations for our eventual return to the Moon and on to Mars, the asteroids, or other strategic deep-space points. These could potentially lead to future manned explorations throughout our solar system later this century and into the next.

International Space Station operations

With the majority of the construction completed, the ISS has finally been able to become the leading scientific research facility it was always intended to be. Expeditions comprising up to six crew members are now utilizing the international facilities and resources to conduct a range of investigations in research fields that have been promoted as the reasons for the existence of large space stations for over 30 years.

The known crewing (as of October 31, 2012) for ISS operations through 2015 are presented in Table 5.1.

The subtle change from constructing the ISS to learning from it was not an overnight event. Indeed, "science" had been conducted on the facility almost from the start of operations, but it was not until the arrival of the U.S. laboratory Destiny in 2001 that "real" science on the ISS could begin. Gradually over the years (and over 30 expeditions), the science program has expanded, apart from the hiatus caused by the tragic loss of Columbia in 2003. The completion of the truss and solar array assembly, the arrival of the European and Japanese science facilities during 2008, and the increased permanent crew size from three to six the following year have enabled the scientific research program to reach its true potential. Regular reports are posted on the internet from the partner agencies, detailing the latest daily operations and activities across the station. Among the most useful sites are

- *http://www.nasa.gov/directorates/heo/reports/iss_reports/*
  (daily reports on ISS activities)
- *http://www.esa.int/SPECIALS/Columbus/SEMBQ84S18H_0.html*
  (regular weekly reports on activities aboard the European Columbus laboratory)
- *http://kibo.jaxa.jp/en/experiment/*
  (latest news on activities on the Japanese Kibo laboratory)
- *http://www.federalspace.ru/main.php?id=2&nid=19641*
  (daily updates on activities aboard the Russian Segment—in Russian).

The journey continues, launch of Soyuz TMA-06M in October 2012.

Table 5.1. Future flight manifest.

Spaceflight	Mission	Launch date	Landing date	Position	Prime crew	Agency	Nationality
Soyuz TMA-08M	ISS-35/36	Mar 2013	Sep 2013	Commander	Vinogradov, Pavel (ISS 36 Cdr)	RSA	Russian
				Flight engineer	Misurkin, Alexander	RSA	Russian
				Flight engineer	Cassidy, Christopher	NASA	American
Soyuz TMA-09M	ISS-36/37	May 2013	Nov 2013	Commander	Yurchikhin, Pavel (ISS 37 Cdr)	RSA	Russian
				Flight engineer	Parmitano, Luca	ESA	Italian
				Flight engineer	Nyberg, Karen	NASA	American
Shenzhou-10	Tiangong 1	Jun 2013	Jun 2013	Commander	?	China	Chinese
				Flight engineer	?	China	Chinese
				Mission engineer	?	China	Chinese
Soyuz TMA-10M	ISS-37/38	Sep 2013	Mar 2014	Commander	Kotov, Oleg (ISS 38 Cdr)	RSA	Russian
				Flight engineer	Ryazansky, Sergei	RSA	Russian
				Flight engineer	Hopkins, Michael	NASA	American
Soyuz TMA-11M	ISS-38/39	Nov 2013	May 2014	Commander	Tyurin, Mikhail	RSA	Russian
				Flight engineer	Wakata, Koichi (ISS 39 Cdr)	JAXA	Japanese
				Flight engineer	Mastracchio, Richard	NASA	American
Soyuz TMA-12M	ISS-39/40	Mar 2014	Sep 2014	Commander	Skvortsov, Alexander	RSA	Russian
				Flight engineer	Artemyev, Oleg	RSA	Russian
				Flight engineer	Swanson, Steven (ISS 40 Cdr)	NASA	American
Soyuz TMA-13M	ISS-40/41	May 2014	Nov 2014	Commander	Surayev, Maksim (ISS-41 Cdr)	RSA	Russian
				Flight engineer	Gerst, Alexander	ESA	German
				Flight engineer	Wiseman, Gregory	NASA	American
Soyuz TMA-14M	ISS-41/42	Sep 2014	Mar 2015	Commander	Samokutyayev, Alexandr	RSA	Russian
				Flight engineer	Serova, Yelena	RSA	Russian
				Flight engineer	Wilmore, Barry (ISS 42 Cdr)	NASA	American

Soyuz TMA-15M	ISS-42/43	Nov 2014	May 2015	Commander Flight engineer Flight engineer	Shkaplerov, Anton (ISS 43 Cdr) Christoforetti, Samantha Virts, Terry	RSA ESA NASA	Russian Italian American
Soyuz MS-01	?	Mar 2015?	Sep 2015?	Commander Flight engineer Flight engineer	TBD Skripochka, Oleg TBD	RSA RSA NASA	Russian Russian American
Soyuz TMA-16M	ISS-43/44	Mar 2015	Oct 2015 Mar 2016 Mar 2016	Commander Flight engineer Flight engineer	Lonchakov, Yuri Korniyenko, Mikhail Kelly, Scott	RSA RSA NASA	Russian Russian American
Soyuz MS-02	?	Sep 2015?		Commander Flight engineer Flight engineer	TBD TBD TBD	RSA NASA ?	Russian American ?
Soyuz TMA-17M	ISS 44/45	May 2015	Nov 2015	Commander Flight engineer Flight engineer	Kononenko, Oleg Yui, Kimiya Lindgren, Kjell	RSA JAXA NASA	Russian Japanese American
Soyuz TMA-18	ISS 45/46	Sep 2015	Mar 2016 Oct 2015?	Commander Flight engineer SF participant	Volkov, Sergei Ovchinin, Aleksei Brightman, Sarah?	RSA RSA U.K.	Russian Russian British
Soyuz TMA-19	ISS 46/47	Nov 2015	TBD	Commander Flight engineer Flight engineer	TBD TBD TBD	? ? ?	? ? ?

Key: RSA, Russian Space Agency; CSA, Canadian Space Agency; NASA, National Aeronautics and Space Administration; ESA, European Space Agency; JAXA, Japanese Space Agency.

Note: The launch of Shenzhou 10 is speculative and awaiting official confirmation. Soyuz MS is an upgraded version of the TMA-M vehicle. All dates and crew assignments are subject to change. All information correct as of December 31, 2012.

Europe's Ariane launch vehicle with an ATV prepares to launch to the ISS with supplies.

Sadly, most of this work does not make the news outside of the partner agencies or "space"-dedicated websites, which makes informing the general public of what the crews are actually doing on the station more difficult. With the growth of social media applications there has been an increase of postings from the ISS over the past few years, but the days of the "right-stuff" headline-grabbing missions are long gone, at least for the foreseeable future.

No space flight can truly be termed "routine" and yet that is exactly what the crews on the ISS are attempting to establish—a regular, smooth running and productive scientific research program to provide genuine advances in space operations, at least in Earth orbit. Quite simply, missions like those being conducted on the station just do not make major headlines, unless something goes wrong.

This "routine" aspect is dependent upon a regular supply of logistics from Earth. Since the retirement of the American Space Shuttle, this has become a little more difficult to achieve using only the smaller vehicles that remain available. The loss of the mass-carrying capability of the Shuttle (in particular for removing trash and redundant equipment from the station) will probably never be replaced in the lifetime of the ISS, even with the proposed new vehicles under consideration (see below). The venerable Russian Soyuz (in operational service since 1967) will continue, in its latest TMA-M version, as the primary resident crew transport ferry and rescue vehicle, but has a relatively small capacity for the return of scientific payload from orbit. The TMA-M could be phased out by 2015 to be replaced by a new variant (Soyuz-MS) which is expected to have further systems upgrades but still resemble the basic Soyuz TMA craft. The unmanned Progress version (now flying as the Progress M-M series), which first flew in 1978 to Salyut 6, continues to provide a regular supply service, ferrying cargo to the station and disposing of trash in destructive atmospheric reentries.

Like the Progress, the European ATV and Japanese HTV unmanned resupply vehicles also have the capability to dispose of waste and unwanted hardware during a destructive reentry, but the return of samples and hardware to Earth is not currently available on these operational craft. This void may, at least partially, be filled by the SpaceX Dragon spacecraft, which, in 2012 completed a successful demonstration mission of its capability to fly to the ISS unmanned, dock and deliver supplies, and then return to an ocean recovery. This was a significant step towards replacing the Shuttle with a (partially) commercially funded vehicle.

There remain a few additional modules to be launched to the station for the Russian segment. These much delayed Russian modules are expected to include

- the Nauka ("Science") Multipurpose Laboratory Module (replacing the Pirs module) and delivery of the European Robotic Arm (ERA); and
- the Nodal Module for the attachment of both the Scientific and Power Producing Module-1 and the Scientific and Power Producing Module-2.

It has been agreed by the international partners to continue flying expeditions to the ISS on Soyuz through to 2020, and possibly as long as 2028, as long as the hardware can support the program. If this happens, and there is no certainty that this will be possible, it would result in almost 30 years of continuous operational expedition activity, doubling the record established on Mir and setting the stage for the next step in orbital exploitation around our planet. Exactly what form that next step will take is also, at present, far from certain.

As this book was about to go to press news began to emerge about a new challenge to be attempted on the International Space Station. On October 5, 2012

NASA announced plans to launch in 2015 one American and one Russian (but without identifying them) on a 1 yr long expedition to the ISS, this being part of the plans to understand how humans adapt to long-duration spaceflight, with an eye to returning to the Moon or journeying out deeper into space. Using the ISS as a base for these experiments will help better prepare for such journeys. In July 2012 ISS veteran Peggy Whitson stepped down as Chief Astronaut to resume space station training and it was suggested she may be one of the NASA candidates for the yearlong mission.

The announcement of a two-person crew for the yearlong mission also generated media suggestions that an opportunity might arise for the third seat on Soyuz to be occupied by a space flight participant on a short-stay mission to the station. Of course, should this be the case, room would have to be made for the tourist on the returning Soyuz with only two outgoing resident crew members coming home instead of three as is now standard. Perhaps partial crew exchange would be possible, but this would alter the current crewing protocol we have seen since 2009 when the last space flight participant, Canadian Guy Laliberté, was launched to the station. On October 10 it was announced in Moscow that British classical crossover artist Sarah Brightman was a candidate for such a flight with Space Adventures, possibly representing UNESCO as an "artist for peace" and to continue her STEM (Science Technology, Engineering, Math) scholarship work for women. It will be interesting to see how the future of the Space Flight Partici- pant Program develops over the rest of this decade and how such flights will, or will not, feature in the longer term planning for the remainder of ISS operations.

On the same day of the yearlong mission announcement (October 5) JAXA astronaut Kimiya Yui had been selected by the Japanese space agency for the Expedition 44/45 mission. A fact overlooked by most of the media, he would serve as a flight rngineer for a 6-month residence commencing around June 2015 and would begin expedition training later that month.

Amidst all this speculation on November 26, 2012, it was announced that NASA had chosen Scott Kelly and Roscosmos had selected Mikhail Korniyenko for the year-long mission. Further crewing assignments through the 50th ISS expedition were expected during the early months of 2013.

Tiangong operations

The news of a yearlong residency on the ISS came a few weeks after closer cooperation between Europe and China was reported. This could, it was sug- gested, develop into the possibility of an ESA astronaut flying aboard a Chinese spacecraft by 2020. Whether this would be to a Tiangong station or the ISS was not clear and remains an open issue to be decided as objections, technical issues, and logistics are debated in the coming years. With the expected reduction or demise of ISS operations after 2020 and the predicted increase in Chinese space station operations from that date, clearly the opportunity to continue and perhaps increase the rate of flying European astronauts on long-duration missions has a certain Eastern promise to it.

The Chinese have also indicated a desire to create their own large space station from which to expand their manned space flight operations, possibly looking towards the Moon and perhaps far beyond. Their first steps were completed between 1999 and 2008 with Shenzhou operations, developing manned space flight capability and the infrastructure to support that effort in launch, orbital operations, and recovery. Their successful maiden flight of one person in 2003 drew upon the experiences (and particularly the design) of the Russian program, giving the Chinese a head start in developing their own program. They were to build upon this experience far quicker than either the Soviets or Americans had been able to in the 1960s. By 2008, the Chinese had demonstrated the capability of flying up to three crew members for several days, as well as EVA capability that could be used to support future space station operations. What had taken the Soviets eight missions to achieve with Vostok/Voskhod and the Americans around 10 missions with Mercury and Gemini, the Chinese accomplished in just three flights.

There were of course significant differences between the 1960s and the 2000s, most notably in the number of missions flown in the 1960s and what other achievements had been accomplished. The Soviets had flown 16 manned missions between April 1961 and June 1970, including the first man in space, first female, first group flight, first crew, first EVA, first manned docking and crew transfer (by EVA) and longest solo manned space flight at 18 days. In contrast, the Soviets had only achieved one manned docking and relatively little spacewalking experience in comparison with the Americans during Gemini and Apollo. The five Apollo missions dispatched to the Moon between December 1968 and April 1970 added very little to the database of low Earth orbit operations, but volumes to explorations away from the planet. It is certain that the Chinese will have studied the lessons learned by the Americans during their unmanned precursor lunar missions and the Apollo experience, and from the Soviet successes and setbacks in both their manned and unmanned lunar exploration program.

Although the Moon may indeed be a future target of Chinese space planners, the immediate focus for the next few years is the creation of a series of space stations leading to the establishment of a large complex. This will be similar to the gradual development of Soviet space station operations at Salyut, Mir, and finally the ISS, but again over a much shorter timescale and with far fewer missions. Once again, the Chinese will be learning from others in order to advance their own program without the need to mount unnecessary and expensive development missions. Official Chinese reports have stated that Tiangong-1 is intended as an experimental test bed, designed to develop the skills of rendezvous and docking that are essential to support a larger space station. The first Tiangong is expected to support three missions, one unmanned (Shenzhou 8 in 2011) and two manned (Shenzhou 9 in 2012 and Shenzhou 10 by 2013). Once these missions are completed, the station will be de-orbited later in 2013, to be replaced by the much larger Tiangong-2 and Tiangong-3 laboratories.

According to the Chinese, Tiangong-2 will be able to support much more sophisticated experiments and research than its pioneering predecessor. Tiangong 3 will be a multimodule design (possibly resembling the Mir configuration) which will be resupplied by Progress-type unmanned freighters. The Chinese goal is to have a fully functional (ISS class) space station in orbit by 2020. If this does occur, it will have taken them less than 10 years, in comparison with the *40* yr period between the first Salyut and completion of the ISS!

COMMERCIAL CARGO AND CREW DEVELOPMENT

Studies, plans, and discussions on what exactly would follow the Space Shuttle had circulated for years before the decision was finally made to retire the vehicles following the loss of Columbia in 2003. During these years the growth of commercial interest in developing a new launch system and spacecraft varied considerably but recently there have been a number of companies who have expressed interest in creating an American launch and crew/cargo transport system independent of NASA.

By 2010, in an effort to replace the Space Shuttle program for the transportation of crews and/or cargo to the ISS, NASA funded Space Act Agreements with five companies. The aim was to develop potential capabilities for launching American astronauts and supporting logistics into space from launch sites within the United States. A sixth company, ATK-EADS, was included as an unsolicited and unfunded proposal in May 2012. The development of a new American crew vehicle is conducted under NASA's Commercial Crew Development (CCDev) Program.

The six were

- Space Exploration Technologies (SpaceX)
- Orbital Sciences Corporation (Orbital)
- Blue Origin
- ATK-EADS
- Sierra Nevada Corporation, and
- The Boeing Company.

In addition, the American space agency signed agreements with Alliant Technologies Inc., Excalibur Almaz Inc., and United Launch Alliance, LLC for the exchange of technical information and expertise.

Space Exploration Technologies (SpaceX)

SpaceX has developed its Dragon (cargo) vehicle to launch on their Falcon 9 launch vehicle. The company is also developing a manned version of the same vehicle. An unmanned Dragon cargo vehicle was successfully flown to the ISS in 2012, becoming the first commercial vehicle to attach itself to the ISS. The Dragon spacecraft can handle both pressurized (up to $14 \, m^3$ or $55 \, ft^3$) and unpressurized (up to $10 \, m^3$ or $39 \, ft^3$) payloads in a fully recoverable capsule with a combined capsule and support trunk up-mass of 2,720 lb (6,000 kg) or 1,360 lb (3,000 kg) for the down-mass of only the capsule. It has an impressive mission duration capability of between one week and two years. The first manned flight is planned for 2015 and the vehicle is designed to carry up to seven astronauts on a wide variety of missions.

Orbital Sciences Corporation (Orbital)

Orbital is developing an unmanned cargo vehicle called Cygnus that will be launched on an Antares launch vehicle. This will be an advanced maneuvering spacecraft designed to support cargo delivery services and is planned to fly eight missions over a 2 yr period (currently 2013 to 2015), delivering approximately 20,000 kg (44,000 lb) of cargo to the ISS and then disposing of unwanted waste in a destructive reentry. Using proven technology, the vehicle comprises a common service module and a pressurized cargo module. The pressurized module is based on the Multi-Purpose Logistics Module developed by Thales Alenia Space.

Blue Origin

Blue Origin is developing a relatively secret crew transportation system to be launched initially on an Atlas V launch vehicle, although it is also developing its own reusable launch system.

ATK-EADS

This proposal was based upon utilizing a modified first stage of the Ariane V as a new second stage, with a Shuttle solid rocket motor as the first stage. Ariane V was to have been the launch vehicle for the canceled European Hermes mini-shuttle. This new design of launch vehicle has been named "Liberty" and would be used to launch a composite crew capsule.

Sierra Nevada Corporation

Sierra Nevada is developing a small lifting body–style crew vehicle called Dream Chaser also for launch on an Atlas V. This fourth-generation design of lifting body is based upon the NASA HL-20 design and is a fully reusable pressurized lifting body spacecraft. Capable of landing on a conventional runway, this design offers cross-range capability and reduced g-forces on descending occupants and payloads.

The Boeing Company

Boeing is developing the Crew Space Transportation (CST-100) crew capsule, initially for launch on an Atlas V. The CST, which can carry a crew of seven, is a cone-shaped capsule resembling the Apollo Command Module, but with a dry-land recovery capability. This new Boeing design is larger than the vehicle which took American astronauts to the Moon between 1968 and 1972, to the Skylab space station in 1973, and docked with a Soviet Soyuz in 1975. However, when compared with the previously proposed Orion deep-space vehicle, the CST-100 is smaller in size.

On August 3, 2012, NASA announced the next step in the development of a new American manned spacecraft by revealing three new partnership agreements with SpaceX ($440 million), Boeing ($460 million), and Sierra Nevada ($212.5 million). As a direct result of Congressional restrictions, the competition was reduced from the original five companies competing for the contract to just two, with a third receiving half funds as an added insurance against unforeseen technical hurdles with either of the other two proposals. SpaceX and Boeing were to develop, test, and mature their designs through to the Critical Design Review (CDR) due in April 2014. This would keep the program on target for its first demonstration flights, which are expected to begin in 2016, achieving operational status from 2017 when the chosen vehicle could be flying crews to the ISS. NASA decided to continue to support the development of Sierra Nevada's Dream Chaser concept as the backup option, and while the concept is not expected to participate in the CDR phase it will add further technical analysis of the design and concept of lifting body designs to the data already gathered over the previous 50 years.

As these programs are still in development and the details likely to change, it is too early to include specific information here. Hopefully, the vehicle that becomes America's next operational manned spacecraft launching crews to the ISS will be in service in time for when the next edition of this log is published.

By 2020, it is also expected that the Boeing Orion spacecraft will be available for crew expeditions into deep space, although its final targets are far from certain at this point.

ORION

In 2004 a concept for a new program to send humans back to the Moon and out to Mars was announced by NASA as part of the Vision for Exploration. Under the label of the Constellation Program, a new Crew Exploration Vehicle was proposed for human crews to meet those objectives and eventually received the name Orion. In 2005 designs were sought from industry and in August 2006 Lockheed Martin Corporation won the contract. Development began on the spacecraft and program as the replacement for the Space Shuttle, but the change of administration in the White House and a new President saw the cancellation of Constellation as originally envisaged. Orion was redesignated the Multi-Purpose Crew Vehicle and is currently undergoing development for a wide range of missions to the Moon, Mars, and the asteroids as well as a backup vehicle for cargo and crews

An artist's impression of the proposed Orion spacecraft.

A future Orion-class spacecraft docks with the ISS.

supporting ISS operations. Numerous ground and atmospheric tests and mock-ups have been developed and though it is expected that the first unmanned flight tests of the vehicle in space will commence around 2014, the first astronauts are not expected to fly on board the MPCV/Orion before 2020.

NEW HORIZONS

Whichever new vehicle design is finally chosen to return American astronauts to space from U.S. launch sites it will need to support programs not only in various types of Earth orbit, but also those planned for journeys far beyond our planet, using far more advanced spacecraft to make the actual journeys. There are a range of options available for future space planners and explorers to aim for.

Low (and other) Earth orbit operations

The ISS will continue to be at the forefront of Earth orbital operations for the remainder of this decade, barring any unforeseen major technical problem or emergency situation. Of course, ISS operations also depend upon adequate funding and continued cooperation between the partners, but hopefully we will see possibly 60 resident crews complete their missions and observations on board the ISS by the 20th anniversary of the permanent manning of the complex. It will also require confidence in the ability of the station to continue to support future crews safely and perform its scientific functions properly if operations are to continue into the 2020s. Studies are currently being conducted in order to qualify the main hardware to support ISS operations into 2028, making it a full 30 years since construction began. The work conducted on board the station over the next 15 years or so will presumably be aimed at supporting plans for whatever follows the ISS in low Earth orbit and for missions beyond our planet.

The question of what follows the ISS is an interesting one, as there are no firm plans or suggestions for a follow-on ISS. So can the ISS remain operational and useful for another 20 years without major issues surfacing? It seems doubtful, as there are already signs that the crews' time is being taken up as much by maintenance, repair, and housekeeping as with pure science research. It is also difficult to imagine the complex supporting more than a crew of six, or perhaps nine without additional resources added. Of course, increasing crew numbers will add to the challenge, as more people means more power, further supplies, and logistics, requiring more investment probably for limited extra returns. Then there is the question of added waste and unwanted materials to dispose of. All of this would require further spacecraft to support expanded operations, thus increasing the operating costs.

Merely adding crew members to work on more activities does not really solve the problem, unless the working environment can be made less reliant on crew input for keeping the vehicle operating. If this were possible, then more time would be available for the crews to perform science or research, but this is probably a step beyond the current and potential capabilities of the station. It would be more likely to be included in next-generation vehicles, especially those intended for deep-space operations (see below).

It will be interesting to witness the development of new manned space vehicles, such as Orion, as they are tested in low Earth orbit. Their level of auto-mated or manual operations, and the amount of crew input required for the tasks assigned will also be critical for their success. With advances in robotics and joint operations with automated space vehicles, the argument for involving the full par-ticipation of a human crew will be something of a challenge. Even in Earth orbit, a blend of human and automated space operations is useful. Robots can venture where it is dangerous for humans to go, while humans can be on hand to offer rational decision-making choices, repair, and servicing skills to a degree not found on fully automated machines. And can humans ever truly give up the need to explore and "be there"? Looking at Mars through the eyes of a rover may be thrilling in its own way, but it cannot possibly compare with taking those first steps ourselves.

As for other nations' involvement in human space flight operations, perhaps a truly international program is the way forward, expanding upon the success of the ISS. China is expected to develop its space station program for the rest of this decade and create a viable infrastructure for bolder ventures farther away from Earth. India has also expressed a desire to place its own citizens in orbit, though recent reports have indicated that the supporting technologies required for such a large commitment are not as advanced as originally thought. The first manned domestic Indian flight is still some years in the future, possibly not before 2020. It is also important not to totally ignore comments from Russia about their desire to rekindle their purely national manned space program, though this will of course depend upon sufficient funding commitments.

Another branch of manned operations in Earth orbit are the commercial program, both suborbital and orbital. The forthcoming Virgin Galactic flights in SpaceShipTwo are expected to increase the popularity of short flights to the fringes of space, but not yet into orbit. Since 2009 space tourist flights to the ISS have been suspended due to the increased size of crews on the station, but could be resumed sometime in the near future if hardware and funding become available and the international partners agree to support them. At the time of writing this does not seem likely before 2015 or 2016. Commercially operated space stations are often discussed and the endorsement of privately developed launch and landing systems for ISS support operations are but a step away from developing commercial orbital space operations, perhaps with further moves toward orbital tourist flights.

The idea of factories, large power platforms, and five-star hotels in space may still be a dream of science fiction, but the time will be right when those ideas come to the forefront of space flight, as we are now seeing with commercial launching agreements. One area which will probably be eagerly fought over will be space salvage, the recovery or repair operations to clear up abandoned or failed satellites, opening up the location to new and updated spacecraft. Exciting devel-

A distant Moon beckons future explorers.

opments in low Earth orbit await future space explorers and investors. The head-line glory may come from venturing outwards, but the long-term investment in Earth orbital infrastructure will allow us to look after our own planet, utilizing the huge investments in space exploration to date to improve the quality of life here on the ground.

There are, of course other types of orbits around our planet yet to be investigated by human crews. Often spoken about in tales of science fiction or yet-to-be-achieved space plans these include polar, synchronous, and geostationary orbits and are primary candidates for human expansion in the future.

Return to the Moon

Should we go back to the Moon first or go straight to Mars? That question has been debated for years and continues to be discussed when trying to determine where we go next away from low Earth orbit. We have been to the Moon before, so it is to an extent familiar territory, but the last Apollo landed 40 years ago. So much has changed since we first stepped on to the lunar surface that returning will be almost like starting over again. A return to the Moon had been debated even before the final journeys of Apollo were completed, and many more times since we stopped going there in 1972.

Curiosity rover on Mars pioneers geological sampling.

Clearly NASA would love to return to the Moon, and soon. Russia never made it in the 1960s so to do so in the near future would give mixed emotions to those still alive who participated in the program to beat the Americans to the surface 50 years ago. It would hurt that they could not have done it sooner, but equally would give them pride that they had finally made it. And then there are the Chinese, who clearly have the Moon in their sights. But why go back?

Proving that it can be done again is one argument, but in this current global climate more is needed than national pride and technological achievement. It did not sustain a long-term program last time, so why would it do so now? Other reasons, such as a scientific research base, mining potential, a remote simulation facility for other extraterrestrial explorations deeper into space, an extensive Earth observation platform, a medical isolation facility for those returning from distant targets, or a launching site for interplanetary probes have all been suggested and all have their merits and disadvantages.

No one really thought it would be so long before we considered going back but the argument remains the same: Why should we return, where on the surface should we aim our seventh landing crew, and for what purpose? Would it be to support other programs or for definite objectives of its own?

What is clear though is that, being the closest celestial neighbor, the Moon will surely not be ignored in our expansion beyond low Earth orbit? Even if we initially fly past it on the way to somewhere else, we will return. The overriding questions of a sustained human return to the Moon are the same as they were in 1972—those of when, where, how, and most importantly why. The added question today is also who?

Mission to Mars

The mysterious Red Planet, a land of imagination, dreams, and hopes, is one still to explore and is well within our reach. Again, the question is not so much whether we will go there, but more one of how do we get there, when would this be, and why would we send humans when robots still have difficulties operating that far from home. Out of all the probes sent to Mars, the failure rate is still higher than the success rate, though the odds are improving. No one ever said exploring space was easy or straightforward. Far from it, as the automated exploration of Mars has demonstrated time and time again since the early 1960s.

Again there are countless reasons for exploring the planet. Not least of these is the familiar desire to find evidence of life as we know it or, more precisely, the chemical evidence of the potential for such life in ancient times. But there are other reasons to go to Mars, including mineral mining, a whole range of scientific studies of a different planet and its environment, and as a staging post for missions farther into the outer reaches of the solar system.

Whatever the next decade or two brings in human space endeavor, it is clear Mars will feature highly in long-range objectives. It is hoped that the "long range" will be shortened somewhat to be achievable in our lifetime.

Asteroids

There have also been studies into sending humans to the asteroids over the decades. More recently, serious thought has been given to making such journeys, possibly using Orion-class spacecraft. NASA has commenced a series of pioneering simulations and evaluations for such a mission, which could be mounted towards the end of the 2020s. This work would be valuable for obvious scientific reasons: to gather a better understanding of these strange small worlds and to help plan both robotic and manned missions to Mars. If flown before manned flights to the Red Planet, these asteroid missions would be the farthest humans have ventured into space, with a proposed 1 yr round trip mission some 3 million miles

Earth-based simulations of new space suits and surface exploration equipment.

from Earth and a stay of up to 30 days at the chosen asteroid. Studies into working on the surface of these objects would prove invaluable if one is discovered to be on a collision course with Earth. Other reasons for visiting asteroids are similar to those for Mars, such as the potential for mining minerals or to use them as staging posts for expeditions to the outer reaches of the solar system.

Lagrange points

Despite no clear commitment to return to the Moon or venture to Mars, there are a number of committed individuals and groups who have produced countless plans and studies for deep-space exploration, hoping for the day when these plans turn into reality. Aside from a return to the Moon, human exploration of Mars and visits to far-flung asteroids, another target for future human exploration often features in these plans—the so-called "gravity parking sites" in space. Called Lagrange (or Lagrangian) points, these are great expanses of space at which the gravitational forces of the Sun and the Earth are equalized, so any spacecraft placed there could remain in place with little effort. Within the Earth–Moon system there are five such points. They are far enough away from the Sun or planetary bodies that they make ideal places to situate observation platforms,

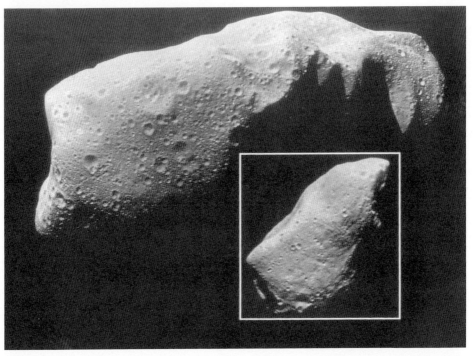

The moons of Mars and asteroids are future targets for robotic and perhaps human exploration.

such as ultra-cold telescopes that measure temperature fluctuations in space. Lagrange points are found around other planets and could be used to site remote operations centers intended to control robotic vehicles to explore the less hospitable places in the solar system more easily.

The L2 point is about 1 million kilometers from Earth and is the target location for the James Webb (infrared) Space Telescope, the replacement for the Hubble Space Telescope. Placing the spacecraft at this point makes servicing from Earth extremely difficult, and until the appearance of Orion or a similar spacecraft it will be impossible for several years to come. Once we have the capability to send crews to these points, they will be able to service and repair the range of telescopes currently being planned to be located there, extending their useful life and expanding their science program as in the case of Hubble. It may also be possible to perform construction tasks with large space structures or spacecraft at these points before sending them to the distant reaches of the solar system.

These locations could provide useful preparation points for trips to Mars and for controlling automated spacecraft on the Moon—a sort of Mission Control in space. With the development of more artificial intelligence spacecraft, operating a control room from deeper into space where communications would be much quicker would clearly be more advantageous than waiting for signals sent between

Underwater simulations help prepare for deep-space planning.

Earth and Mars that would need a 40 min round trip. As Gemini was a step for Apollo to the Moon, perhaps Lagrange points will be a stepping stone to deep-space human exploration missions.

TO BOLDLY GO ...

It was just over five decades ago that a young Soviet air force pilot was sitting strapped to an ejection seat in the confined compartment of a new type of vehicle called a spacecraft. After a rocket-boosted flight of a few minutes, he found himself high above the Earth in the vacuum of space and for just one orbit, becoming the only living human not to be on Earth or within its atmosphere. With this short mission, Yuri Gagarin became history's first explorer of the cosmos. In the decades since that bold leap, over 500 individuals have followed in his trail, creating news pages in history along the way.

Who will be the 13th person on the Moon?

Sadly over the past five decades, we have lost many of the pioneers from the early days of the space program, both those who made the journey from Earth and those who made such missions possible, from administrators and managers, to flight controllers, launch technicians, spacecraft designers and engineers, and so many more. As these pages were being written, two more pioneers were lost in the space of one month: on July 23, 2012, Sally Ride, the first American woman to fly in space, lost her battle with cancer at the age of 61. This was followed on August 25 by the death of Neil Armstrong, the first man to step on to the Moon, following complications after heart surgery at the age of 82. Their contribution, along with their colleagues and fellow workers in the first 50 years of the global space program, will never be forgotten. No matter how far humans may venture or what marvels they may encounter in the exploration of space, their achievements will have been built upon the foundations laid by pioneers such as these. The exploits of those who created and flew the first missions from Earth are recounted in documents such as this log.

Appendix A

Manned Spaceflight Log Book 1961–2012

World space flight sequence	Country of origin/ sequence	Earth orbit flight sequence	Mission designation	Prime crew members	Launch to landing dates (dd/mm/yy)	Crew duration (dd:hh:mm:ss)	Notes
1961							
1	U.S.S.R. 1	1	Vostok	Gagarin	04/12/61	000:01:48:00	
2	U.S.A. 1	Suborbital	Mercury 3	Shepard	05/05/61	000:00:15:28	
3	U.S.A. 2	Suborbital	Mercury 4	Grissom	07/21/61	000:00:15:37	
4	U.S.S.R. 3	2	Vostok 2	Titov G.	08/06/61–08/07/61	001:01:18:00	
1962							
5	U.S.A. 3	3	Mercury 6	Glenn	02/20/62	000:04:55:23	
6	U.S.A. 4	4	Mercury 7	Carpenter	05/24/62	000:04:56:05	
—	U.S.A. –	Astro-flight	X-15-3-62	White R.	07/17/62	000:00:10:00	
7	U.S.S.R. 3	5	Vostok 3	Nikolayev	08/11/62–08/15/62	003:22:22:00	
8	U.S.S.R. 4	6	Vostok 4	Popovich	08/12/62–08/15/62	002:22:57:00	
9	U.S.A. 5	7	Mercury 8	Schirra	10/03/62	000:09:13:11	

(continued)

World space flight sequence	Country of origin/ sequence	Earth orbit flight sequence	Mission designation	Prime crew members	Launch to landing dates (dd/mm/yy)	Crew duration (dd:hh:mm:ss)	Notes
1963							
—	U.S.A. —	Astro-flight		Walker J.	01/17/63	000:00:10:00	
10	U.S.A. 6	8	Mercury 9	Cooper	05/15/63–05/16/63	001:10:19:49	
11	U.S.S.R. 5	9	Vostok 5	Bykovsky	06/14/63–06/19/63	004:23:06:00	
12	U.S.S.R. 6	10	Vostok 6	Tereshkova	06/16/63–06/19/63	002:22:50:00	
—	U.S.A. —	Astro-flight	X-15-3-87	Rushworth	06/27/63	000:00:10:00	
—	U.S.A. —	Astro-flight	X-15-3-90	Walker J.	07/19/63	000:00:10:00	
—	U.S.A. —	Astro-flight	X-15-3-91	Walker J.	08/22/63	000:00:10:00	
1964							
13	U.S.S.R. 7	11	Voskhod	Komarov/Feoktistov/Yegorov	10/12/64–10/13/64	001:00:17:03	
1965							
14	U.S.S.R. 8	12	Voskhod 2	Belyayev/Leonov	03/18/65–03/19/65	001:02:02:17	
15	U.S.A. 7	13	Gemini 3	Grissom/Young	03/23/65	000:04:52:51	
16	U.S.A. 8	14	Gemini 4	McDivitt/White	06/03/65–06/07/65	004:01:56:12	
—	U.S.A. —	Astro-flight	X-15-3-138	Engle	06/29/65	000:00:10:00	
—	U.S.A. —	Astro-flight	X-15-1-143	Engle	08/10/65	000:00:10:00	
17	U.S.A. 9	15	Gemini 5	Cooper/Conrad	08/21/65–08/29/65	007:22:55:14	
—	U.S.A. —	Astro-flight	X-15-3-150	McKay	09/29/65	000:00:10:00	
—	U.S.A. —	Astro-flight	X-15-1-153	Engle	10/14/65	000:00:10:00	
18	U.S.A. 10	16	Gemini 7	Borman/Lovell	12/04/65–12/16/65	013:18:35:01	
19	U.S.A. 11	17	Gemini 6	Schirra/Stafford	12/15/65–12/16/65	001:01:51:54	

1966							
20	U.S.A. 12	18	Gemini 8	Armstrong/Scott	03/16/66	000:010:41:26	
21	U.S.A. 13	19	Gemini 9	Stafford/Cernan	06/03/66–06/06/66	003:00:20:50	
22	U.S.A. 14	20	Gemini 10	Young/Collins	07/18/66–07/21/66	002:22:46:39	
23	U.S.A. 15	21	Gemini 11	Conrad/Gordon	09/12/66–09/15/66	002:23:17:08	
—	U.S.A. —	Astro-flight	X-15-3-174	Dana	11/01/66	000:00:10:00	
24	U.S.A. 16	22	Gemini 12	Lovell/Aldrin	11/11/66–11/16/66	003:22:34:31	
1967							
—	U.S.A. —	Pre-launch pad fire	Apollo 1	Grissom/White/Chaffee	01/27/67		Fatal pad fire prior to launch date
25	U.S.S.R. 9	23	Soyuz 1	Komarov	04/23/67–24/24/67	001:02:47:52	Komarov died in crash landing
—	U.S.A. —	Astro-flight	X-15-3-190	Knight	10/17/67	000:00:10:00	
—	U.S.A. —	Astro-flight	X-15-3-191	Adams	11/15/67	000:00:10:00	
1968							
—	U.S.A. —	Astro-flight	X-15-1-197	Knight	08/21/68	000:00:10:00	
26	U.S.A. 17	24	Apollo 7	Schirra/Eisele/Cunningham	10/11/68–10/22/68	010:20:09:03	
27	U.S.S.R. 10	25	Soyuz 3	Beregovoi	10/26/68–10/30/68	003:22:50:45	
28	U.S.A. 18	26	Apollo 8	Borman/Lovell/Anders	12/21/68–12/28/68	006:03:00:42	
1969							
29	U.S.S.R. 11	27	Soyuz 4	Shatalov	01/14/69–01/17/69	002:23:20:47	Volynov
30	U.S.S.R. 12	28	Soyuz 5	Volynov/Yeliseyev/Khrunov	01/15/69–01/18/69	003:00:54:15 / 001:23:45:50	Yeliseyev/Khrunov (down on Soyuz 4)
31	U.S.A. 19	29	Apollo 9	McDivitt/Scott/Schweickart	03/03/69–03/13/69	010:01:00:54	

(continued)

World space flight sequence	Country of origin/ sequence	Earth orbit flight sequence	Mission designation	Prime crew members	Launch to landing dates (dd/mm/yy)	Crew duration (dd:hh:mm:ss)	Notes
1969	(cont.)						
32	U.S.A. 20	30	Apollo 10	Stafford/Young/Cernan	05/18/69–05/26/69	008:00:03:23	
33	U.S.A. 21	31	Apollo 11	Armstrong/Collins M./Aldrin	07/16/69–07/24/69	008:03:18:35	
34	U.S.S.R. 13	32	Soyuz 6	Shonin/Kubasov	10/11/69–10/16/69	004:22:42:47	
35	U.S.S.R. 14	33	Soyuz 7	Filipchenko/Gorbatko/Volkov V.	10/12/69–10/17/69	004:22:40:23	
36	U.S.S.R. 15	34	Soyuz 8	Shatalov/Yeliseyev	10/13/69–10/18/69	004:22:50:49	
37	U.S.A. 22	35	Apollo 12	Conrad/Gordon/Bean	11/14/69–11/24/69	010:04:36:25	
1970							
38	U.S.A. 23	36	Apollo 13	Lovell/Swigert/Haise	04/11/70–04/17/70	005:22:54:41	
39	U.S.S.R. 16	37	Soyuz 9	Nikolayev/Sevastyanov	06/01/70–06/19/70	017:16:58:55	
1971							
40	U.S.A. 24	38	Apollo 14	Shepard/Roosa/Mitchell	01/31/71–02/09/71	009:00:01:57	
41	U.S.S.R. 17	39	Soyuz 10	Shatalov/Yeliseyev/Rukavishnikov	04/23/71–04/25/71	001:23:45:54	
42	U.S.S.R. 18	40	Soyuz 11	Dobrovolsky/Volkov V./Patsayev	06/06/71–06/30/71	023:18:21:43	
43	U.S.A. 25	41	Apollo 15	Scott D./Worden/Irwin	07/26/71–08/07/71	012:07:11:53	
1972							
44	U.S.A. 26	42	Apollo 16	Young/Mattingly/Duke	04/16/72–04/27/72	011:01:51:25	
45	U.S.A. 27	43	Apollo 17	Cernan/Evans/Schmitt	12/06/72–12/19/72	012:13:51:59	

1973							
46	U.S.A. 28	44	Skylab 2	Conrad/Kerwin/Weitz	05/25/73–06/22/73	028:00:49:49	
47	U.S.A. 29	45	Skylab 3	Bean/Garriott/Lousma	07/28/73–09/25/73	059:11:09:04	
48	U.S.S.R. 19	46	Soyuz 12	Lazarev/Makarov	09/27/73–09/29/73	001:23:15:32	
49	U.S.A. 30	47	Skylab 4	Carr/Gibson E./Pogue	01/15/73–02/08/74	084:01:15:37	
50	U.S.S.R. 20	48	Soyuz 13	Klimuk/Lebedev	12/18/73–12/26/73	007:20:55:35	
1974							
51	U.S.S.R. 21	49	Soyuz 14	Popovich/Artyukhin	07/03/74–07/19/74	015:17:30:28	
52	U.S.S.R. 22	50	Soyuz 15	Sarafanov/Demin	08/26/74–08/28/74	002:00:12:11	
53	U.S.S.R. 23	51	Soyuz 16	Filipchenko/Rukavishnikov	12/02/74–12/08/74	005:22:23:35	
1975							
54	U.S.S.R. 24	52	Soyuz 17	Gubarev/Grechko	01/11/75–12/09/75	029:13:19:45	
55	U.S.S.R. 25	Launch abort	Soyuz 18-1	Lazarev/Makarov	04/05/75	000:00:21:27	
56	U.S.S.R. 26	53	Soyuz 18	Klimuk/Sevastyanov	05/24/75–06/26/75	062:23:20:08	
57	U.S.S.R. 27	54	Soyuz 19	Leonov/Kubasov	07/15/75–07/21/75	005:22:30:51	
58	U.S.A. 31	55	Apollo 18	Stafford/Brand/Slayton	07/15/75–07/24/75	09:01:28:24	
1976							
59	U.S.S.R. 28	56	Soyuz 21	Volynov/Zholobov	07/06/76–08/24/76	049:06:23:32	
60	U.S.S.R. 29	57	Soyuz 22	Bykovsky/Aksenov	09/15/76–09/23/76	007:21:52:17	
61	U.S.S.R. 30	58	Soyuz 23	Zudov/Rozhdestvensky	10/14/76–10/16/76	002:00:06:35	
1977							
62	U.S.S.R. 31	59	Soyuz 24	Gorbatko/Glazkov	02/07/77–02/25/77	017:17:25:58	
63	U.S.S.R. 32	60	Soyuz 25	Kovalenok/Ryumin	10/09/77–10/11/77	002:00:44:45	
64	U.S.S.R. 33	61	Soyuz 26	Romanenko/Grechko	12/10/77–03/16/78	096:10:00:07	Down on Soyuz 27

(continued)

World space flight sequence	Country of origin/ sequence	Earth orbit flight sequence	Mission designation	Prime crew members	Launch to landing dates (dd/mm/yy)	Crew duration (dd:hh:mm:ss)	Notes
1978							
65	U.S.S.R. 34	62	Soyuz 27	Dzhanibekov/Makarov	01/10/78–01/16/78	005:22:58:58	Down on Soyuz 26
66	U.S.S.R. 35	63	Soyuz 28	Gubarev/Remek	03/02/78–03/10/78	007:22:16:00	
67	U.S.S.R. 36	64	Soyuz 29	Kovalenok/Ivanchenko	07/15/78–11/02/78	139:14:47:32	Down on Soyuz 31
68	U.S.S.R. 37	65	Soyuz 30	Klimuk/Hermaszewski	06/27/78–07/05/78	007:22:02:59	
69	U.S.S.R. 38	66	Soyuz 31	Bykovsky/Jaehn	08/26/78–09/03/78	007:20:49:04	Down on Soyuz 29
1979							
70	U.S.S.R. 39	67	Soyuz 32	Lyakhov/Ryumin	02/25/79–08/19/79	175:00:35:37	Down on Soyuz 34
71	U.S.S.R. 40	68	Soyuz 33	Rukavishnikov/Ivanov G.	04/10/79–04/12/79	001:23:01:06	
1980							
72	U.S.S.R. 41	69	Soyuz 35	Popov/Ryumin	04/09/80–10/11/80	184:20:11:35	Down on Soyuz 37
73	U.S.S.R. 42	70	Soyuz 36	Kubasov/Farkas	05/26/80–06/03/80	007:20:45:44	Down on Soyuz 35
74	U.S.S.R. 43	71	Soyuz T-2	Malyshev/Aksenov	06/05/80–06/09/80	003:22:19:30	
75	U.S.S.R. 44	72	Soyuz 37	Gorbatko/Pham Tuan	07/23/80–07/31/80	007:20:42:00	
76	U.S.S.R. 45	73	Soyuz 38	Romanenko/Tamayo-Mendez	09/18/80–09/26/80	007:20:43:24	
77	U.S.S.R. 46	74	Soyuz T-3	Kizim/Makarov/Strekalov	11/27/80–12/10/80	012:19:07:42	Down on Soyuz 36
1981							
78	U.S.S.R. 47	75	Soyuz T-4	Kovalenok/Savinykh	03/13/81–05/26/81	074:17:37:23	
79	U.S.S.R. 48	76	Soyuz 39	Dzhanibekov/Gurragcha	03/22/81–03/30/81	007:20:42:03	
80	U.S.A. 32	77	STS-1	Young/Crippen	04/12/81–04/14/81	002:06:20:53	
81	U.S.S.R. 49	78	Soyuz 40	Popov/Prunariu	05/14/81–05/22/81	007:20:41:52	
82	U.S.A. 33	79	STS-2	Engle/Truly	11/12/81–11/14/81	002:06:13:13	

Year	No.	Country	Flt	Flight	Crew	Date	Duration	Notes
1982	83	U.S.A. 34	80	STS-3	Lousma/Fullerton	03/22/82–03/30/82	008:00:04:45	
	84	U.S.S.R. 50	81	Soyuz T-5	Berezovoy/Lebedev	05/13/82–12/10/82	211:09:04:32	Down on Soyuz T-7
	85	U.S.S.R. 51	82	Soyuz T-6	Dzhanibekov/Ivanchenko/Chretien	06/24/82–07/02/82	007:21:50:52	
	86	U.S.A. 35	83	STS-4	Mattingly/Hartsfield	06/27/82–07/02/82	007:01:09:31	
	87	U.S.S.R. 52	84	Soyuz T-7	Popov/Serebrov/Savitskaya	08/19/82–08/24/82	007:21:52:24	Down on Soyuz T-5
	88	U.S.A. 36	85	STS-5	Brand/Overmyer/Allen J./Lenoir	11/11/82–11/16/82	005:02:14:26	
1983	89	U.S.A. 37	86	STS-6	Weitz/Bobko/Musgrave/Peterson	04/04/83–04/09/83	005:00:23:42	
	90	U.S.S.R. 53	87	Soyuz T-8	Titov V./Strekalov/Serebrov	04/20/83–04/22/83	002:00:17:48	
	91	U.S.A. 38	88	STS-7	Crippen/Hauck/Fabian/Ride/Thagard	06/18/83–06/24/83	006:02:23:59	
	92	U.S.S.R. 54	89	Soyuz T-9	Lyakhov/Alexandrov	06/27/83–11/23/83	149:10:46:01	
	93	U.S.A. 39	90	STS-8	Truly/Brandenstein/Bluford/Gardner/Thornton W.	08/30/83–09/05/83	006:01:08:43	
	—	Pad Abort		Soyuz T 10-1	Titov V./Strekalov	09/26/83	N/A	Launchpad abort prior to liftoff
	94	U.S.A. 40	91	STS-9	Young/Shaw/Garriott/Parker/Lichtenberg/Merbold	11/28/83–12/08/83	010:07:47:23	
1984	95	U.S.A. 41	92	STS-41-B	Brand/Gibson R./McNair/Stewart/McCandless	02/03/84–02/11/84	007:23:15:55	
	96	U.S.S.R. 55	93	Soyuz T-10	Kizim/Solovyov V./Atkov	02/08/84–10/02/84	236:22:49:04	Down on Soyuz T-11
	97	U.S.S.R. 56	94	Soyuz T-11	Malyshev/Strekalov/Sharma	04/03/84–04/11/84	007:21:40:06	Down on Soyuz T-10

(continued)

World space flight sequence	Country of origin/ sequence	Earth orbit flight sequence	Mission designation	Prime crew members	Launch to landing dates (dd/mm/yy)	Crew duration (dd:hh:mm:ss)	Notes
1984	(cont.)						
98	U.S.A. 42	95	STS-41-C	Crippen/Scobee/Hart/ Van Hoften/Nelson G.	04/06/84–04/13/84	006:23:40:06	
99	U.S.S.R. 57	96	Soyuz T-12	Dzhanibekov/Savitskaya/Volk	07/17/84–07/29/84	011:19:14:36	
100	U.S.A. 43	97	STS-41-D	Hartsfield/Coats/Mullane/ Hawley/Resnik/Walker C.	08/30/84–09/05/84	006:00:56:04	
101	U.S.A. 44	98	STS-41-G	Crippen/McBride/Sullivan/ Ride/Leestma/Scully-Power/ Garneau	10/05/84–10/13/84	008:05:23:38	
102	U.S.A. 45	99	STS-51-A	Hauck/Walker D./Allen J./ Fisher A./ Gardner D.	11/08/84–11/16/84	007:23:44:56	
1985							
103	U.S.A. 46	100	STS-51-C	Mattingly/Shriver/Onizuka/ Buchli/Payton	01/24/85–01/27/85	003:01:23:23	
104	U.S.A. 47	101	STS-51-D	Bobko/Williams/Griggs/ Hoffman/Seddon/Garn/ Walker C.	04/12/85–04/19/85	006:23:55:23	
105	U.S.A. 48	102	STS-51-B	Overmyer/Gregory F./Lind/ Thagard/Thornton W./Wang/ Van den Berg	04/29/85–05/06/85	007:00:08:46	
106	U.S.S.R. 58	103	Soyuz T-13	Dzhanibekov/Savinykh	06/06/85–09/26/85	112:03:12:06 / 168:03:51:00	Dzhanibekov Savinykh (down on Soyuz T-14)
107	U.S.A. 49	104	STS-51-G	Brandenstein/Creighton/ Fabian/Nagel/Lucid/Baudry/ Al-Saud	06/17/85–06/24/85	007:01:38:52	

				Crew	Dates	Duration	Notes
108	U.S.A. 50	105	STS-51-F	Fullerton/Bridges/Henize/Musgrave/England/Acton/Bartoe	07/29/85–08/06/85	007:22:45:26	
109	U.S.A. 51	106	STS-51-I	Engle/Covey/Van Hoften/Lounge/Fisher W.	08/27/85–09/03/85	007:02:17:42	
110	U.S.S.R. 59	107	Soyuz T-14	Vasyutin/Grechko/Volkov A.	09/17/85–11/21/85	064:21:52:08 / 008:21:13:06	Vasyutin/Volkov A. Grechko (down on Soyuz T-13)
111	U.S.A. 52	108	STS-51-J	Bobko/Grabe/Hilmers/Stewart/Pailes	10/03/85–10/07/85	004:01:44:38	
112	U.S.A. 53	109	STS-61-A	Hartsfield/Nagel/Dunbar/Buchli/Bluford/Furrer/Messerschmid/Ockels	10/30/85–11/06/85	007:00:44:53	
113	U.S.A. 54	110	STS-61-B	Shaw/O'Connor/Ross/Cleve/Spring/Walker C./Neri-Vela	11/26/85–12/03/85	006:21:04:49	
1986							
114	U.S.A. 55	111	STS-61-C	Gibson R./Bolden/Nelson G./Hawley/Chang-Diaz/Cenker/Nelson B.	01/12/86–01/18/86	006:02:03:51	
—		Launch accident	STS-51-L	Scobee/Smith M./Onizuka/Resnik/McNair/Jarvis/McAuliffeascent	01/28/86	000:00:01:13	Fatal explosion during the ascent
115	U.S.S.R. 60	112	Soyuz T-15	Kizim/Solovyov V.	03/13/86–07/16/86	125:00:00:56	
1987							
116	U.S.S.R. 61	113	Soyuz TM-2	Romanenko/Laveikin	02/06/87–12/29/87	326:11:37:57 / 174:03:25:56	Romanenko (down on Soyuz TM-3) Laveikin (down on TM-2)

(continued)

World space flight sequence	Country of origin/ sequence	Earth orbit flight sequence	Mission designation	Prime crew members	Launch to landing dates (dd/mm/yy)	Crew duration (dd:hh:mm:ss)	Notes
1987	(cont.)						
117	U.S.S.R. 62	114	Soyuz TM-3	Viktorenko/Alexandrov/Faris	07/22/87–07/30/87	007:23:04:55 160:07:16:58	Viktorenko/Faris (down on Soyuz TM-2) Alexandrov (down on Soyuz TM-3)
118	U.S.S.R. 63	115	Soyuz TM-4	Titov V./Manarov/Levchenko	12/21/87–12/21/88	365:22:38:57 007:21:58:12	Titov V./Manarov (down on Soyuz TM-TM6) Levchenko (down on Soyuz TM-3)
1988							
119	U.S.S.R. 64	116	Soyuz TM-5	Solovyov A./Savinykh/Alexandrov	06/07/88–06/17/88	009:20:09:19	
120	U.S.S.R. 65	117	Soyuz TM-6	Lyakhov/Polyakov/Mohmand	08/29/88–07/07/88	008:20:26:27 240:22:34:47	Lyakhov/Mohmand (down on Soyuz TM-5) Polyakov (down on Soyuz TM-7)
121	U.S.A. 56	118	STS-26	Hauck/Covey/Lounge/Hilmers/Nelson G.	09/29/88–10/03/88	004:01:00:11	
122	U.S.S.R. 66	119	Soyuz TM-7	Volkov A./Krikalev/Chretien	11/26/88–04/26/89	151:11:08:23 024:18:07:25	Volkov A./Krikalev (down on Soyuz TM-7) Chretien (down on Soyuz TM-6)
123	U.S.A. 57	120	STS-27	Gibson R./Gardner G./Mullane/Ross/Shepherd	12/02/88–12/06/88	004:09:05:35	

1989

124	U.S.A. 58	121	STS-29	Coats/Blaha/Buchli/Springer/Bagian	03/13/89–03/18/89	004:23:38:50
125	U.S.A. 59	122	STS-30	Walker D./Grabe/Thagard/Cleave/Lee	05/04/89–05/08/89	004:00:56:27
126	U.S.A. 60	123	STS-28	Shaw/Richards R./Leestma/Adamson/Brown M.	08/08/89–08/13/89	005:01:00:09
127	U.S.S.R. 67	124	Soyuz TM-8	Viktorenko/Serebrov	09/06/89–02/19/90	166:06:58:16
128	U.S.A. 61	125	STS-34	Williams D./McCulley/Lucid/Chang-Diaz/Baker E.	10/18/89–10/23/89	004:23:39:21
129	U.S.A. 62	126	STS-33	Gregory F./Blaha/Carter/Musgrave/Thornton K.	11/22/89–11/27/89	005:00:06:48

1990

130	U.S.A. 63	127	STS-32	Brandenstein/Wetherbee/Dunbar/Ivins/Low	01/09/90–01/20/90	010:21:00:36
131	U.S.S.R. 68	128	Soyuz TM-9	Solovyov A./Balandin	02/11/90–08/09/90	179:01:17:57
132	U.S.A. 64	129	STS-36	Creighton/Casper/Hilmers/Mullane/Thuot	02/28/90–03/04/90	004:10:18:22
133	U.S.A. 65	130	STS-31	Shriver/Bolden/McCandless/Hawley/Sullivan	04/24/90–04/29/90	005:01:16:06
134	U.S.S.R. 69	131	Soyuz TM-10	Manakov/Strekalov	08/01/90–12/10/90	130:20:35:51
135	U.S.A. 66	132	STS-41	Richards/Cabana/Melnick/Shepherd/Akers	10/06/90–10/10/90	004:02:10:04
136	U.S.A. 67	133	STS-38	Covey/Culbertson/Springer/Meade/Gemar	11/15/90–11/20/90	004:21:54:31
137	U.S.A. 68	134	STS-35	Brand/Gardner G./Hoffman/Lounge/Parker/Durrance/Parise	12/02/90–12/10/90	008:23:05:08

(continued)

World space flight sequence	Country of origin/ sequence	Earth orbit flight sequence	Mission designation	Prime crew members	Launch to landing dates (dd/mm/yy)	Crew duration (dd:hh:mm:ss)	Notes
1990	(cont.)						
138	U.S.S.R. 70	135	Soyuz TM-11	Afanasyev/Manarov/Akiyama	12/02/90–05/26/91	175:01:51:42 007:21:54:40	Afanasyev/Manarov (down on Soyuz TM-11) Akiyama (down on Soyuz TM-10)
1991							
139	U.S.A. 69	136	STS-37	Nagel/Cameron/Godwin/ Ross/Apt	04/05/91–04/11/91	005:23:32:44	
140	U.S.A. 70	137	STS-39	Coats/Hammond/Harbaugh/ McMonagle/Bluford/Veach/ Hieb	04/28/91–05/06/91	008:07:22:23	
141	U.S.S.R. 71	138	Soyuz TM-12	Artsebarsky/Krikalev/ Sharman	05/18/91–10/10/91	144:15:21:50 007:21:14:20 311:20:01:54	Artsebarsky (down on Soyuz TM-12) Sharman (down on Soyuz TM-11) Krikalev (down on Soyuz TM-13)
142	U.S.A. 71	139	STS-40	O'Connor/Gutierrez/Bagian/ Jernigan/Seddon/Gaffney/ Hughes-Fulford	06/05/91–06/14/91	009:02:14:20	
143	U.S.A. 72	140	STS-43	Blaha/Baker M./Lucid/Low/ Adamson	08/02/91–08/11/91	008:21:21:25	
144	U.S.A. 73	141	STS-48	Creighton/Reightler/Gemar/ Buchli/Brown M.	09/12/91–09/18/91	005:08:27:38	

145	U.S.S.R. 72	142	Soyuz TM-13	Volkov A./Aubakirov/ Viehbock	10/02/91–03/25/92	175:02:52:43 007:22:12:59	Volkov A. (down on Soyuz TM-13) Aubakirov/Viehbock (down on Soyuz TM12)
146	U.S.A. 74	143	STS-44	Gregory F./Henricks/ Voss J.S./Musgrave/Runco/ Hennen	11/24/91–12/01/91	006:22:50:44	
1992						
147	U.S.A. 75	144	STS-42	Grabe/Oswald/Thagard/ Readdy/Hilmers/Bondar/ Merbold	01/22/92–01/30/92	008:01:14:44	
148	Russia 73	145	Soyuz TM-14	Viktorenko/Kaleri/Flade	03/17/92–08/10/92	145:14:10:32 007:21:56:52	Viktorenko/Kaleri (down on Soyuz TM-14) Flade (down on Soyuz TM-13)
149	U.S.A. 76	146	STS-45	Bolden/Duffy/Sullivan/ Leestma/Foale/Frimout/ Lichtenberg	03/24/92–04/02/92	008:22:09:28	
150	U.S.A. 77	147	STS-49	Brandenstein/Chilton/Hieb/ Melnick/Thuot/Thornton K./ Akers	05/07/92–05/16/92	008:21:17:38	
151	U.S.A. 78	148	STS-50	Richards R./Bowersox/ Dunbar/Baker E./Meade/ DeLucas/Trinh	06/25/92–07/09/92	013:19:30:04	
152	Russia 74	149	Soyuz TM-15	Solovyov A./Avdeyev/Tognini	07/27/92–02/01/93	188:21:41:15 013:18:56:14	Solovyov A./Avdeyev (down on Soyuz TM-15) Tognini (down on Soyuz TM-14)
153	U.S.A. 79	150	STS-46	Shriver/Allen A./Nicollier/ Ivins/Hoffman/Chang-Diaz/ Malerba	07/31/92–08/08/92	007:23:15:03	

(continued)

World space flight sequence	Country of origin/sequence	Earth orbit flight sequence	Mission designation	Prime crew members	Launch to landing dates (dd/mm/yy)	Crew duration (dd:hh:mm:ss)	Notes
1992	*(cont.)*						
154	U.S.A. 80	151	STS-47	Gibson R./Brown C./Lee/Apt/Davis/Jemison/Mohri	09/12/92–09/20/92	007:22:30:23	
155	U.S.A. 81	152	STS-52	Wetherbee/Baker M./Veach/Shepherd/Jernigan/MacLean	10/22/92–11/01/92	009:20:56:13	
156	U.S.A. 82	153	STS-53	Walker D./Cabana/Bluford/Voss/Clifford	12/02/92–12/09/92	007:07:19:47	
1993							
157	U.S.A. 83	154	STS-54	Casper/McMonagle/Harbaugh/Runco/Helms	01/13/93–01/19/93	005:23:38:19	
158	Russia 75	155	Soyuz TM-16	Manakov/Poleschuk	01/24/93–07/02/93	179:00:43:46	
159	U.S.A. 84	156	STS-56	Cameron/Oswald/Foale/Cockrell/Ochoa	04/07/93–04/17/93	009:06:08:24	
160	U.S.A. 85	157	STS-55	Nagel/Henricks/Ross/Precourt/Harris/Walter/Schlegel	04/26/93–05/06/93	009:23:39:59	
161	U.S.A. 86	158	STS-57	Grabe/Duffy/Low/Sherlock/Wisoff/Voss J.E.	06/21/93–07/02/93	009:23:44:54	
162	Russia 76	159	Soyuz TM-17	Tsibliyev/Serebrov/Haigneré J-P	07/01/93–01/14/94	196:17:45:22 / 020:16:08:52	Tsibliyev/Serebrov Haigneré (down on Soyuz TM-16)
163	U.S.A. 87	160	STS-51	Culbertson/Readdy/Newman/Bursch/Walz	09/12/93–09/22/93	009:20:11:11	

164	U.S.A. 88	STS-58	Blaha/Searfoss/Seddon/McArthur W./Lucid/Wolf/Fettman	10/18/93–11/01/93	014:00:12:32	
165	U.S.A. 89	STS-61	Covey/Bowersox/Thornton K./Nicollier/Hoffman/Musgrave/Akers	12/02/93–12/12/93	010:19:58:37	
1994						
166	Russia 77	Soyuz TM-18	Afanasyev/Usachev/Polyakov	01/08/94–07/09/94	182:00:27:02 437:17:58:31	Afanasyev/Usachev (down on Soyuz TM-18) Polyakov (down on Soyuz TM-20)
167	U.S.A. 90	STS-60	Bolden/Reightler/Davis/Sega/Chang-Diaz/Krikalev	02/03/94–02/11/94	008:07:09:22	
168	U.S.A. 91	STS-62	Casper/Allen A./Thuot/Gemar/Ivins	03/04/94–03/18/94	013:23:16:41	
169	U.S.A. 92	STS-59	Gutierrez/Chilton/Apt/Clifford/Godwin/Jones T.	04/09/94–04/20/94	011:05:49:30	
170	Russia 78	Soyuz TM-19	Malenchenko/Musabayev	07/01/94–11/04/94	125:22:53:36	
171	U.S.A. 93	STS-65	Cabana/Halsell/Heib/Walz/Thomas D./Chiao/Mukai	07/08/94–07/23/94	014:17:55:00	
172	U.S.A. 94	STS-64	Richards R./Hammond/Linenger/Helms/Meade/Lee	09/09/94–09/20/94	010:22:49:57	
173	U.S.A. 95	STS-68	Baker M./Wilcutt/Smith S./Bursch/Wisoff/Jones T.	09/30/94–10/11/94	011:05:46:08	
174	Russia 79	Soyuz TM-20	Viktorenko/Kondakova/Merbold	10/04/94–03/22/95	169:05:21:35 031:12:35:56	Viktorenko/Kondakova (down on Soyuz TM-20) Merbold (down on Soyuz TM-19)
175	U.S.A. 96	STS-66	McMonagle/Brown C./Ochoa/Tanner/Clervoy/Parazynski	11/03/94–11/14/94	010:22:34:02	

(continued)

World space flight sequence	Country of origin/ sequence	Earth orbit flight sequence	Mission designation	Prime crew members	Launch to landing dates (dd/mm/yy)	Crew duration (dd:hh:mm:ss)	Notes
1995							
176	U.S.A. 97	173	STS-63	Wetherbee/Collins E./Harris/ Foale/Voss J E ./Titov V.	02/02/95–02/11/95	008:06:28:15	
177	U.S.A. 98	174	STS-67	Oswald/Gregory W./ Grunsfeld/Lawrence/Jernigan/ Durrance/Parise	03/02/95–03/18/95	016:15:08:48	
178	Russia 80	175	Soyuz TM-21	Dezhurov/Strekalov/Thagard	03/14/95–09/11/95	115:08:43:02	Down on STS-71
179	U.S.A. 99	176	STS-71	Gibson R./Precourt/Baker E./ Harbaugh/Dunbar/ Solovyov A./Budarin	06/27/95–07/07/95	009:19:22:17	STS-71 orbiter crew only
—	—	Launched on STS-71	Mir EO-19	Solovyov A./Budarin (launched on STS-71)	06/27/95–09/11/95	075:11:20:21	Down on Soyuz TM-21 Down on Soyuz TM-21
180	U.S.A. 100	177	STS-70	Henricks/Kregel/Thomas D./ Currie/Weber	07/13/95–07/22/95	008:22:20:05	
181	Russia 81	178	Soyuz TM-22	Gidzenko/Avdeyev/Reiter	09/03/95–02/29/96	179:01:41:46	
182	U.S.A. 101	179	STS-69	Walker D./Cockrell/Voss J.S./ Newman/Gernhardt	09/07/95–09/18/95	010:20:28:56	
183	U.S.A. 102	180	STS-73	Bowersox/Rominger/Coleman/ López-Alegría/Thornton K./ Leslie/Sacco	10/20/95–11/05/95	015:21:52:28	
184	U.S.A. 103	181	STS-74	Cameron/Halsell/Hadfield/ Ross/McArthur W.	11/12/95–11/20/95	008:04:30:44	

1996							
185	U.S.A. 104	182	STS-72	Duffy/Jett/Chiao/Scott W./Wakata/Barry	01/11/96–01/20/96	008:22:01:47	
186	Russia 82	183	Soyuz TM-23	Onufriyenko/Usachev	02/23/96–09/02/96	193:19:07:35	
187	U.S.A. 105	184	STS-75	Allen A./Horowitz/Hoffman/Cheli/Nicollier/Chang-Diaz/Guidoni	02/22/96–03/09/96	015:17:40:21	
188	U.S.A. 106	185	STS-76	Chilton/Searfoss/Sega/Clifford/Godwin/Lucid	03/22/96–03/31/96	009:05:15:53 188:04:00:11	STS-76 orbiter crew Lucid (down on STS-79)
189	U.S.A. 107	186	STS-77	Casper/Brown C./Thomas A./Bursch/Runco/Garneau	05/19/96–05/29/96	010:00:39:18	
190	U.S.A. 108	187	STS-78	Henricks/Kregel/Linnehan/Helms/Brady/Favier/Thirsk	06/20/96–07/07/96	016:21:47:45	
191	Russia 83	188	Soyuz TM-24	Korzun/Kaleri/Andre-Deshays	08/17/96–03/02/97	196:17:26:13 015:18:23:37	Korzun/Kaleri (down on Soyuz TM-24) Andre-Deshays (down on Soyuz TM-23)
192	U.S.A. 109	189	STS-79	Readdy/Wilcutt/Apt/Akers/Walz/Blaha	09/16/96–09/26/96	010:03:18:26 128:05:27:55	STS-79 orbiter crew Blaha (down on STS-81)
193	U.S.A. 110	190	STS-80	Cockrell/Rominger/Jernigan/Jones T./Musgrave	11/19/96–12/17/96	017:15:53:18	
1997							
194	U.S.A. 111	191	STS-81	Baker M./Jett/Wisoff/Grunsfeld/Ivins/Linenger	01/12/97–01/22/97	010:04:55:21 132:04:00:21	STS-81 orbiter crew Linenger (down on STS-84)
195	Russia 84	192	Soyuz TM-25	Tsibliyev/Lazutkin/Ewald	02/10/97–08/14/97	184:22:07:41 019:16:34:46	Tsibliyev/Lazutkin (down on Soyuz TM-25) Ewald (down on Soyuz TM-24)

(continued)

World space flight sequence	Country of origin/ sequence	Earth orbit flight sequence	Mission designation	Prime crew members	Launch to landing dates (dd/mm/yy)	Crew duration (dd:hh:mm:ss)	Notes
1997	*(cont.)*						
196	U.S.A. 112	193	STS-82	Bowersox/Horowitz/Tanner/ Hawley/Harbaugh/Lee/ Smith S.	02/11/97–02/21/97	009:23:37:09	
197	U.S.A. 113	194	STS-83	Halsell/Still/Voss J.E./ Gernhardt/Thomas D./ Crouch/Linteris	04/04/97–04/08/97	003:23:12:39	
198	U.S.A. 114	195	STS-84	Precourt/Collins E./Clervoy/ Nicollier/Lu/Kondakova/Foale	05/15/97–05/24/97	009:05:19:56 144:13:47:21	STS-84 orbiter crew Foale (down on STS-86)
199	U.S.A. 115	196	STS-94	Halsell/Still/Voss J.E.	07/01/97–07/17/97	015:16:34:04	
200	Russia 85	197	Soyuz TM-26	Solovyov A./Vinogradov	08/05/97–02/19/98	197:17:34:36	
201	U.S.A. 116	198	STS-85	Brown C./Rominger/Davis/ Curbeam/Robinson/ Tryggvason	08/17/97–08/19/97	011:20:26:59	
202	U.S.A. 117	199	STS-86	Wetherbee/Bloomfield/ Titov V./Parazynski/Chretien/ Lawrence/Wolf	09/25/97–10/06/97	010:19:20:50 127:20:00:50	STS-86 orbiter crew Wolf (down on STS-86)
203	U.S.A. 118	200	STS-87	Kregel/Lindsey/Chawla/ Scott W./Doi/Kadenyuk	11/19/97–12/05/97	015:16:34:04	
1998							
204	U.S.A. 119	201	STS-89	Wilcutt/Edwards/Reilly/ Anderson/Dunbar/Sharipov/ Thomas A.	01/22/98–01/31/98	008:19:46:54 140:15:12:06	STS-89 orbiter crew Thomas (down on STS-91)

				Crew	Dates	Duration	Notes
205	Russia 86	202	Soyuz TM-27	Musabayev/Budarin/Eyharts	01/29/98–08/25/98	207:12:51:02	Musabayev/Budarin (down on Soyuz TM-27) Eyharts (down on Soyuz TM-26)
						020:16:36:48	
206	U.S.A. 120	203	STS-90	Searfoss/Altman/Linnehan/Hire/Williams D./Buckley/Pawelczyk	04/17/98–08/28/98	015:21:49:59	
207	U.S.A. 121	204	STS-91	Precourt/Gorie/Kavandi/Lawrence/Chang-Diaz/Ryumin	06/02/98–06/12/98	009:19:53:54	
208	Russia 87	205	Soyuz TM-28	Padalka/Avdeyev/Baturin	08/13/98–02/08/99	198:16:31:20	Padalka (down on Soyuz TM-28) Avdeyev (down on Soyuz TM-29) Baturin (down on Soyuz TM-27)
						379:14:51:10	
						011:19:41:33	
209	U.S.A. 122	206	STS-95	Brown C./Lindsey/Robinson/Parazynski/Duque/Mukai/Glenn	10/29/98–11/17/98	008:21:43:56	
210	U.S.A. 123	207	STS-88	Cabana/Sturckow/Ross/Currie/Newman/Krikalev	12/04/98–12/15/98	011:19:17:57	
1999							
211	Russia 88	208	Soyuz TM-29	Afanasyev/Haigneré J-P./Bella	02/20/99–08/28/99	188:20:16:19	Afanasyev/Haigneré J-P. (down on Soyuz TM-29) Bella (down on Soyuz TM-28)
						007:21:56:29	
212	U.S.A. 124	209	STS-96	Rominger/Husband/Jernigan/Ochoa/Barry/Payette/Tokarev	05/27/99–06/06/99	009:19:13:57	
213	U.S.A. 125	210	STS-93	Collins E./Ashby/Hawley/Coleman/Tognini	07/23/99–07/27/99	004:02:49:37	
214	U.S.A. 126	211	STS-103	Brown C./Kelly S./Grunsfeld/Smith S./Foale/Nicollier/Clervoy	15/19/99–12/27/99	007:23:10:47	

(continued)

World space flight sequence	Country of origin sequence	Earth orbit flight sequence	Mission designation	Prime crew members	Launch to landing dates (dd/mm/yy)	Crew duration (dd:hh:mm:ss)	Notes
2000							
215	U.S.A. 127	212	STS-99	Kregel/Gorie/Thiele/Kavandi/Voss J.E./Mohri	02/11/00–02/22/00	011:05:39:41	
216	Russia 89	213	Soyuz TM-30	Zaletin/Kaleri	04/04/00–06/16/00	072:19:42:16	
217	U.S.A. 128	214	STS-101	Halsell/Horowitz/Weber/Williams J./Voss J.S./Helms/Usachev	05/19/00–05/09/00	009:21:10:10	
218	U.S.A. 129	215	STS-106	Wilcutt/Altman/Lu/Mastracchio/Burbank/Malenchenko/Morukov	09/08/00–09/19/00	011:19:12:15	
219	U.S.A. 130	216	STS-92	Duffy/Melroy/López-Alegria/Wisoff/McArthur W./Chiao/Wakata	10/11/00–10/24/00	012:21:43:47	
220	Russia 90	217	Soyuz TM-31	Shepherd/Gidzenko/Krikalev (ISS-1)	10/31/00–03/21/01	140:23:38:55	Down on STS-102
221	U.S.A. 131	218	STS-97	Jett/Bloomfield/Tanner/Garneau/Noriega	11/30/00–12/11/00	010:19:58:20	
2001							
222	U.S.A. 132	219	STS-98	Cockrell/Polansky/Curbeam/Jones T./Ivins	02/07/01–02/20/01	012:21:21:00	
223	U.S.A. 133	220	STS-102	Wetherbee/Kelly J.M./Richards P./Thomas A.	03/08/01–03/21/01	012:19:51:57	
—	—	Launched on STS-102	ISS-2	Usachev/Voss J.S./Helms	03/08/01–08/22/01	167:06:40:49	Down on STS-105

224	U.S.A. 134	221	STS-100	Rominger/Ashby/Hadfield/Parazynski/Guidoni/Phillips/Lonchakov	04/19/01–05/01/01	011:21:31:14	Down on Soyuz TM-31
225	Russia 91	222	Soyuz TM-32	Musabayev/Baturin/Tito	04/28/01–05/06/01	007:22:04:08	
226	U.S.A. 135	223	STS-104	Lindsey/Hobaugh/Gernhardt/Reilly/Kavandi	07/12/01–07/23/01	012:18:36:39	
227	U.S.A. 136	224	STS-105	Horowitz/Sturckow/Barry/Forrester	08/10/01–08/22/01	011:21:13:52	
—	—	Launched on STS-105	ISS-3	Culbertson/Dezhurov/Tyurin	08/10/01–12/17/01	128:20:44:56	Down on STS-108
228	Russia 92	225	Soyuz TM-33	Afanasyev/Haigneré C./Kozeev	10/21/01–10/31/01	009:20:00:25	Down on Soyuz TM-32
229	U.S.A. 137	226	STS-108	Gorie/Kelly M./Godwin/Tani	12/05/01–12/17/01	011:19:36:45	
—	—	Launched on STS-108	ISS-4	Onufriyenko/Bursch/Walz	12/05/01–09/17/02	195:19:38:12	Down on STS-113
2002							
230	U.S.A. 138	227	STS-109	Altman/Carey/Currie/Grunsfeld/Linnehan/Newman	03/01/02–03/12/02	010:22:11:09	
231	U.S.A. 139	228	STS-110	Bloomfield/Frick/Walheim/Ochoa/Smith S./Morin/Ross	04/08/02–04/19/02	010:19:43:48	
232	Russia 93	229	Soyuz TM-34	Gidzenko/Vittori/Shuttleworth	04/25/02–05/05/02	009:21:25:18	Down on Soyuz TM-33
233	U.S.A. 140	230	STS-111	Cockrell/Lockhart/Chang-Diaz/Perrin	06/05/02–06/19/02	013:20:35:56	
—	—	Launched on STS-111	ISS-5	Korzun/Whitson/Treschev	06/05/02–12/07/02	184:22:14:23	Down on STS-113
234	U.S.A. 141	231	STS-112	Ashby/Melroy/Wolf/Magnus/Sellers/Yurchikhin	10/07/02–10/18/02	010:19:58:44	
235	Russia 94	232	Soyuz TMA-1	Zaletin/De Winne/Lonchakov	10/30/02–11/10/02	010:20:53:09	Down on Soyuz TM-34
236	U.S.A. 142	233	STS-113	Wetherbee/Lockhart/López-Alegría/Herrington	11/23/02–12/07/02	013:18:48:38	
—	—	Launched on STS-113	ISS-6	Bowersox/Budarin/Petit	11/23/02–05/03/03	161:01:14:38	Down on Soyuz TMA-1

(continued)

World space flight sequence	Country of origin/ sequence	Earth orbit flight sequence	Mission designation	Prime crew members	Launch to landing dates (dd/mm/yy)	Crew duration (dd:hh:mm:ss)	Notes
2003							
237	U.S.A. 143	234	STS-107	Husband/McCool/Brown D./Chawla/Anderson/Clark/Ramon	01/16/03–02/01/03	015:22:20:22	Fatal breakup of vehicle during entry/landing phase
238	Russia 95	235	Soyuz TMA-2	Malenchenko/Lu (ISS-7)	04/26/03–10/27/03	184:22:46:09	
239	China 1	236	Shenzhou 5	Yang	10/15/03–10/16/03	021:26:00	
240	Russia 96	237	Soyuz TMA-3	Foale/Kaleri (ISS-8)/Duque	01/18/03–04/30/04	194:18:23:43 009:21:01:58	Foale/Kaleri Duque (down on Soyuz TMA-2)
2004							
241	Russia 97	238	Soyuz TMA-4	Padalka/Finke (ISS-9)/Kuipers	04/19/04–10/24/04	187:21:16:09 010:20:52:46	Padalka/Finke Kuipers (down on Soyuz TMA-3)
—	U.S.A.	Astro-flight	Spaceship 1-60	Melvill	06/21/04	000:00:24:00	X-Prize flight
—	U.S.A.	Astro-flight	Spaceship 1-65	Melvill	09/29/04	000:00:24:00	X-Prize flight
—	U.S.A.	Astro-flight	Spaceship 1-66	Binnie	10/04/04	000:00:24:00	X-Prize flight
242	Russia 98	239	Soyuz TMA-5	Sharipov/Chiao (ISS-10)/Shargin	10/14/04–04/24/05	192:19:00:59 009:21:29:00	Sharipov/Chiao Shargin (down on Soyuz TMA-4)

					Crew		Notes
2005							
243	Russia 99	240	Soyuz TMA-6	Krikalev/Phillips (ISS-11)/Vittori	04/14/05–10/11/05	179:00:23:00 009:21:21:02	Krikalev/Phillips Vittori (down on Soyuz TMA-5)
244	U.S.A. 144	241	STS-114	Collins E./Kelly J.M./Noguchi/Robinson/Thomas A./Lawrence/Camarda	07/26/05–08/09/05	013:21:32:48	
245	Russia 100	242	Soyuz TMA-7	McArthur W./Tokarev (ISS-12)/Olsen	10/01/05–04/09/06	189:19:53:00 009:21:15:00	McArthur W./Tokarev Olsen (down on Soyuz TMA-6)
246	China 2	243	Shenzhou 6	Fei/Nie	10/12/05–10/16/05	4:19:33:00	
2006							
247	Russia 101	244	Soyuz TMA-8	Vinogradov/Williams J. (ISS-13)/Pontes (VC-10)	03/30/06–09/29/06	182:22:43:00 009:21:17:00	Vinogradov/Williams J. Pontes (down on Soyuz TMA-7)
248	U.S.A. 145	245	STS-121	Lindsey/Kelly M./Fossum/Nowak/Wilson/Sellers/Reiter (ISS FE)	07/04/06–07/17/06	012:18:37:54 171:03:54:05	STS-121 orbiter crew Reiter (down on STS-116)
249	U.S.A. 146	246	STS-115	Jett/Ferguson/Tanner/Burbank/Stefanyshyn-Piper/MacLean	09/09/06–09/21/06	011:19:07:24	
250	Russia 102	247	Soyuz TMA-9	Lopez-Algeria/Tyurin (ISS-14)/Ansari (VC-11)	09/19/06–04/21/07	215:08:22:48 010:21:05:00	López-Algería/Tyurin Ansari (down on Soyuz TMA-8)
251	U.S.A. 147	248	STS-116	Polansky/Oefelein/Patrick/Curbeam/Fuglesang/Higginbotham/Williams S. (ISS FE)	12/09/06–12/2206	012:20:45:16 194:18:58:00	STS-116 orbiter crew Williams down on STS-117

(continued)

World space flight sequence	Country of origin/ sequence	Earth orbit flight sequence	Mission designation	Prime crew members	Launch to landing dates (dd/mm/yy)	Crew duration (dd:hh:mm:ss)	Notes
2007							
252	Russia 103	249	Soyuz TMA-10	Yurchikhin/Kotov (ISS-15)/ Simonyi (VC-12)	04/07/07–10/21/07	196:17:04:35 013:18:59:50	Yurchikhin/Kotov Simonyi (down on Soyuz TMA-9)
253	U.S.A. 148	250	STS-117	Sturckow/Archambault/ Forrester/Swanson/Olivas/ Reilly/Anderson C. (ISS FE)	06/08/07–06/22/07	013:20:12.44 151:18:24:09	STS-117 orbiter crew Anderson (down on STS-120)
254	U.S.A. 149	251	STS-118	Kelly S./Hobaugh/Caldwell/ Mastracchio/Williams D./ Morgan/Drew	08/08/07–08/21/07	012:17:55:34	
255	Russia 104	252	Soyuz TMA-11	Malenchenko/ Whitson (ISS-16)/ Muszaphar (VC-13)	10/10/07–04/19/08	191:19:07:05 010:21:14:00	Malenchenko/Whitson Muszaphar (down on Soyuz TMA-10)
256	U.S.A. 150	253	STS-120	Melroy/Zamka/Parazynski/ Wilson/Wheelock/Nespoli/ Tani (ISS FE)	10/23/07–11/07/07	015:02:24:02 119:21:29:01	STS-120 orbiter crew Tani (down on STS-122)
2008							
257	U.S.A. 151	254	STS-122	Frick/Poindexter/Melvin/ Walheim/Schlegel/Love/ Eyharts (ISS FE)	02/07/08–02/20/08	012:18:21:50 048:04:53:38	STS-122 orbiter crew Eyharts (down on STS-123)
258	U.S.A. 152	255	STS-123	Gorie/Johnson G./Behnken/ Foreman/Doi/Linnehan/ Reisman (ISS FE)	03/11/08–03/26/08	015:18:10:54 095:08:47:05	STS-123 orbiter crew Reisman (down on STS-124)

No.	Country	Mission	Crew	Dates	Duration	Notes
259	Russia 105	Soyuz TMA-12	Volkov S./Kononenko (ISS-17)/Yi (VC-14)	04/08/08–10/24/08	198:16:20:31 / 010:21:19:21	Volkov S./Kononenko Yi (down on Soyuz TMA-11)
260	U.S.A. 153	STS-124	Kelly M./Ham/Nyberg/Garan/Fossum/Hoshide/Chamitoff (ISS)	05/31/08–06/14/08	013:18:13:07 / 183:00:22:54	STS-124 orbiter crew Chamitoff (down on STS-126)
261	China 3	Shenzhou 7	Zhai/Liu/Jing	09/25/08–09/27/08	002:02:27:35	
262	Russia 106	Soyuz TMA-13	Lonchakov/Fincke (ISS-18)/Garriott R. (VC-15)	10/12/08–04/08/09	178:00:13:38 / 011:20:35:37	Lonchakov/Fincke Garriott (down on Soyuz TMA-12)
263	U.S.A. 154	STS-126	Ferguson/Boe/Petit/Bowen/Stefanyshyn-Piper/Kimbrough/Magnus (ISS FE)	11/14/08–11/30/08	015:20:29:27 / 133:18:17:38	STS-126 orbiter crew Magnus (down on STS-119)
2009						
264	U.S.A. 155	STS-119	Archambault/Antonelli/Acaba/Swanson/Arnold/Phillips/Wakata (ISS FE)	03/15/09–03/28/09	012:19:29:33 / 137:15:04:23	STS-119 orbiter crew Wakata (down on STS-127)
265	Russia 107	Soyuz TMA-14	Padalka/Barratt (ISS-19/20)/Simonyi (VC-16)	03/26/09–10/11/09	198:16:42:22 / 012:19:25:52	Padalka/Barratt Simonyi (down on Soyuz TMA-13)
266	U.S.A. 156	STS-125	Altman/Johnson G.C./Good/McArthur/Grunsfeld/Massimino/Feustel	05/11/09–05/24/09	012:21:38:09	
267	Russia 108	Soyuz TMA-15	Romanenko R./De Winne/Thirsk (ISS-20/21)	05/27/09–12/01/09	187:20:41:38	
268	U.S.A. 157	STS-127	Polansky/Hurley/Wolf/Cassidy/Payette/Marshburn/Kopra (ISS FE)	07/15/09–07/31/09	015:16:44:57 / 058:02:50:10	STS-127 orbiter crew Kopra (down on STS-128)
269	U.S.A. 158	STS-128	Sturckow/Ford/Forrester/Hernandez/Olivas/Fuglesang/Stott (ISS FE)	08/28/09–09/12/09	013:20:53:43 / 090:10:44:43	STS-128 orbiter crew Stott (down on STS-129)

(continued)

World space flight sequence	Country of origin/ sequence	Earth orbit flight sequence	Mission designation	Prime crew members	Launch to landing dates (dd/mm/yy)	Crew duration (dd:hh:mm:ss)	Notes
2009	*(cont.)*						
270	Russia 109	267	Soyuz TMA-16	Surayev/ Williams J. (ISS-21/22)/ Laliberté (VC-17)	09/30/09–03/18/10	169:04:09:37 010:21:16:55	Surayev/Williams J. Laliberté (down on Soyuz TMA-14)
271	U.S.A. 159	268	STS-129	Hobaugh/Wilmore/Melvin/ Bresnik/Foreman/Satcher	11/16/09–11/27/09	010:19:16:13	
272	Russia 110	269	Soyuz TMA-17	Kotov/Noguchi/ Creamer (ISS22/23)	12/21/09–06/02/10	163:05:32:32	
2010							
273	U.S.A. 160	270	STS-130	Zamka/Virts/Hire/Robinson/ Patrick/Behnken	02/08/10–02/21/10	013:18:06:22	
274	Russia 111	271	Soyuz TMA-18	Skvortsov/Kornienko/ Caldwell-Dyson (ISS23/24)	04/02/10–09/25/10	176:01:18:38	
275	U.S.A. 161	272	STS-131	Poindexter/Dutton/ Mastracchio/ Metcalf-Lindenburger/Wilson/ Yamazaki/Anderson C.	04/05/10–04/20/10	015:02:47:10	
276	U.S.A. 162	273	STS-132	Ham/Antonelli/Reisman/ Good/Bowen/Sellers	05/14/10–05/26/10	011:18:29:09	
277	Russia 112	274	Soyuz TMA-19	Yurchikhin/Walker S./ Wheelock (ISS24/25)	06/16/10–11/26/10	163:07:10:47	
278	Russia 113	275	Soyuz TMA-M	Kaleri/Skripochka/ Kelly S (ISS-25/26)	10/08/10–03/16/11	159:08:43:05	
279	Russia 114	276	Soyuz TMA-20	Kondratyev/Coleman/ Nespoli (ISS 26/27)	12/15/10–05/24/11	159:08:17:15	

2011				
280	U.S.A. 163	277	STS-133	Lindsey/Boe/Drew/Bowen/Barratt/Stott	02/24/11–03/09/11	012:19:03:51
281	Russia 115	278	Soyuz TMA-21	Samokutyaev/Borisenko/Garan	04/04/11–09/16/11	164:05:41:19
282	U.S.A. 164	279	STS-134	Kelly M./Johnson G H./Fincke/Vittori/Feustel/Chamitoff	05/16/11–06/01/11	015:17:38:22
283	Russia 116	280	Soyuz TMA-02M	Volkov S./Furukawa/Fossum (ISS-28/29)	07/06/11–22/11/11	167:06:12:05
284	U.S.A. 165	281	STS-135	Ferguson/Hurley/Magnus/Walheim	07/08/11–07/21/11	012:18:27:52
285	Russia 117	282	Soyuz TMA-22	Shkaplerov/Ivanishin/Burbank (ISS-29/30)	11/14/11–04/27/12	165:07:31:34
286	Russia 118	283	Soyuz TMA-03M	Kononenko/Kuipers/Pettit (ISS-30/31)	12/21/11–07/01/12	192:18:58:21
2012				
287	Russia 119	284	Soyuz TMA-04M	Padalka/Revin/Acaba (ISS-31/32)	05/15/12–09/17/12	124:23:51:30
288	China 4	285	Shenzhou 9	Jing/Liu W/Liu Y (Tiangong-1 1st crew)	06/16/12–06/29/12	012:15:24:00
289	Russia 120	286	Soyuz TMA-05M	Malenchenko/Williams S./Hoshide (ISS-32/33)	07/15/12–11/19/12	126:23:13:27
290	Russia 121	287	Soyuz TMA-06M	Novitsyky/Tarekin/Ford (ISS-33/34)	10/23/12–	In space
291	Russia 122	288	Soyuz TMA-07M	Romanenko R./Hadfield/Marshburn (ISS-34/35)	12/19/12	In space

(continued)

World space flight sequence	Country of origin/ sequence	Earth orbit flight sequence	Mission designation	Prime crew members	Launch to landing dates	Crew duration	Notes
					(dd/mm/yy)	(dd:hh:mm:ss)	
2013							

Appendix B

Cumulative space flight experience (order of most experience)

Name	Country	Total space flights	Time in space (dd:hh:mm:ss)
Sergei K. Krikalev	U.S.S.R./Russia	6	804:08:17:52
↑ *800 days* ↑			
Alexandr Y. Kaleri	Russia	5	769:06:34:44
Sergei V. Avdeyev	Russia	3	746:06:34:44
Gennady I. Padalka	Russia	4	710:06:21:50
↑ *700 days* ↑			
Valery V. Polyakov	U.S.S.R./Russia	2	678:16:33:18
Anatoly Y. Solovyov	U.S.S.R./Russia	5	651:00:03:28
Yuri I. Malenchenko	Russia	5	641:11:12:27
↑ *600 days* ↑			
Viktor M. Afanasyev	U.S.S.R./Russia	4	555:18:35:28
Yuri V. Usachev	Russia	4	552:23:25:36
Musa K. Manarov	U.S.S.R.	2	541:00:31:39
↑ *500 days* ↑			
Alexandr A. Viktorenko	U.S.S.R./Russia	4	489:01:34:18
Nikolai V. Budarin	Russia	3	444:01:26:01
Yuri V. Romanenko	U.S.S.R.	3	430:18:21:30
↑ *400 days* ↑			
Alexandr A. Volkov	U.S.S.R.	3	391:11:53:14
Oleg D. Kononenko	Russia	2	391:11:18:52
Yuri I. Onufriyenko	Russia	2	389:14:45:47

(continued)

Name	Country	Total space flights	Time in space (dd:hh:mm:ss)
Vladimir G. Titov	U.S.S.R./Russia	4	386:22:48:36
Vasily V. Tsibliyev	Russia	2	381:15:53:03
Valery G. Korzun	Russia	2	381:15:40:36
E. Michael Finke	U.S.A.	3	381:15:09:51
Pavel V. Vinogradov	Russia	2	380:17:18:36
Peggy A. Whitson	U.S.A.	2	376:17:21:23
Leonid D. Kizim	U.S.S.R.	3	374:17:57:42
C. Michael Foale	U.S.A.	6	373:18:31:00
Alexandr A. Serebrov	U.S.S.R./Russia	4	372:22:53:50
Valery V. Ryumin	U.S.S.R./Russia	4	371:17:27:01
Fyodor N Yurchikhin	Russia	3	369:20:14:44
Donald R. Pettit	U.S.A.	3	369:16:42:36
Sergei A. Volkov	Russia	2	365:22:32:36
→ *365 days–1 year accumulated experience* ←			
Jeffrey N. Williams	U.S.A.	3	362:01:03:17
Vladimir A. Solovyov	U.S.S.R.	2	361:22:50:00
Oleg V. Kotov	Russia	2	359:23:37:00
Thomas A. Reiter	Germany	2	350:05:35:44
Mikhail V. Tyurin	Russia	2	344:15:07:44
Talgat A. Musabayev	Kazakhstan/ Russia	3	341:09:48:46
Vladimir A. Lyakhov	U.S.S.R.	3	333:07:48:05
Yuri P. Gidzenko	Russia	3	329:22:45:59
Sunita Williams	U.S.A.	2	321:18:11:27
Gennady M. Manakov	U.S.S.R./Russia	2	309:21:19:37
Alexandr P. Alexandrov	U.S.S.R.	2	309:18:02:59
↑ *300 days* ↑			
Gennady M. Strekalov	U.S.S.R./Russia	5	268:22:24:29
Michael E. López-Alegría	U.S.A.	3	258:08:48:31
Viktor P. Savinykh	U.S.S.R.	3	252:17:37:42
Vladimir N. Dezhurov	Russia	2	244:05:27:58
Oleg Y. Atkov	U.S.S.R.	1	236:22:49:01
Carl E. Walz	U.S.A.	4	230:13:06:06
Leroy Chiao	U.S.A.	4	229:08:42:44
Daniel W. Bursch	U.S.A.	4	226:22:17:39
William S. McArthur	U.S.A.	4	224:22:22:04
Shannon W. Lucid	U.S.A.	5	223:02:56:03
Valentin V. Lebedev	U.S.S.R.	2	219:06:00:07
Vladimir V. Kovalenok	U.S.S.R.	3	216:09:09:40
Kenneth D. Bowersox	U.S.A.	5	211:14:35:51
Michael R. Barratt	U.S.A.	2	211:14:35:27
Anatoly N. Berezovoi	U.S.S.R.	1	211:09:04:32

Name	Country	Total space flights	Time in space (dd:hh:mm:ss)
Susan J. Helms	U.S.A.	5	211:00:09:38
Jean-Pierre Haigneré	France	2	209:14:25:11
Edward Tsang Lu	U.S.A.	3	205:23:20:12
Robert B. Thirsk	Canada	2	204:18:28:33
John L. Phillips	U.S.A.	3	204:16:02:15
André Kuipers	Netherlands	2	203:15:51:07
Salizhan S. Sharipov	Russia	2	201:14:49:05
Yuri V. Lonchakov	Russia	3	200:18:39:23
Leonid I. Popov	U.S.S.R.	3	200:14:45:51
↑ *200 days* ↑			
Valery I. Tokarev	Russia	2	199:15:06:57
James S. Voss	U.S.A.	5	199:06:32:23
Gregory G. Chamitoff	U.S.A.	2	198:18:01:05
Frank De Winne	Belgium	2	198:17:32:49
Michael E. Fossum	U.S.A.	3	193:19:03:06
Daniel C. Burbank	U.S.A.	3	188:22:51:13
Tracy E. Caldwell-Dyson	U.S.A.	2	188:19:14:34
↑ *Roman Y. Romanenko* ↑	*Russian*	2	*187:20:41:38 in space*
Sergei V. Treshchev	Russia	1	184:22:14:23
Alexandr I. Lazutkin	Russia	1	184:20:07:41
Catherine G. Coleman	U.S.A.	3	180:04:00:36
Scott J. Kelly	U.S.A.	3	180:01:49:08
Alexandr N. Balandin	U.S.S.R.	1	179:01:17:57
Alexandr F. Poleshchuk	Russia	1	179:00:43:46
Yelena V. Kondakova	Russia	2	178:10:42:23
Ronald J. Garan	U.S.A.	2	177:23:54:26
Douglas H. Wheelock	U.S.A.	2	177:09:35:02
Andrew S. W. Thomas	U.S.A.	4	177:09:17:07
Alexandr I. Laveikin	U.S.S.R.	1	174:03:25:56
Soichi Noguchi	Japan	2	177:03:05:48
Mikhail B. Korniyenko	Russia	1	176:01:19:00
Aleksandr A. Skvortsov	Russia	1	176:01:19:00
Paolo A. Nespoli	Italy	2	174:10:41:02
Maxim V. Surayev	Russia	1	169:04:09:37
David A. Wolf	U.S.A.	4	168:08:58:05
Satoshi Furukawa	Japan	1	167:06:12:05
Clayton C. Anderson	U.S.A.	2	165:21:10:10
Anton N. Shkaplerov	Russia	1	165:07:31:34
Anatoli A. Ivanishin	Russia	1	165:07:31:34
Andrei I. Borisenko	Russia	1	164:05:41:19
Aleksandr M. Samokutyaev	Russia	1	164:05:41:19
Timothy J. Creamer	U.S.A.	1	163:06:32:00

(*continued*)

Name	Country	Total space flights	Time in space (dd:hh:mm:ss)
Shannon Walker	U.S.A.	1	162:07:11:00
John E. Blaha	U.S.A.	5	161:02:51:20
Oleg I. Skripochka	Russia	1	159:08:42:00
Dmitri Kondratyev	Russia	1	159:08:17:00
William M. Shepherd	U.S.A.	4	159:07:53:22
Alexandr S. Ivanchenkov	U.S.S.R.	2	147:03:38:36
Vladimir A. Dzhanibekov	U.S.S.R.	5	145:15:58:36
Anatoly P. Artsebarsky	U.S.S.R.	1	144:15:21:50
Frank L. Culbertson, Jr.	U.S.A.	3	143:14:52:20
Jerry M. Linenger	U.S.A.	2	143:02:51:19
Akihiko Hoshide	Japan	2	140:17:26:34
Norman E. Thagard	U.S.A.	5	140:13:31:40
Joseph M. Acaba	U.S.A.	2	137:19:22:31
Koichi Wakata	Japan	3	137:15:04:00
Georgi M. Grechko	U.S.S.R.	3	135:20:32:58
Daniel M. Tani	U.S.A.	2	131:18:05:48
Sergei N. Revin	Russia	1	124:23:51:30
Garrett E. Reisman	U.S.A.	2	107:03:16:09
Nicole P. Stott	U.S.A.	2	103:14:49:50
Sandra H. Magnus	U.S.A.	3	107:10:08:34
↑ *100 days* ↑			
Gerald P. Carr	U.S.A.	1	84:01:15:37
Edward G. Gibson	U.S.A.	1	84:01:15:37
William R. Pogue	U.S.A.	1	84:01:15:37
Sergei V. Zaletin	Russia	2	83:16:35:25
Vital I. Sevastyanov	U.S.S.R.	2	80:16:19:03
Pyotr I. Klimuk	U.S.S.R.	3	78:18:18:42
Owen K. Garriott	U.S.A.	2	69:18:57:21
Alan L. Bean	U.S.A.	2	69:15:45:29
Léopold Eyharts	France	2	68:21:30:48
Jack R. Lousma	U.S.A.	2	67:11:15:15
Kent V. Rominger	U.S.A.	5	67:03:01:04
Franklin R. Chang-Diaz	U.S.A.	7	66:18:23:51
James D. Wetherbee	U.S.A.	6	66:10:31:20
Vladimir V. Vasyutin	U.S.S.R.	1	64:21:52:08
Kenneth D. Cockrell	U.S.A.	5	64:12:30:40
Tamara E. Jernigan	U.S.A.	5	63:01:30:42
Steven W. Lindsey	U.S.A.	5	62:22:39:18
Curtis L. Brown, Jr.	U.S.A.	6	61:05:28:25
Richard M. Linnehan	U.S.A.	4	59:12:00:38
Boris V. Volynov	U.S.S.R.	2	59:07:17:47
Marsha S. Ivins	U.S.A.	5	58:21:52:49
Jerry L. Ross	U.S.A.	7	58:18:30:30

Name	Country	Total space flights	Time in space (dd:hh:mm:ss)
John M. Grunsfeld	U.S.A.	5	58:15:06:09
Timothy L. Kopra	U.S.A.	1	58:02:50:00
Scott L. Parazynski	U.S.A.	5	57:15:37:21
F. Story Musgrave	U.S.A.	6	53:09:05:12
Thomas D. Jones	U.S.A.	4	53:00:53:00
Kevin R. Kregel	U.S.A.	4	52:18:14:18
James D. Halsell, Jr.	U.S.A.	5	52:05:37:13
Scott D. Altman	U.S.A.	4	51:12:51:29
Frederick W. Sturckow	U.S.A.	4	51:09:37:58
Wendy B. Lawrence	U.S.A.	4	51:03:59:56
Jeffrey A. Hoffman	U.S.A.	5	50:11:59:38
Bonnie J. Dunbar	U.S.A.	5	50:18:29:45
↑ *50 days* ↑			
Ulf D. Merbold	Germany	3	49:21:39:55
Vitaly M. Zholobov	U.S.S.R.	1	49:06:23:32
Charles Conrad, Jr.	U.S.A.	4	49:03:38:36
Dominic L. Gorie	U.S.A.	4	48:15:11:30
Janice E. Voss	U.S.A.	5	48:11:34:27
Stephen K. Robinson	U.S.A.	4	48:09:52:17
Scott J. Horowitz	U.S.A.	4	47:11:43:26
Joseph R. Tanner	U.S.A.	4	44:12:49:07
Peter J. K. Wisoff	U.S.A.	4	44:08:13:24
Jean-Loup J. M. Chrétien	France	3	43:11:20:33
James H. Newman	U.S.A.	4	43:10:11:53
Donald A. Thomas	U.S.A.	4	43:02:16:24
Michael L. Gernhardt	U.S.A.	4	43:01:05:46
Thomas T. Henricks	U.S.A.	4	42:18:43:06
Brent W. Jett Jr.	U.S.A.	4	42:17:34:32
Claude Nicollier	Switzerland	4	42:12:08:36
Mark E. Kelly	U.S.A.	4	42:07:06:37
Terrence W. Wilcutt	U.S.A.	4	42:00:07:16
Stephanie D. Wilson	U.S.A.	3	41:23:49:06
Nancy J. (Sherlock) Currie	U.S.A.	4	41:15:37:00
Mark Polansky	U.S.A.	3	41:10:51:12
Ellen L. Ochoa	U.S.A.	4	40:19:40:57
Brian Duffy	U.S.A.	4	40:17:42:00
Kathryn C. Thornton	U.S.A.	4	40:15:19:14
Christopher J. Ferguson	U.S.A.	3	40:10:05:51
Michael A. Baker	U.S.A.	4	40:03:03:17
Stephen L. Smith	U.S.A.	4	40:00:42:44

(continued)

Name	Country	Total space flights	Time in space (dd:hh:mm:ss)
↑ *40 days* ↑			
Patrick G. Forrester	U.S.A.	3	39:14:19:37
Rex J. Walheim	U.S.A.	3	39:08:21:34
Richard A. Searfoss	U.S.A.	3	39:03:21:17
Charles J. Precourt	U.S.A.	4	38:20:20:03
Pamela A. Melroy	U.S.A.	3	38:20:06:33
Richard A. Mastracchio	U.S.A.	3	38:15:54:59
Linda M. Godwin	U.S.A.	4	38:06:17:36
Andrew M. Allen	U.S.A.	3	37:16:15:08
Robert L. Curbeam, Jr.	U.S.A.	3	37:14:34:25
Alexei A. Gubarev	U.S.S.R.	2	37:11:35:45
Robert D. Cabana	U.S.A.	4	37:04:08:16
Eileen M. Collins	U.S.A.	4	36:08:13:30
Charles O. Hobaugh	U.S.A.	3	36:07:48:26
Robert L. Gibson	U.S.A.	5	36:04:22:35
Roberto Vittori	Italy	3	35:13:05:11
Piers J. Sellers	U.S.A.	3	35:08:49:37
Jay Apt	U.S.A.	4	35:07:15:01
John W. Young	U.S.A.	6	34:19:43:26
James F. Reilly, II	U.S.A.	3	34:13:30:54
John H. Casper	U.S.A.	4	34:09:56:15
Gregory J. Harbaugh	U.S.A.	4	34:02:03:35
Stephen S. Oswald	U.S.A.	3	33:22:34:53
Thomas D. Akers	U.S.A.	4	33:22:48:44
Richard N. Richards	U.S.A.	4	33:21:33:45
Janet L. Kavandi	U.S.A.	3	33:10:01:24
Paul J. Weitz	U.S.A.	2	33:01:14:25
Mark C. Lee	U.S.A.	4	32:22:27:40
Daniel C. Brandenstein	U.S.A.	4	32:21:09:29
Michael J. Bloomfield	U.S.A.	3	32:11:03:19
Steven A. Hawley	U.S.A.	5	32:03:47:21
Gregory H. Johnson	U.S.A.	2	31:11:48:51
Takao Doi	Japan	2	31:10:44:59
Vance D. Brand	U.S.A.	4	31:02:06:59
Daniel T. Barry	U.S.A.	3	30:14:29:39
Viktor V. Gorbatko	U.S.S.R.	3	30:12:48:21
Margaret R. Seddon	U.S.A.	3	30:02:25:18
Steven R. Nagel	U.S.A.	4	30:01:40:09

Name	Country	Total space flights	Time in space (dd:hh:mm:ss)
↑ *30 days* ↑			
G. David Low	U.S.A.	3	29:18:10:00
Guion S. Bluford, Jr.	U.S.A.	4	29:16:39:38
Carl J. Meade	U.S.A.	3	29:16:17:23
Kathryn P. Hire	U.S.A.	2	29:15:57:20
Kevin P. Chilton	U.S.A.	3	29:08:25:48
J. J. Marc Garneau	Canada	3	29:02:03:09
James A. Lovell, Jr.	U.S.A.	4	28:23:04:55
William F. Readdy	U.S.A.	3	28:23:04:55
George D. Zamka	U.S.A.	2	28:20:30:26
Vladislav N. Volkov	U.S.S.R.	2	28:17:02:06
Dafydd R. Williams	Canada	2	28:15:46:30
Eric A. Boe	U.S.A.	2	28:15:33:30
Andrew J. Fuestal	U.S.A.	2	28:15:14:56
Ellen S. Baker	U.S.A.	3	28:14:34:27
Douglas G. Hurley	U.S.A.	2	28:11:13:48
Charles F. Bolden	U.S.A.	4	28:08:42:36
Jean-François A. Clervoy	France	3	28:03:11:34
N. Jan Davis	U.S.A.	3	28:02:10:35
Joseph P. Kerwin	U.S.A.	1	28:00:49:49
→ *28 days–1 month accumulated experience* ←			
Michael R. U. Clifford	U.S.A.	3	27:18:28:12
John D. Olivas	U.S.A.	2	27:17:05:19
Heidemarie M. Stefanyshyn-Piper	U.S.A.	2	27:15:37:01
Paul S. Lockhart	U.S.A.	2	27:15:24:34
Umberto Guidoni	Italy	2	27:15:12:39
Stephen G. Bowen	U.S.A.	3	27:14:58:46
Jeffrey S. Ashby	U.S.A.	3	27:14:20:16
Robert L. Benkhen	U.S.A.	2	27:12:19:31
Pierre J. Thuot	U.S.A.	3	27:06:55:27
Richard O. Covey	U.S.A.	4	27:05:19:42
James M. Kelly	U.S.A.	2	27:02:24:45
Alan G. Poindexter	U.S.A.	2	26:21:08:56
Kalpana Chawla	U.S.A.	2	26:17:42:15
Ronald J. Grabe	U.S.A.	4	26:03:44:57
Christer Fuglesang	U.S.A.	2	26:17:39:01
Lee J. Archambault	U.S.A.	2	26:15:43:01
Steven R. Swanson	U.S.A.	2	26:15:42:35
Nicholas J. M. Patrick	U.S.A.	2	26:14:57:40
Charles Simonyi	U.S.A.	2	26:14:25.02
Michael J. Foreman	U.S.A.	2	26:13:26:13

(*continued*)

Name	Country	Total space flights	Time in space (dd:hh:mm:ss)
Ronald J. Grabe	U.S.A.	4	26:03:44:57
Hans W. Schlegel	Germany	2	25:18:51:00
Richard D. Husband	U.S.A.	2	25:17:33:57
Claudi (Deshays) Haigneré	France	2	25:15:09:02
Samuel T. Durrance	U.S.A.	2	25:14:15:54
Ronald A. Parise	U.S.A.	2	25:14:15:54
Julie Payette	Canada	2	25:11:58:55
Donald R. McMonagle	U.S.A.	3	25:05:37:19
Michael P. Anderson	U.S.A.	2	24:18:08:06
Michael T. Good	U.S.A.	2	24:16:06:18
Winston E. Scott	U.S.A.	2	24:14:36:49
Eugene A. Cernan	U.S.A.	3	24:14:16:12
Dominic A. Antonelli	U.S.A.	2	24:14:00:10
Charles D. Gemar	U.S.A.	3	24:05:41:26
Georgi T. Dobrovolsky	U.S.S.R.	1	23:18:21:43
Viktor I. Patsayev	U.S.S.R.	1	23:18:21:43
Chiaki Mukai	Japan	2	23:15:50:04
Stephen N. Frick	U.S.A.	2	23:14:04:28
Robert L. Crippen	U.S.A.	4	23:13:52:52
Leland D. Melvin	U.S.A.	2	23:13:37:57
Kenneth L. Cameron	U.S.A.	3	23:10:14:45
Michael J. Massimino	U.S.A.	2	23:05:48:18
Mario Runco, Jr.	U.S.A.	3	22:23:11:52
David R. Scott	U.S.A.	3	22:18:54:13
Brewster H. Shaw, Jr.	U.S.A.	3	22:05:55:21
Kathryn D. Sullivan	U.S.A.	3	22:04:52:04
David C. Leestma	U.S.A.	3	22:04:35:49
Richard J. Hieb	U.S.A.	3	21:22:38:01
Steven G. MacLean	Canada	2	21:16:04:40
Andrian G. Nikolayev	U.S.S.R.	2	21:15:20:55
Thomas K. Mattingly, II	U.S.A.	3	21:04:36:41
Thomas P. Stafford	U.S.A.	4	21:03:24:31
Valery F. Bykovsky	U.S.S.R.	3	20:17:47:21
Oleg G. Makarov	U.S.S.R.	4	20:17:43:39
David C. Hilmers	U.S.A.	4	20:15:21:40
Kenneth T. Ham	U.S.A.	2	20:12:42:16
James F. Buchli	U.S.A.	4	20:10:18:12
Henry W. Hartsfield, Jr.	U.S.A.	3	20:02:53:51
John M. Lounge	U.S.A.	3	20:02:23:01
Sidney M. Gutierrez	U.S.A.	2	20:02:02:58
↑ *Chris A. Hadfield* ↑	*Canada*	2	*20:02:02:58 in space*
Carlos I. Noriega	U.S.A.	2	20:01:19:08

Name	Country	Total space flights	Time in space (dd:hh:mm:ss)
↑ *20 days* ↑			
Charles D. Walker	U.S.A.	3	19:21:59:48
Frank F. Borman, II	U.S.A.	2	19:21:35:43
Yuri M. Baturin	Russian	2	19:17:45:41
Svetlana Y. Savitskaya	U.S.S.R.	2	19:17:07:00
Reinhold Ewald	Germany	1	19:16:34:46
Michael L. Coats	U.S.A.	3	19:12:00:01
Roger K. Crouch	U.S.A.	2	19:09:59:11
Gregory T. Linteris	U.S.A	.2	19:09:59:11
Susan L. (Kilrain) Still	U.S.A.	2	19:09:59:11
David M. Walker	U.S.A.	4	19:08:04:29
L. Blaine Hammond, Jr.	U.S.A.	2	19:06:14:15
Robert A. R. Parker	U.S.A.	2	19:06:54:22
Byron K. Lichtenberg	U.S.A.	2	19:05:58:41
Mamoru M. Mohri	Japan	2	19:03:00:54
Frederick D. Gregory	U.S.A.	3	18:23:10:11
Mary Ellen Weber	U.S.A.	2	18:19:31:15
Pedro F. Duque	Spain	2	18:18:45:54
Valery N. Kubasov	U.S.S.R.	3	18:17:59:22
Michel A. C. Tognini	France	2	18:17:46:32
Pavel R. Popovich	U.S.S.R.	2	18:16:27:28
Bernard A. Harris, Jr.	U.S.A.	2	18:06:10:36
Charles L. Veach	U.S.A.	2	18:04:20:33
Frederick H. Hauck	U.S.A.	3	18:03:12:29
Yuri N. Glazkov	U.S.S.R.	1	17:17:25:58
Ronald M. Sega	U.S.A.	2	17:12:27:01
George D. Nelson	U.S.A.	3	17:02:46:44
Charles E. Brady, Jr.	U.S.A.	1	16:21:48:33
Jean-Jacques Favier	France	1	16:21:48:33
John O. Creighton	U.S.A.	3	16:20:27:24
William G. Gregory	U.S.A.	1	16:15:09:49
Karol J. Bobko	U.S.A.	3	16:03:06:53
Loren J. Shriver	U.S.A.	3	16:02:07:32
Bryan D. O'Connor	U.S.A.	2	15:23:21:26
Charles G. Fullerton	U.S.A.	2	15:22:52:33
William C. McCool	U.S.A.	1	15:22:20:00
David M. Brown	U.S.A.	1	15:22:20:00
Laurel B. S. Clark	U.S.A.	1	15:22:20:00
Ilan Ramon	Israel	1	15:22:20:00
Albert Sacco, Jr.	U.S.A.	1	15:21:53:18
Jay C. Buckey	U.S.A.	1	15:21:50:56
James A. Pawelczyk	U.S.A.	1	15:21:50:56
Maurizio Cheli	Italy	1	15:17:41:25

(*continued*)

Name	Country	Total space flights	Time in space (dd:hh:mm:ss)
Yuri P. Artyukhin	U.S.S.R.	1	15:17:30:28
Christopher J. Cassidy	U.S.A.	1	15:16:44:58
↑ *Thomas H. Marshburn* ↑	*U.S.A.*	*1*	*15:16:44:58 in space*
Leonid K. Kadenyuk	Ukraine	1	15:16:34:59
Fred W. Leslie	U.S.A.	1	15:12:53:18
James P. Bagian	U.S.A.	2	15:01:54:53
Richard M. Mullane	U.S.A.	3	14:20:22:38
Jing Haipeng	China	2	14:17:51:35
Sally K. Ride	U.S.A.	2	14:07:50:06
James A. McDivitt	U.S.A.	2	14:02:57:06
James P. Dutton	U.S.A.	1	14:02:47:10
Metcalf Lindenberger	U.S.A.	1	14:02:47:10
Naoko (Sumino) Yamazaki	Japan	1	14:02:47:10
James D. A. Van Hoften	U.S.A.	2	14:01:59:26
Dale A. Gardner	U.S.A.	2	14:00:55:32
Martin J. Fettman	U.S.A.	1	14:00:13:33
→ *14 days (2 weeks)* ←			
James C. Adamson	U.S.A.	2	13:22:23:15
Charles J. Camarda	U.S.A.	1	13:21:32:48
↑ *Kevin A. Ford* ↑	*U.S.A.*	*2*	*13:20:53:45+ in space*
Philippe Perrin	France	1	13:20:35:56
Lawrence J. DeLucas	U.S.A.	1	13:19:31:02
Eugene H. Trinh	U.S.A.	1	13:19:31:02
John B. Herrington	U.S.A.	1	13:18:48:38
Karen L. Nyberg	U.S.A.	1	13:18:13:07
Terry W. Virts	U.S.A.	1	13:18:06:24
Kenneth S. Reightler, Jr.	U.S.A.	2	13:15:38:40
Guy S. Gardner, Jr.	U.S.A.	2	13:08:12:23
John M. Fabian	U.S.A.	2	13:04:05:15
Richard F. Gordon, Jr.	U.S.A.	2	13:03:53:33
Joseph P. Allen, IV	U.S.A.	2	13:02:10:28
William E. Thornton	U.S.A.	2	13:01:19:26
Bruce McCandless, II	U.S.A.	2	13:00:34:08
Bruce E. Melnick	U.S.A.	2	12:23:29:29
Gregory C. Johnson	U.S.A.	1	12:21:37:39
K. Megan McArthur	U.S.A.	1	12:21:37:09
Joan Higginbotham	U.S.A.	1	12:20:45:16
William A. Oefelein	U.S.A.	1	12:20:45:16
Paul W. Richards	U.S.A.	1	12:19:51:57
Richard R. Arnold	U.S.A.	1	12:19:31:01
José M. Hernández	U.S.A.	1	12:19:04:50
Lisa M. Nowak	U.S.A.	1	12:18:37:54
Stanley G. Love	U.S.A.	1	12:18:21:44

Name	Country	Total space flights	Time in space (dd:hh:mm:ss)
Barbara R. Morgan	U.S.A.	1	12:17:55:54
Liu Wang	China	1	12:15:24:00
Liu Yang	China	1	12:15:24:00
Ronald E. Evans, Jr.	U.S.A.	1	12:13:51:59
Harrison H. Schmitt	U.S.A.	1	12:13:51:59
Walter M. Schirra, Jr.	U.S.A.	3	12:07:14:08
James B. Irwin	U.S.A.	1	12:07:11:53
Alfred M. Worden, Jr.	U.S.A.	1	12:07:11:53
Robert F. Overmyer	U.S.A.	2	12:02:25:22
Robert L. Stewart	U.S.A.	2	12:02:02:47
Buzz (Edwin) E. Aldrin	U.S.A.	2	12:01:53:06
Donald E. Williams	U.S.A.	2	11:23:36:45
Richard A. Garriott	U.S.A.	1	11:20:35:37
Bjarni V. Tryggvason	Canada	1	11:20:28:09
Vladimir V. Aksyonov	U.S.S.R.	2	11:20:11:47
Yuri V. Malyshev	U.S.S.R.	2	11:19:59:36
Igor P. Volk	U.S.S.R.	1	11:19:14:36
Boris V. Morukov	Russian	1	11:19:12:15
Gerhard P. J. Thiele	German	1	11:05:29:41
John M. Lounge	U.S.A.	2	11:03:49:28
Michael Collins	U.S.A.	2	11:02:05:14
Charles M. Duke, Jr.	U.S.A.	1	11:01:51:25
Duane G. Carey	U.S.A.	1	10:22:11:01
Mary L. Cleave	U.S.A.	2	10:22:03:42
Guy Laliberté	Canada	1	10:21:17:40
Muszaphar Shuker Al Masrie	Malaysia	1	10:21:14:00
So Yeon Yi	South Korea	1	10:21:13:00
Anousheh Ansari	U.S.A.	1	10:21:05:00
Anatoly V. Filipchenko	U.S.S.R.	2	10:21:03:58
R. Walter Cunningham	U.S.A.	1	10:20:09:03
Donn F. Eisele	U.S.A.	1	10:20:09:03
Lee M. E. Morin	U.S.A.	1	10:19:42:44
Randolph J. Bresnik	U.S.A.	1	10:19:16:13
Robert L. Satcher	U.S.A.	1	10:19:16:13
Barry E. Wilmore	U.S.A.	1	10:19:16:13
Mark N. Brown	U.S.A.	2	10:09:29:19
Russell L. Schweickart	U.S.A.	1	10:01:00:54
↑ *10 days* ↑			
Ulrich Walter	Germany	1	09:23:41:00
Vladimir A. Shatalov	U.S.S.R.	3	09:21:57:30
Robert C. Springer	U.S.A.	2	09:21:35:02
Yuri G. Shargin	Russian	1	09:21:29:00
Mark R. Shuttleworth	South Africa	1	09:21:25:16

(*continued*)

Name	Country	Total space flights	Time in space (dd:hh:mm:ss)
Marcos C. Pontes	Brazil	1	09:21:17:00
Gregory H. Olsen	U.S.A.	1	09:21:15:00
Nikolai N. Rukavishnikov	U.S.S.R.	3	09:21:10:35
Alexandr P. Alexandrov	Bulgaria	1	09:20:09:19
Konstantin M. Kozeev	Russia	1	09:20:00:25
Joe H. Engle	U.S.A.	2	09:09:03:78
L. Gordon Cooper, Jr.	U.S.A.	2	09:07:15:04
John H. Glenn, Jr.	U.S.A.	2	09:20:40:19
F. Andrew Gaffney	U.S.A.	1	09:02:15:14
Millie E. Hughes-Fulford	U.S.A.	1	09:02:15:14
Donald K. Slayton	U.S.A.	1	09:01:28:24
Alan B. Shepard, Jr.	U.S.A.	2	09:00:16:25
Edgar D. Mitchell	U.S.A.	1	09:00:01:57
Stuart A. Roosa	U.S.A.	1	09:00:01:57
Alexei S. Yeliseyev	U.S.S.R.	3	08:22:22:33
Dirk D.D.D. Frimout	Belgium	1	08:22:10:24
Abdul Ahad Mohmand	Afghan	1	08:20:26:27
Joe F. Edwards. Jr.	U.S.A.	1	08:19:48:06
Neil A. Armstrong	U.S.A.	2	08:14:00:01
Richard H. Truly	U.S.A.	2	08:07:24:40
Jon A. McBride	U.S.A.	1	08:05:24:33
Paul D. Scully-Power	U.S.A.	1	08:05:24:33
Roberta K. Bondar	Canada	1	08:01:15:42
Ronald E. McNair	U.S.A.	2	07:23:18:14
Anna L. Fisher	U.S.A.	1	07:23:45:59
Franco E. Malerba	Italy	1	07:23:16:07
Mohammed Ahmed Faris	Syria	1	07:23:04:55
Loren J. Acton	U.S.A.	1	07:22:46:22
John-David F. Bartoe	U.S.A.	1	07:22:46:22
Roy D. Bridges	U.S.A.	1	07:22:46:22
Anthony W. England	U.S.A.	1	07:22:46:22
Karl G. Henize	U.S.A.	1	07:22:46:22
Mae C. Jemison	U.S.A.	1	07:22:31:13
Vladimir Remek	Czechoslovakia	1	07:22:16:00
Toktar O. Aubakirov	U.S.S.R.	1	07:22:15:59
Franz A. Viehbock	Austria	1	07:22:15:59
Dennis Tito	U.S.A.	1	07:22:04:08
Miroslaw Hermaszewski	Poland	1	07:22:02:59
Anatoly S. Levchenko	U.S.S.R.	1	07:21:58:12
Klaus-Dietrich Flade	Germany	1	07:21:56:52
Ivan Bella	Slovakia	1	07:21:56:29
Toyohiro Akiyama	Japan	1	07:21:54:40
Rakesh Sharma	India	1	07:21:40:16
Helen P. Sharman	U.K.	1	07:21:14:20

Name	Country	Total space flights	Time in space (dd:hh:mm:ss)
Sigmund W. P. Jähn	Germany (GDR)	1	07:20:49:04
Bertalan Farkas	Hungary	1	07:20:45:44
Arnoldo Tamayo Méndez	Cuba	1	07:20:43:24
Jugderdemidin Gurragcha	Mongolia	1	07:20:42:03
Pham Tuan	Vietnam	1	07:20:42:00
Dumitru D. Prunariu	Romania	1	07:20:41:52
William F. Fisher	U.S.A.	1	07:02:18:31
Sultan bin Salman al-Saud	Saudi Arabia	1	07:01:39:42
Patrick P. R. Baudry	France	1	07:01:39:42
Wubbo J. Ockels	Netherlands	1	07:00:45:48
Rheinhard A. Furrer	Germany	1	07:00:45:48
Ernst W. Messerschmid	Germany	1	07:00:45:48
Alexei A. Leonov	U.S.S.R.	2	07:00:33:08
Don L. Lind	U.S.A.	1	07:00:09:53
Lodewijk van den Berg	U.S.A.	1	07:00:09:53
Taylor G. Wang	U.S.A.	1	07:00:09:53
→ 7 days (1 week) ←			
S. David Griggs	U.S.A.	1	06:23:56:31
Edwin J. Garn	U.S.A.	1	06:23:56:31
F. Richard Scobee	U.S.A.	2	06:23:42:08
Terry J. Hart	U.S.A.	1	06:23:40:55
Thomas J. Hennen	U.S.A.	1	06:22:52:28
Rudolfo Neri Vela	Mexico	1	06:21:06:12
Sherwood C. Spring	U.S.A.	1	06:21:06:12
William A. Anders	U.S.A.	1	06:03:00:42
Robert J. Cenker	U.S.A.	1	06:02:04:52
C. William Nelson, Jr.	U.S.A.	1	06:02:04:52
Judith A. Resnik	U.S.A.	2	06:00:59:18
Fred W. Haise, Jr.	U.S.A.	1	05:22:54:41
John L. Swigert, Jr.	U.S.A.	1	05:22:54:41
William B. Lenoir	U.S.A.	1	05:02:15:29
Donald H. Peterson	U.S.A.	1	05:00:24:36
Manley L. Carter, Jr.	U.S.A.	1	05:00:07:50
Michael J. McCulley	U.S.A.	1	04:23:40:14
Georgi S. Shonin	U.S.S.R.	1	04:22:42:47
Fei Junlong	China	1	04:19:33:00
Nie Haisheng	China	1	04:19:33:00
William A. Pailes	U.S.A.	1	04:02:45:46
Edward H. White, II	U.S.A.	1	04:01:56:12
Georgi T. Beregovoi	U.S.S.R.	1	03:22:50:45
Ellison S. Onizuka	U.S.A.	2	03:01:35:31
Gary E. Payton	U.S.A.	1	03:01:34:18

(continued)

Name	Country	Total space flights	Time in space (dd:hh:mm:ss)
Valentina V. Tereshkova	U.S.S.R.	1	02:22:50:00
Vladimir M. Komarov	U.S.S.R.	1	02:03:04:55
Liu Boming	China	1	02:02:27:35
Zhai Zhigang	China	1	02:02:27:35
Lev S. Dyomin	U.S.S.R.	1	02:00:12:11
Gennady V. Sarafanov	U.S.S.R.	1	02:00:12:11
Valery I. Rozhdestvensky	U.S.S.R.	1	02:00:06:35
Vyacheslav D. Zudov	U.S.S.R.	1	02:00:06:35
Yevgeny V. Khrunov	U.S.S.R.	1	01:23:45:50
Vasily G. Lazarev	U.S.S.R.	2	01:23:36:59
Georgi I. Ivanov	Bulgaria	1	01:23:01:06
Pavel I. Belyayev	U.S.S.R.	1	01:02:02:17
Gherman S. Titov	U.S.S.R.	1	01:01:18:00
Konstantin P. Feoktistov	U.S.S.R.	1	01:00:17:03
Boris B. Yegorov	U.S.S.R.	1	01:00:17:03 *

↑ *1 day (24 hours)* ↑			
Yang Liwei	China	1	00:21:26:00
Virgil I. Grissom	U.S.A.	2	00:05:08:28
M. Scott Carpenter	U.S.A.	1	00:04:56:05
Yuri A. Gagarin	U.S.S.R.	1	00:01:48:00
↑ *Oleg V. Novitsky* ↑	*Russia*	1	*In space*
↑ *Yevgeni Tarekin* ↑	*Russia*	1	*In space*

↓ *Flights in the quest for space* ↓ *Astro-flight experience: X-15 (1961–1968) and SpaceShipOne (2004)*			
Mike Melvill (SS-1)	U.S.A.	2	00:00:48:00
Joseph A. Walker (X-15)	U.S.A.	3	00:00:30:00
Brian Binnie (SS-1)	U.S.A.	1	00:00:24:00
William J. Knight (X-15)	U.S.A.	2	00:00:20:00
Robert A. Rushworth (X-15)	U.S.A.	1	00:00:10:00
John B. McKay (X-15)	U.S.A.	1	00:00:10:00
William H. Dana (X-15)	U.S.A.	1	00:00:10:00
Michael J. Adams (X-15)	U.S.A.	1	00:00:10:00

Mission in progress: STS 51-L (25) *Challenger launch accident (January 28, 1986)*			
Gregory B. Jarvis	U.S.A.	1	00:00:01:13
S. Christa McAuliffe	U.S.A.	1	00:00:01:13

Note: As of December 31, 2012 Kevin Ford, Oleg Novitskiy, Evgeny Tarelkin (Kazbek crew/Soyuz TMA-06M) and Chris Hadfield, Roman Romanenko, Tom Marshburn (Parus crew/Soyuz TMA-07M) were in space aboard the International Space Station serving as the ISS-34 expedition).

Appendix C

EVA log 1965–2012

EVA	Start date of EVA	Time (h:min)	EVA crew members	Mission/Spacecraft	Notes
001	1965 Mar 18	0:12	Leonov	Voskhod 2	First person (Soviet) to perform EVA; experienced difficulty in returning to the air lock
002	1965 Jun 4	0:21	White	Gemini 4	First U.S. EVA; used zip gun to maneuver; McDivitt exposed to space vacuum; did not exit (IVA)
003	1966 Jun 5	2:10	Cernan	Gemini 9	Tried to put on AMU, but failed due to overheating of suit; Stafford on IVA; first EVA to exceed 1 hour
004	1966 Jul 19	0:39	Collins	Gemini 10	Stand-up EVA; photographs of stellar background; cut short due to impurities in air supply; Young on IVA
005	1966 Jul 20	0:50	Collins	Gemini 10	Removed Agena 8 experiment package; used HHMU; lost stills camera from open hatch; Young on IVA
006	1966 Sep 13	0:33	Gordon	Gemini 11	Straddled neck of Gemini to attach 100 ft tether to Agena; dubbed "Space Cowboy"; Conrad on IVA
007	1966 Sep 14	2:08	Gordon	Gemini 11	Stand-up EVA for photography of stellar background in UV; Conrad on IVA

(continued)

EVA	Start date of EVA	Time (h:min)	EVA crew members	Mission/Spacecraft	Notes
008	1966 Nov 12	2:29	Aldrin	Gemini 12	Stand-up EVA for astronomical photography; equipment prep for later EVAs; Lovell on IVA; EVA record
009	1966 Nov 13	2:09	Aldrin	Gemini 12	Evaluation of restraints and tethers during completion of EVA work task tests; Lovell on IVA
010	1966 Nov 14	0:59	Aldrin	Gemini 12	Final Gemini EVA; completed astronomical photography objectives; Lovell on IVA
011	1969 Jan 16	0:37	Khrunov/Yeliseyev	Soyuz 5/4	First Soviet EVA for 4 years; first EVA crew transfer between two spacecraft; related to lunar program
012	1969 Mar 6	1:07	Schweickart/Scott D.	Apollo 9/LM/CM	First test of Apollo lunar EVA suit in Earth orbit; Scott conducted stand-up EVA from CM hatch; Schweickart demonstrated exit on to LM porch and partial crew transfer to LM; McDivitt remained inside LM on IVA
013	1969 Jul 20	2:32	Armstrong/Aldrin	Apollo 11/LM	First moonwalk; EVA record; set up experiments and took samples; Tranquility Base EVA world's most televised event to date
014	1969 Nov 19	3:56	Conrad/Bean	Apollo 12/LM	Second moonwalk; set up first ALSEP; geological sampling near landing site; EVA record
015	1969 Nov 20	3:49	Conrad/Bean	Apollo 12/LM	Third moonwalk; geological field trip on foot; retrieved parts from unmanned Surveyor III
016	1971 Feb 5	4:48	Shepard/Mitchell	Apollo 14/LM	Fourth moonwalk; deployed second ALSEP and completed short geological field trip on foot; EVA record.
017	1971 Feb 6	4:35	Shepard/Mitchell	Apollo 14/LM	Fifth moonwalk; geological field trip on foot to Cone Crater using MET to carry equipment.

018	1971 Jul 30	0:27	Scott, D.	Apollo 15/LM	Stand-up EVA from LM top hatch to photographically survey Hadley Rille landing site; Irwin remained in LM on IVA
019	1971 Jul 31	3:33	Scott, D./Irwin	Apollo 15/LM	Sixth moonwalk; set up third ALSEP geological traverse on first LRV set a new EVA record.
020	1971 Aug 1	7:12	Scott, D./Irwin	Apollo 15/LM	Seventh moonwalk; geological traverse on LRV; round trip to Hadley–Apennine Front; new EVA record
021	1971 Aug 2	4:50	Scott, D/Irwin	Apollo 15/LM	Eighth moonwalk; geological traverse on LRV to Hadley Rille
022	1971 Aug 5	0:39	Worden/Irwin	Apollo 15/CM	First deep-space EVA; Worden retrieved SIM bay film cassettes; Irwin filmed and assisted during stand-up EVA in CM hatch; Scott exposed to vacuum, but did not exit CM
023	1972 Apr 21	7:11	Young/Duke	Apollo 16/LM	Ninth moonwalk; deployed fourth ALSEP; drove second LRV on geological traverse to Flag Crater
024	1972 Apr 22	7:23	Young/Duke	Apollo 16/LM	Tenth moonwalk; LRV geological traverse to Stone Mountain; new EVA record
025	1972 Apr 23	5:40	Young/Duke	Apollo 16/LM	Eleventh moonwalk; LRV geological traverse to House Rock; largest boulder visited by Apollo crews
026	1972 Apr 25	1:24	Mattingly/Duke	Apollo 16/CM	Second deep-space EVA; Mattingly retrieved SIM bay film cassettes from SM; Duke filmed and assisted during stand-up EVA in CM hatch; Young exposed to vacuum but did not exit CM
027	1972 Dec 11	7:12	Cernan/Schmitt	Apollo 17/LM	Twelfth moonwalk; deployed fifth and final ALSEP; drove third LRV on short geological traverse
028	1972 Dec 12	7:37	Cernan/Schmitt	Apollo 17/LM	Thirteenth moonwalk; LRV geological traverse; discovered "orange soil"; set new EVA record

(continued)

EVA	Start date of EVA	Time (h:min)	EVA crew members	Mission/Spacecraft	Notes
029	1972 Dec 13	7:15	Cernan/Schmitt	Apollo 17/LM	Fourteenth and last moonwalk; LRV geological traverse; performed ceremonies marking end of Apollo
030	1972 Dec 17	1:06	Evans/Schmitt	Apollo 17/CM	Third deep-space EVA; Evans retrieved SIM bay film cassettes; Schmitt filmed and assisted during stand-up EVA in CM hatch; Cernan exposed to vacuum but did not exit CM; final Apollo EVA
031	1973 May 25	0:37	Weitz	Skylab 2/CM	Stand-up EVA from the open CM hatch; attempts to deploy stuck solar array; assisted from inside CM by Kerwin and Conrad on IVA
032	1973 Jun 7	3:25	Conrad/Kerwin	Skylab 2/OWS	Successfully deployed stuck solar wing, saving station; tethered Conrad catapulted into space by motion of array
033	1973 Jun 19	1:44	Conrad/Weitz	Skylab 2/OWS	Retrieval and replacement of ATM film cassettes; inspected solar array and previously deployed parasol sunshade
034	1973 Aug 6	6:31	Garriott/Lousma	Skylab 3/OWS	Erected twin-pole assembly over parasol to improve thermal conditions inside Skylab OWS
035	1973 Aug 24	4:30	Garriott/Lousma	Skylab 3/OWS	ATM film cassette retrieval and replacement
036	1973 Sep 22	2:45	Bean/Garriott	Skylab 3/OWS	ATM film cassette retrieval and replacement
037	1973 Nov 22	6:33	Gibson/Pogue	Skylab 4/OWS	ATM film cassette retrieval and replacement; routine repair tasks
038	1973 Dec 25	7:01	Carr/Pogue	Skylab 4/OWS	ATM film cassette retrieval and replacement; first EVA on Christmas Day; observed Comet Kohoutek
039	1973 Dec 29	3:38	Carr/Gibson	Skylab 4/OWS	ATM film cassette retrieval and replacement; further observations of Kohoutek

040	1974 Feb 3	5:19	Carr/Gibson	Skylab 4/OWS	Final ATM film cassette retrieval; retrieved experiment packages from exterior of Skylab; final Skylab EVA
041	1977 Dec 20	1:28	Grechko	EO-1/Salyut 6	Stand-up EVA examined forward docking port after failure of Soyuz 25 to dock; Romanenko conducted IVA; first Soviet EVA in almost 9 years; first EVA from a Salyut space station
042	1978 Jul 29	2:05	Ivanchenko/Kovalenok	EO-2/Salyut 6	Removed and replaced samples from exterior of Salyut 6; Kovalenok performed stand-up EVA
043	1979 Aug 15	1:23	Ryumin/Lyakhov	EO-3/Salyut 6	Unscheduled EVA to free KRT-10 telescope antenna from aft docking port
044	1982 Jul 30	2:33	Lebedev/Berezovoi	EO-1/Salyut 7	Collected and replaced samples on exterior of Salyut
045	1983 Apr 7	4:17	Musgrave/Peterson	STS-6/OV-099	First Shuttle demonstration EVA; evaluated new EVA suits and restraint system
046	1983 Nov 1	2:50	Alexandrov/Lyakhov	EO-2/Salyut 7	First in a series of EVAs adding extra panels to solar arrays; installed additional panels to central array
047	1983 Nov 3	2:55	Alexandrov/Lyakhov	EO-2/Salyut 7	Added second panel to central array
048	1984 Feb 7	5:55	McCandless/Stewart	STS-41B/OV-099	First untethered EVAs; McCandless flew first MMU 300 feet from orbiter; Stewart also test-flew MMU
049	1984 Feb 9	6:17	McCandless/Stewart	STS-41B/OV-099	Further MMU flights; also evaluated procedures for satellite repairs planned for later missions
050	1984 Apr 8	2:57	Nelson/Van Hoften	STS-41C/OV-099	Nelson attempts to capture Solar Max by flying MMU; unsuccessful; later captured by Shuttle RMS
051	1984 Apr 11	6:16	Nelson/Van Hoften	STS-41C/OV-099	Repaired Solar Max in payload bay; redeployed; Van Hoften flew untethered MMU in payload bay

(continued)

EVA	Start date of EVA	Time (h:min)	EVA crew members	Mission/Spacecraft	Notes
052	1984 Apr 23	4:15	Kizim/Solovyov, V.	EO-3/Salyut 7	First of a series of six EVAs; transported ladder and EVA tools to work area, and prepared work site
053	1984 Apr 26	4:56	Kizim/Solovyov, V.	EO-3/ Salyut 7	Installed new propellant valve by cutting into station's skin
054	1984 Apr 29	2:45	Kizim/Solovyov, V.	EO-3/Salyut 7	Installation of new conduit; replacement of thermal covering of station
055	1984 May 3	2:45	Kizim/Solovyov, V.	EO-3/ Salyut 7	Installation of second conduit; verification of both conduits; fuel leak pinpointed
056 array	1984 May 18	3:05	Kizim/Solovyov, V.	EO-3/ Salyut 7	Installation of a second set of solar array extensions to main
057	1984 Jul 25	3:35	Savitskaya/Dzhanibekov	T12/ Salyut 7	First female to perform EVA; tested multipurpose welding gun
058	1984 Aug 8	5:00	Kizim/Solovyov, V.	EO-3/Salyut 7	Leaking fuel pipe sealed; retrieved samples from solar arrays for evaluation on Earth
059	1984 Oct 11	3:27	Leestma/Sullivan	STS-41G/OV-099	First U.S. female to perform EVA; completed satellite-refueling demonstration in payload bay of Shuttle
060	1984 Nov 12	6:00	Allen/Gardner	STS-51A/OV-103	Used MMU to retrieve rogue Palapa Comsat
061	1984 Nov 14	5:42	Allen/Garner	STS-51A/OV-103	Used MMU to retrieve rogue Westar Comsat
062	1985 Apr 16	3:00	Hoffman/Griggs	STS-51D/OV-103	First unscheduled (contingency) U.S. EVA; crew attached "Flyswatter" device to RMS attempting to activate Leasat
063	1985 Aug 2	5:00	Savinykh/Dzhanibekov	EO-4/ Salyut 7	Attached third and final set of additional solar panels to main arrays; evaluated new Orlan EVA suits

064	1985 Aug 31	7:08	Van Hoften/Fisher	STS-51I/OV-103	Manual capture of Leasat deployed on 51D; commenced repairs in payload bay
065	1985 Sep 1	4:26	Van Hoften/Fisher	STS-51I/OV-103	Completed Leasat repair and redeployed by hand from end of RMS
066	1985 Nov 29	5:30	Spring/Ross	STS-61B/OV-104	Space construction tests using EASE and ACCESS in payload bay for future space station activities
067	1985 Dec 1	6:30	Spring/Ross	STS-61B/OV-104	Space construction tests with EASE and ACCESS related to Space Station Freedom program
068	1986 May 28	3:50	Kizim/Solovyov, V.	EO-5/ Salyut 7	Collected experiments from exterior of station; evaluated a beam builder for future space construction tasks
069	1986 May 31	5:00	Kizim/Solovyov, V.	EO-5/Salyut 7	Completed additional space construction test; used improved URI welding gun; final EVA from Salyut station
070	1987 Apr 11	3:40	Romanenko/Laveikin	EO-2/Mir node	Unscheduled EVA to remove a foreign object from rear Mir docking port which prevented hard docking of Kvant astrophysical module; first EVA from Mir
071	1987 Jun 12	1:53	Romanenko/Laveikin	EO-2/Mir node	Added an extra set of solar panels to the Mir exterior
072	1987 Jun 16	3:15	Romanenko/Laveikin	EO-2/Mir node	Installed a second set of additional solar panels to exterior of Mir
073	1988 Feb 26	4:25	Titov, V./Manarov	EO-3/Mir node	Replaced elements of solar array panels erected by EO-2 crew
074	1988 Jun 30	5:10	Titov, V./Manarov	EO-3/Mir node	Attempted repair of the X-ray telescope on Kvant-1; terminated due to broken wrench
075	1988 Oct 20	4:12	Titov, V./Manarov	EO-3/Mir node	Completed repair of the TTM telescope on Kvant
076	1988 Dec 9	5:57	Volkov, A./Chrétien	EO-4/Mir node	First non-U.S. non-U.S.S.R. EVA; first French EVA (Chrétien); erected French ERA structure (by kicking the container); carried out French experiments

(continued)

EVA	Start date of EVA	Time (h:min)	EVA crew members	Mission/Spacecraft	Notes
077	1990 Jan 8	2:56	Viktorenko/Serebrov	EO-5/Mir node	Deployment of two star sensors on exterior of Mir and retrieved samples from hull of station
078	1990 Jan 11	2:54	Viktorenko/Serebrov	EO-5/Mir node	Retrieved French experiments deployed in December 1988 and installed new experiment packages
079	1990 Jan 26	3:02	Viktorenko/Serebrov	EO-5/Kvant-2	First use of specialized air lock on Kvant-2 module; prepared docking device for use with Soviet MMU; removed Kurs antenna; installed new TV system; tested improved EVA suit
080	1990 Feb 1	4:59	Viktorenko/Serebrov	EO-5/Kvant-2	First flight of Soviet MMU; Serebrov flew (tethered) up to 30 meters from Mir
081	1990 Feb 5	3:45	Viktorenko/Serebrov	EO-5/Kvant-2	Viktorenko flies (tethered) MMU 45 meters from Mir and performs a victory roll in celebration
082	1990 Jul 17	7:00	Solovyov, A./Balandin	EO-6/Kvant-2	Attempted repair of damaged Soyuz thermal blankets; a damaged outer Kvant-2 hatch meant using a backup method to reenter Mir
083	1990 Jul 26	3:31	Solovyov, A./Balandin	EO-6/Kvant-2	Stowed external ladders on Mir for future use; completed temporary repairs to damaged Kvant-2 outer hatch
084	1990 Oct 30	3:45	Manakov/Strekalov	EO-7/Kvant-2	Completed a partially successful repair of Kvant-2 outer hatch
085	1991 Jan 7	5:18	Afanasyev/Manarov	EO-8/Kvant-2	Completed repairs to Kvant-2 outer hatch; installed support structure for a crane to relocate solar arrays from Kristall to Kvant-1 module
086	1991 Jan 23	5:33	Afanasyev/Manarov	EO-8/Kvant-2	Installation of first crane jib near the forward multiple docking adapter node on Mir core module

087	1991 Jan 26	6:20	Afanasyev/Manarov	EO-8/Kvant-2	Installation of second crane jib on Kvant-1 for future relocation of solar arrays
088	1991 Apr 7	4:38	Ross/Apt	STS-37/OV-104	First U.S. EVA for 64 months; unscheduled EVA to repair stuck Gamma Ray Observatory high-gain antenna
089	1991 Apr 8	6:11	Ross/Apt	STS-37/OV104	Completed EVA experiments related to future space station construction, including CETA mobility tests
090	1991 Apr 25	3:34	Afanasyev/Manarov	EO-8/Kvant-2	Inspection of faulty Kurs antenna discovering missing receiver dish; replacement of exterior TV camera
091	1991 Jun 25	4:58	Artsebarsky/Krikalev	EO-9/Kvant-2	Replacement of damaged Kurs antenna
092	1991 Jun 28	3:24	Artsebarsky/Krikalev	EO-9/Kvant-2	Installation of U.S. cosmic ray detector experiment on exterior of Mir
093	1991 Jul 15	5:55	Artsebarsky/Krikalev	EO-9/ Kvant-2	Commenced construction of 15 m Sofora girder on the exterior of Kvant-1
094	1991 Jul 19	6:20	Artsebarsky/Krikalev	EO-9/Kvant-2	Continued construction of Sofora
095	1991 Jul 23	5:34	Artsebarsky/Krikalev	EO-9/Kvant-2	Continued construction of Sofora
096	1991 Jul 27	6:49	Artsebarsky/Krikalev	EO-9/Kvant-2	Completed construction of Sofora; Artsebarsky's suit overheats and he is blinded by perspiration; he is guided back to the hatch by Krikalev
097	1992 Feb 20	4:12	Volkov, A./Viktorenko	EO-10/Kvant-2	Installation of new equipment on exterior of Kvant-2; Volkov's suit experienced problems that forced an early termination of the EVA
098	1992 May 10	3:43	Thuot/Hieb	STS-49/OV-105	An attempt to capture Intelsat VI (satellite fails)

(continued)

EVA	Start date of EVA	Time (h:min)	EVA crew members	Mission/Spacecraft	Notes
099	1992 May 12	5:30	Thuot/Hieb	STS-49/OV-105	Second attempt to capture Intelsat VI satellite fails
100	1992 May 13	8:29	Thuot/Hieb/Akers	STS-49/OV-105	First three-person EVA; trio captured Intelsat VI by hand, then attached a new kick motor and redeployed it; the hundredth EVA was also the longest in EVA history
101	1992 May 14	7:45	Thornton, K./Akers	STS-49/OV-105	Tested Space Station Freedom construction techniques and crew rescue procedures; Thornton set a new EVA duration record for a female
102	1992 Jul 8	2:05	Viktorenko/Kaleri	EO-11/Kvant-2	Completed external repairs to the Kvant-2 module
103	1992 Sep 3	3:56	Solovyov, A./Avdeyev	EO-12/Kvant-2	Commenced work to install new VDU propulsion system onto the Sofora girder
104	1992 Sep 7	5:08	Solovyov, A./Avdeyev	EO-12/Kvant-2	Continued installation of VDU on Sofora girder; lowered U.S.S.R. flag from outside Mir
105	1992 Sep 11	5:44	Solovyov, A./Avdeyev	EO-12/Kvant-2	Continued installation of VDU on Sofora girder
106	1992 Sep 15	3:33	Solovyov, A./Avdeyev	EO-12/Kvant-2	Antenna attached to Kristall module in order to assist planned Buran and U.S. Shuttle docking radar
107	1993 Jan 17	4:28	Harbaugh/Runco	STS-54/OV-105	Demonstration and tests of techniques under development for construction of Space Station Freedom and HST repair activities
108	1993 Apr 19	5:25	Manakov/Poleschuk	EO-13/Kvant-2	Commenced process of transferring solar arrays on exterior of Mir
109	1993 Jun 18	4:18	Manakov/Poleschuk	EO-13Kvant-2	Commenced configuration of Mir exterior for a series of EVAs by next main crew

	Date	Duration	Crew	Mission	Description
110	1993 Jun 25	5:50	Low/Wisoff	STS-57/OV-105	Crew attached EURECA antenna; evaluated Space Station Freedom EVA tasks and HST repair methods
111	1993 Sep 16	4:18	Tsibliyev/Serebrov	EO-14/Kvant-2	First of three EVAs to install Rapana mast on exterior of Mir; also checked hull for Perseid meteoroid damage
112	1993 Sep 16	7:05	Walz/Newman	STS-51/OV-103	Further tests of HST repair tools and techniques
113	1993 Sep 20	3:13	Tsibliyev/Serebrov	EO-14/Kvant-2	Continued deployment of Rapana mast; also installed sample packages for later retrieval
114	1993 Sep 28	1:52	Tsibliyev/Serebrov	EO-14/Kvant-2	Planned EVA to complete installation of Rapana and film exterior of station; the EVA was shortened when Tsibliyev's suit overheated
115	1993 Oct 22	0:38	Tsibliyev/Serebrov	EO-14/Kvant-2	Installation of new instrument blocks on Kvant-2; completed exterior filming of station
116	1993 Oct 29	4:12	Tsibliyev/Serebrov	EO-14/Kvant-2	Inspected solar arrays and exterior antenna; checked Sofora mount; retrieved materials samples in order to determine future operational lifetime of Mir
117	1993 Dec 4	7:54	Hoffman/Musgrave	STS-61/OV-105	First HST service mission; first service EVA included the replacement of malfunctioning gyroscopes
118	1993 Dec 5	6:36	Thornton, K./Akers	STS-61/OV-105	Second HST service EVA to remove old solar arrays and install new ones
119	1993 Dec 6	6:47	Hoffman/Musgrave	STS-61/OV-105	Third HST service EVA; installed new camera
120	1993 Dec 7	6:50	Thornton, K./Akers	STS-61/OV-105	Fourth HST service EVA; installed COSTAR and a new computer; cut loose solar panel
121	1993 Dec 8	7:21	Hoffman/Musgrave	STS-61/ OV-105	Fifth HST service EVA; new control systems installed

(continued)

EVA	Start date of EVA	Time (h:min)	EVA crew members	Mission/Spacecraft	Notes
122	1994 Sep 9	5:06	Malenchenko/Musabayev	EO-16/Kvant-2	Inspected docking port after collision by Progress M-24; repaired thermal blanket torn by Soyuz TM-17; attached new solar panels; configured Mir for planned U.S. Shuttle docking.
123	1994 Sep 13	6:01	Malenchenko/Musabayev	EO-16/Kvant-2	Samples retrieved from Rapana; maintenance work completed on Sofora truss; maintenance on exterior of Kvant-2 and solar panels
124	1994 Sep 16	6:51	Lee/Meade	STS-61/OV-103	First tests of SAFER, free-flying astronaut rescue jet pack (smaller MMU)
125	1995 Feb 9	4:39	Harris/Foale	STS-63/OV-103	Evaluation of the astronaut's ability to translate large objects; related to future Alpha Space Station tasks (EDFT-03); unsuccessful test of cold temperature EVA gloves.
126	1995 May 12	6:15	Dezhurov/Strekalov	EO-18/Kvant-2	Exterior of Mir was prepared for the transfer of solar panels allowing for docking of the U.S. Shuttle
127	1995 May 17	6:30	Dezhurov/Strekalov	EO-18/Kvant-2	Commenced moving solar panels but failed to complete initial move
128	1995 May 22	5:15	Dezhurov/Strekalov	EO-18/Kvant-2	Completed the move of the first solar array started on previous EVA
129	1995 May 28	0:21	Dezhurov/Strekalov	EO-18/Mir node	Internal EVA (first of several at Mir) wearing full EVA suits in forward transfer compartment to relocate Konus equipment to allow docking of Spektr module on June 1
130	1995 Jun 1	0:24	Dezhurov/Strekalov	EO-18/Mir node	Second internal EVA returning the Konus docking cone to its pre-May 28 location

131	1995 Jul 14	5:34	Solovyov, A./Budarin	EO-19/Kvant-2	Inspection of a leaky docking collar; moved two solar arrays allowing later transfer of Kristall module
132	1995 Jul 19	3:08	Solovyov, A./Budarin	EO-19/Kvant-2	Budarin commences installation of MIRAS infrared spectrometer; Solovyov unable to exit hatch due to a suit problem and remains in transfer compartment
133	1995 Jul 21	5:50	Solovyov, A./ Budarin	EO-19/Kvant-2	Completed installation of MIRAS spectrometer
134	1995 Sep 15	6:46	Voss/Gernhardt	STS-69/OV-105	Tests of EVA thermal gear; continued tests on Space Station Alpha techniques (EDFT-02)
135	1995 Oct 20	5:16	Avdeyev/Reiter	EO-20/Kvant-2	ESA astronaut Reiter (first German to perform EVA) erects European experiment on outside of Mir
136	1995 Dec 8	0:29	Gidzenko/Avdeyev	EO-20/Mir node	Third Mir internal EVA; relocation of −Z docking cone to +Z port for Priroda module docking
137	1996 Jan 15	6:09	Chiao/Barry	STS-72/OV-105	Space station hardware evaluation, the third EVA Development Flight Test (EDFT-03) exercise; this time umbilical lines, utility boxes and, work platforms were used
138	1996 Jan 17	6:54	Chiao/Scott, W.	STS-72/OV-105	Continuation of the EDFT program begun on previous EVA; in addition, a 30 min cold soak of Scott's EMU was completed to test its thermal properties.
139	1996 Feb 8	3:06	Gidzenko/Reiter	EO-20/Kvant-2	Retrieved the exposure facility ESEF that was deployed during EO-20 EVA 1 on October 20, 1995; the crew also moved the Ikarus MMU; second German EVA
140	1996 Mar 15	5:51	Onufriyenko/Usachev	EO-21/Kvant-2	Installation of a second Strela crane on the Mir base blocks; they also set cables ready for new solar panels on Kvant

(continued)

EVA	Start date of EVA	Time (h:min)	EVA crew members	Mission/Spacecraft	Notes
141	1996 Mar 27	6:02	Godwin/Clifford	STS-76/OV-104	First Shuttle-based EVA by American astronauts docked with Mir; the crew attached MEEP dust collectors to the exterior of Mir docking module but did not traverse over to Mir; they also evaluated common foot restraints and tether hooks intended for ISS EVA operations
142	1996 May 20	5:20	Onufriyenko/Usachev	EO-21/Kvant-2	Relocated solar battery from exterior of Mir docking module to exterior of Kvant-1
143	1996 May 24	5:43	Onufriyenko/Usachev	EO-21/Kvant-2	Installed Russian/American solar panel on to the exterior of Kvant-1
144	1996 May 30	4:20	Onufriyenko/Usachev	EO-21/Kvant-2	Installation of MOMS-2 camera and an EVA handrail to the exterior of Mir
145	1996 Jun 6	3:34	Onufriyenko/Usachev	EO-21/Kvant-2	Installed the SKK-11 experiment on the outside of Mir and replaced the Komza experiment package. They also filmed part one of a sponsored Pepsi commercial
146	1996 Jun 13	5:46	Onufriyenko/Usachev	EO-21/Kvant-2	Crew installed and deployed Ferma-3 girder and repaired the Travers Antenna; they also completed filming the second part of the Pepsi commercial
147	1996 Dec 2	5:57	Korzun/Kaleri	EO-22/Kvant-2	Linked power cables from solar battery to the main electrical bus on Mir
148	1996 Dec 9	6:36	Korzun/Kaleri	EO-22/Kvant-2	Completed the linkage of power cables begun on previous EVA
149	1997 Feb 13	6:42	Lee/Smith	STS-82/OV-103	Sixth HST-servicing EVA; replacement of older High Resolution Spectrograph with the new Space Telescope Imaging Spectrograph (STIS) and Near Infrared Camera and Multi-Object Spectrometer (NICMOS)

No.	Date	Duration	Crew	Mission	Description
150	1997 Feb 14	7:27	Harbaugh/Tanner	STS-82/OV-103	Seventh HST-servicing EVA; replacement of the Far Guidance Sensor (FGS) and out-of-date recorders; installation of the Optical Control Electronics Enhancement Kit (OCE-EK); the crew also noted insulation damage on the telescope
151	1997 Feb 15	7:11	Lee/Smith	STS-82/OV-103	Eighth HST-servicing EVA; replacement of the Data Interface unit (DIU); installation of a new solid state data recorder
152	1997 Feb 16	6:34	Harbaugh/Tanne	STS-82/OV-103	Ninth HST-servicing EVA; replacement of Solar Array Drive electronics (SADE); installation of covers for magnetometers; commenced the repair of insulation noted on February 14 EVA
153	1997 Feb 17	5:17	Lee/Smith	STS-82/OV-103	Tenth HST-servicing EVA; this was an additional EVA added to the flight plan in order to attach thermal insulation blankets to the exterior of the telescope
154	1997 Apr 29	4:48	Tsibliyev/Linenger	EO-23/Kvant-2	First U.S./Russian EVA; first American to use a Russian EVA suit (Orlan-M); retrieval of experiment packages from exterior of Mir
155	1997 Aug 22	3:16	Solovyov, A./Vinogradov	EO-24/Mir node	Fourth Mir IVA; connected power cables from the damaged Spektr module to the Mir base block
156	1997 Sep 6	6:00	Solovyov, A./Foale	EO-24/Kvant-2	Second U.S./Russian EVA; crew searched for evidence of puncture in hull of Spektr but found none; completed a manual realignment of solar arrays
157	1997 Oct 1	5:01	Parazynski/Titov, V.	STS-86/OV-104	Third U.S./Russian EVA; first from Shuttle; retrieval of MEEP from exterior of Mir
158	1997 Oct 20	6:38	Solovyov, A./Vinogradov	EO-24/Mir node	Fifth Mir IVA; completed connections of Spektr power cables
159	1997 Nov 3	6:04	Solovyov, A./Vinogradov	EO-24/Kvant-2	Disconnected the old Kvant solar array and replaced it; hand-launched mini replica of Sputnik, marking the 40th anniversary of the Space Age (October 4)

(continued)

EVA	Start date of EVA	Time (h:min)	EVA crew members	Mission/Spacecraft	Notes
160	1997 Nov 6	6:17	Solovyov, A./Vinogradov	EO-24/Kvant-2	Transferred solar panel; temporarily installed a cap for possible Spektr leak repair; a hatch leak delayed EVA closeout procedures
161	1997 Nov 24	7:43	Scott, W./Doi	STS-87/OV-102	First Japanese (Doi) to perform EVA; unplanned Spartan retrieval; further tests of ISS EVA hardware
162	1997 Dec 3	4:59	Scott, W./Doi	STS-87/OV-102	Second Japanese EVA; continued the originally planned ISS EVA hardware/procedures tests delayed from EVA 1
163	1998 Jan 8	4:04	Solovyov, A./Vinogradov	EO-24/Kvant-2	Attempted repair of leaking air lock hatch
164	1998 Jan 14	6:38	Solovyov, A./Wolf	EO-24/Kvant-2	Fourth U.S./Russian EVA; inspection of Mir exterior and repair of the faulty EVA air lock hatch
165	1998 Apr 1	6:40	Musabayev/Budarin	EO-2/Kvant-2	Installation of handrails on the outside of Mir in preparation for a planned repair of Spektr solar array mounting
166	1998 Apr 6	4:23	Musabayev/Budarin	EO-25/Kvant-2	Strengthening of Spektr solar array mounting; the EVA was shortened by mission control error in regard to the orientation system on Mir
167	1998 Apr 11	6:25	Musabayev/Budarin	EO-25/Kvant-2	Crew dismantled and discarded the exterior control engine (VDU) unit
168	1998 Apr 17	6:32	Musabayev/Budarin	EO-25/Kvant-2	Dismantled the Strela boom and T3; commenced the installation of a new VDU unit
169	1998 Apr 22	6:21	Musabayev/Budarin	EO-25/Kvant-2	Completion of the installation of the new VDU unit on top of a tower on the exterior of Mir; 100th space station–related EVA
170	1998 Sep 15	0:30	Avdeyev/Padalka	EO-26/Kvant-2	Set up six new external experiments; conducted a program of maintenance and repair; and hand-launched second replica Sputnik satellite

#	Date	Duration	Crew	Mission	Description
171	1998 Nov 10	5:54	Avdeyev/Padalka	EO-26/Kvant-2	Set up six new external experiments; conducted maintenance and repair program; hand-launched second replica mini Sputnik satellite
172	1998 Dec 7	7:21	Ross/Newman	STS-88/OV-105	First ISS-related EVA (Shuttle Orbiter based); connected electrical cables from Zarya to Unity; installation of handrails and support equipment for future EVAs
173	1998 Dec 9	7:02	Ross/Newman	STS-88/OV-105	Installation of two antennae on Unity and removal of launch restraint pins on Node hatches; also deployment of stuck TORU antenna on Zarya
174	1998 Dec 12	6:59	Ross/Newman	STS-88/OV-105	Stowed bag of EVA tools on Unity for future EVA crews; freed second stuck TORU antenna on Zarya; conducted a survey of exterior of station hull
175	1999 Apr 16	6:19	Afanasyev/Haigneré	EO-26/Kvant-2	Second French EVA; performed CNES experiments; deployment by hand of third mini-replica of Sputnik
176	1999 May 28	7:55	Jernigan/Barry	STS-96/OV-103	Relocation of stowed U.S. Orbital Transfer Device and Russian Strela cranes from Shuttle payload bay to Unity; attached a further set of EVA tools to the station; longest EVA for a female
177	1999 Jul 23	6:07	Afanasyev/Avdeyev	EO-26/Kvant-2	An attempt to deploy Russian/Georgian reflector antenna
178	1999 Jul 28	5:22	Afanasyev/Avdeyev	EO-26/Kvant-2	Deployment of Russian/Georgian reflector; deployed and retrieved exposed sample cassettes
179	1999 Dec 22	8:16	Smith/Grunsfeld	STS-103/OV-103	Eleventh HST-servicing EVA
180	1999 Dec 23	8:10	Foale/Nicollier	STS-103/OV-103	Twelfth HST-servicing EVA; first Swiss EVA
181	1999 Dec 24	8:08	Smith/Grunsfeld	STS-103/OV-103	Thirteenth HST-servicing EVA

(continued)

EVA	Start date of EVA	Time (h:min)	EVA crew members	Mission/Spacecraft	Notes
182	2000 May 12	4:52	Zaletin/Kaleri	EO-28/Kvant-2	The final EVA from the Mir complex; used Germatisator Sealing Experiment; completed a panoramic inspection; examined a failed solar battery
183	2000 May 21	6:44	Williams/Voss	STS-101/OV-104	Secured OTD; installation of final elements of Strela crane and replaced antenna on Unity; completed a number of get-ahead tasks
184	2000 Sep 10	6:14	Lu/Malenchenko	STS-106/OV-104	Fifth U.S./Russian EVA; connection of electrical and communication cables between Zvezda and Zarya; also installed magnetometer
185	2000 Oct 15	6:28	Chiao/McArthur	STS-92/OV-103	Deployed two antenna assemblies and installed EVA toolbox; electrical cable connection for Z1 truss
186	2000 Oct 16	7:07	Wisoff/López-Alegría	STS-92/OV-103	Preparation of Z1 truss for attachment of solar arrays; released launch latches securing PMA-3
187	2000 Oct 17	6:48	Chiao/McArthur	STS-92/OV-103	Installed second EVA toolbox, continued program of reconfigurations and connections of electrical cables; installation of two DC converter units on Z1 truss
188	2000 Oct 18	6:56	Wisoff/López-Alegría	STS-92/OV-103	Tested SAFER backpack; completed EVA wrap-up tasks
189	2000 Dec 3	7:33	Tanner/Noriega	STS-97/OV-105	Installed P6 truss segment and deployed two solar arrays (after one stuck); connected power and data cables
190	2000 Dec 5	6:37	Tanner/Noriega	STS-97/OV-105	Inspection of partially deployed solar array; connected power and data cables and coolant lines from P6; replacement of S-band assembly at P6

191	2000 Dec 7	5:10	Tanner/Noriega	STS-97/OV-105	Repaired P6; installation of a small antenna and a sensor on a radiator; also installed a Floating Potential Probe and a centerline camera
192	2001 Feb 10	7:34	Jones/Curbeam	STS-98/OV-105	Monitored relocation of PMA-2; connected power and data cables; Curbeam's suit was contaminated with leaking ammonia requiring a "bake-out" in sunlight to clean
193	2001 Feb 12	6:50	Jones/Curbeam	STS-98/OV-104	EVA crew assisted in repositioning of PMA-2 on Destiny Lab; installed insulation covers in payload bay of Discovery; installed SSRMS base
194	2001 Feb 14	5:25	Jones/Curbeam	STS-98/OV-105	Attachment of S-band antenna; released cooling radiator; inspected connections; test-flew SAFER; this marked the 100th U.S. EVA since Gemini 4 in June 1965
195	2001 Mar 11	8:56	Voss/Helms (ISS EO-2)	STS-102/OV-103	Relocated Early Communications antenna from Unity to PMA attachment; installation of Lab Cradle Assembly on Destiny; installed SSRMS cable tray; set new EVA endurance record
196	2001 Mar 13	6:21	Thomas/Richards	STS-102/OV-103	Configuration of stowage platform; completed the connection of cables commenced in previous EVA; made minor adjustments to solar array brace
197	2001 Apr 22	7:10	Parazynski/Hadfield	STS-100/OV-105	First Canadian EVA (Hadfield); installation of UHF antenna and video command and power cables between the SSRMS and Destiny Lab robotic workstation; installation of Canadarm
198	2001 Apr 24	7:40	Parazynski/Hadfield	STS-100/OV-105	Second Canadian EVA; rewired and rerouted power and data cables for Canadarm2
199	2001 Jun 8	0:19	Usachev/Voss	EO-2/Zvezda	First IVA at ISS; sixth U.S./Russian EVA; replacement of flat plate in Zvezda nadir docking port with a docking cone; first IVA at ISS without a Shuttle docked

(continued)

EVA	Start date of EVA	Time (h:min)	EVA crew members	Mission/Spacecraft	Notes
200	2001 Jul 14	5:59	Gernhardt/Reilly	STS-104/OV-104	Configuration of Quest air lock module for transfer to Unity module; connection of heater cables
201	2001 Jul 17	6:29	Gernhardt/Reilly	STS-104/OV-104	Transfer of two oxygen tanks and one nitrogen storage tank from Atlantis payload bay to exterior of Quest
202	2001 Jul 20	4:02	Gernhardt/Reilly	STS-104/Quest	First EVA from Quest air lock; transferred one nitrogen tank from Atlantis payload to exterior of Quest; examined gimbal assembly on top of solar array truss; inauguration of Quest fell on 32nd anniversary of first moonwalk during Apollo 11 (1969)
203	2001 Aug 16	6:16	Barry/Forrester	STS-105/OV-103	Crew installed Early Ammonia Servicer (EAS) on P6 and the Material ISS Experiment (MISSE) on Quest
204	2001 Aug 18	5:29	Barry/Forrester	STS-105/OV-103	Installation of six EVA handrails; relocation of a further two on Destiny; strung heater cables for future S0 truss
205	2001 Oct 8	4:58	Dezhurov/Tyurin	EO-3/Pirs	First from Pirs and first all-Russian ISS EVA; erected antenna and docking targets on exterior of Pirs module; first EVA from ISS without Shuttle docked with it; 100th Soviet/Russian EVA since Voskhod 2 in March 1965.
206	2001 Oct 15	5:52	Dezhurov/Tyurin	EO-3/ Pirs	Placed Kravka sample detector on Zvezda; placed MPAC and SEED experiments on Zvezda; revoked Russian flag and installed commercial logos
207	2001 Nov 12	15:04	Dezhurov/Culbertson	EO-3/ Pirs	Routed and fixed cables from Pirs to ISS interior; checked and photographed small section of solar battery on Zvezda; installed handrails on Pirs; tested extension of Strela cargo crane

208	2001 Dec 3	2:46	Dezhurov/Tyurin	EO-3/Pirs	First unplanned EVA at ISS; removed rubber O-ring obstruction from Pirs docking equipment left behind during November 22 docking of Progress MI-7 which was achieved later on December 3
209	2001 Dec 11	4:12	Godwin/Tani	STS-108/OV-105	Installed insulation blankets on beta gimbal assembly located on the top of the P6 truss; unsuccessful attempt to free stuck solar array cable
210	2002 Jan 14	6:03	Onufriyenko/Walz	EO-4/Pirs	Installed ham radio antenna; completed assembly of the Strela unit
211	2002 Jan 25	5:59	Onufriyenko/Bursch	EO-4/Pirs	Installation of six thruster plume deflectors; installed four new experiment packages and retrieved one older one; attached tether guides and a ham radio antenna
212	2002 Feb 20	5:47	Walz/Bursch	EO-4/Quest	Connected cables from Destiny to Z1 truss; removed tools and handrails that had been used on earlier EVAs; this EVA was conducted on the 40th anniversary of John Glenn's Mercury 6 flight
213	2002 Mar 4	7:01	Grunsfeld/Linnehan	STS-109/OV-102	Fourteenth HST-servicing EVA; replacement of one of the telescope's two second-generation solar arrays; replaced a Diode Box assembly; completed other prep work for later EVAs on this mission
214	2002 Mar 5	7:16	Newman/Massimino	STS-109/OV-102	Fifteenth HST-servicing EVA; replacement of second array with a new unit and its Diode Box Assembly; replacement of Reaction Wheel Assembly-1
215	2002 Mar 6	6:48	Grunsfeld/Linnehan	STS-109/OV-102	Sixteenth HST-servicing EVA; replacement of the telescope's Power Control Unit in Bay 4; Linnehan conducted an inspection of HST's exterior handrails to be used during the fourth and fifth EVAs

(continued)

EVA	Start date of EVA	Time (h:min)	EVA crew members	Mission/Spacecraft	Notes
216	2002 Mar 7	7:30	Newman/Massimino	STS-109/OV-102	Seventeenth HST-servicing EVA; replacement of the Faint Object Camera with the new Advanced Camera for Surveys; completed Power Control Unit cleanup tasks
217	2002 Mar 8	7:20	Grunsfeld/Linnehan	STS-109/OV-102	Eighteenth HST-servicing EVA; installed the Cryogenic Cooler and its cooling system radiator around the NICMOS experiment
218	2002 Apr 11	7:48	Smith, S./Walheim	STS-110/OV-104	Transfer and attachment of S0 trusses
219	2002 Apr 13	7:30	Ross/Morin	STS-110/OV-104	Completed attachment of S0 trusses; attached redundant power cable
220	2002 Apr 14	6:27	Smith, S./Walheim	STS-110/OV-104	Connection of mobile transporter; routed power connections to SSRMS through S0 truss
221	2002 Apr 16	6:37	Ross/Morin	STS-110/OV-104	Installed work lights and air lock spur for future EVA work on S0 truss
222	2002 Jun 9	7:14	Chang-Diaz/Perrin	STS-111/OV-105	Third French EVA; transfer of power/data/grapple fixture to solar array; attached space debris shields; preparation of Mobile Base System (MBS) for installation
223	2002 Jun 11	5:00	Chang-Diaz/Perrin	STS-111/OV-105	Fourth French EVA; hard mate of the MBS and connected power, data, and electronics cables
224	2002 Jun 13	7:17	Chang-Diaz/Perrin	STS-111/OV-105	Fifth French EVA; replacement of failed SSRMS wrist roll joint; preparation of P6 truss for future relocation
225	2002 Aug 16	4:25	Korzun/Whitson	EO-5/ Pirs	Installed first 6 (of planned 23) debris panels onto Zvezda; installation of Russian Kromka experiment postponed due to late start of EVA

226	2002 Aug 26	5:21	Korzun/Treschev	EO-5/Pirs	Installation of exterior frame on Zarya to house components for future EVA assembly tasks; installed new materials samples on a pair of NASDA (Japanese) experiments housed on outside of Zvezda; installed devices to simplify routing of tethers in future assembly EVAs; installed two ham radio antennae on Zvezda; installed Kromka hardware to measure residual emissions from Zvezda jet thrusters
227	2002 Oct 10	7:01	Wolf/Sellers	STS-112/Quest	Attached power, data, and fluid lines between S0 and newly installed S1 truss; deployed second S-band comm system; installed first of two external camera systems; released launch restraints on the truss's mobile EVA workstation (CETA) and released the launch lock holding S1's radiators in place for launch
228	2002 Oct 12	6:04	Wolf/Sellers	STS-112/Quest	Forty-fifth ISS EVA; set up the second external camera system and released further radiator beam launch locks; removed insulation on quick disconnect fittings; installed new Z1 and P6 junctions to install Spool Positioning Devices; released starboard side launch restraints on CETA; attached Ammonia Tank Assembly cables
229	2002 Oct 14	6:36	Wolf/Sellers	STS-112/Quest	Replaced Interface Umbilical Assembly on Mobile Transporter; installed two jumpers to allow flow of coolant between S1 and S0 trusses; released the large metal rod used as a launch restraint for S1 and stowed it (a drag link); installed Spool Positioning Devices on ammonia lines
230	2002 Nov 26	6:45	López-Alegría/Herrington	STS-113/Quest	Forty-seventh ISS EVA; made connections between P1 and S0 trusses; released launch restraint on CETA; installed Spool Positioning Devices onto ISS; removed drag link on P1; installed wireless video system on External Transceiver Assembly on to Unity node to support EVA helmet camera operations

(continued)

EVA	Start date of EVA	Time (h:min)	EVA crew members	Mission/Spacecraft	Notes
231	2002 Nov 28	6:10	López-Alegría/Herrington	STS-113/Quest	Installed fluid jumpers at S0/P1 attachment point; removed P1's starboard keel pin; installed a second wireless video system External Transceiver Assembly on to the P1; removed port keel pin; relocated CETA cart from P1 to S1 truss allowing Mobile Transporter to move along P1 to assist in future missions
232	2002 Nov 30	7:00	López-Alegría/Herrington	STS-113/Quest	Installed further SPDs (total for the three EVAs over 10); reconfigured external electrical harnesses that route power through Main Bus Switching Units; attached Ammonia Tank Assembly lines
233	2003 Jan 15	6:51	Bowersox/Pettit	EO-6/Quest	Continued the outfitting and activation of P1 truss; released remaining radiator launch locks allowing full deployment; removal of debris on sealing ring of Unity's Earth-facing docking port; tested P6 truss ammonia reserve; unable to complete installation of a light fixture on CETA (reassigned to a future EVA); cut away thermal cover strap that was interfering with the rotation of Quest air lock hatch and delayed start of this EVA
234	2003 Apr 8	6:26	Bowersox/Pettit	EO-6/Quest	Reconfigures ISS power system to provide a secondary power source for one of the CMG's secured thermal control system quick disconnect fittings; released stuck latch on the CETA cart light support system
235	2004 Feb 26	3:55	Foale/Kaleri	EO-8/ Pirs	Replacement of Russian microgravity experiment cassette containers; removal of JAXA micrometer impact experiment due to a malfunction in the cooling system of Kaleri's pressure suit
236	2004 Jun 24	0.40	Padalka/Fincke	EO-9/Pirs	A pressure fault in the primary oxygen supply tank in Fincke's suit resulted in a rescheduling of the planned EVA to June 30

#	Date	Duration	Crew	Station/Vehicle	Description
237	2004 Jun 30	5:40	Padalka/Fincke	EO-9/Pirs	Replacement of a Remote Power Controller (RPC) which had failed during late April resulting in the loss of power in the No. 2 Control Moment Gyroscope; it was on this EVA that primary control of the EVA was transferred from Moscow to Houston
238	2004 Aug 3	4:30	Padalka/Fincke	EO-9/Pirs	In preparation for the arrival of the first ATV vehicle laser retroreflectors were removed from the assembly compartment on Zvezda; they were replaced with updated units and supplemented with an internal video meter target; two additional antennae were installed and Kromka experiment packages were removed and replaced
239	2004 Sep 3	5:21	Padalka/Fincke	EO-9/Pirs	Further preparatory work completed in advance of the arrival of ATV; installed four safety tethers on Zarya's handrails and three communications antennae; removed the covers on those antennae; replaced the Zarya Control Module flow control panel
240	2005 Jan 26	5:28	Chiao/Sharipov	EO-10/Pirs	Installation of the Universal Work Platform is completed; European commercial experiment ROKVISS (Robotic Component Verification on the ISS) is installed along with its antenna; relocated a JAXA exposure experiment; installed the Russian Biorisk experiment
241	2005 Mar 28	4:30	Chiao/Sharipov	EO-10/Pirs	Deployment of a small (11lb/5 kg) Russian nanosatellite; additional navigational and communication equipment installed in preparation for ATV dockings
242	2005 Jun 30	6:50	Robinson/Noguchi	STS-114/OV-103	At station a base and cabling for an External Stowage Platform were installed and power to CMG-2 was rerouted; two exposure experiments were retrieved and a faulty GPS antenna was replaced; in support of Shuttle return-to-flight requirements TPS repair techniques were demonstrated in the Shuttle payload bay

(continued)

EVA	Start date of EVA	Time (h:min)	EVA crew members	Mission/Spacecraft	Notes
243	2005 Aug 1	7:14	Robinson/Noguchi	STS-114/OV-103	Removal of the faulty CMG-1 from the Z1 truss to the Discovery payload bay; installed a new CMG-1 on the Z1 truss
244	2005 Aug 3	6:01	Robinson/Noguchi	STS-114/OV-103	Installed PCSat2 ham radio satellite; photo-documentation and visual inspection of orbiter heat shield and the manual extraction of two protruding gap fillers from between tiles located on the forward underside of Discovery
245	2005 Aug 18	4:58	Krikalev/Phillips	EO-11/Pirs	Retrieval of one of three Biorisk experiment canisters; from Zvezda several experiments were removed and a docking TV camera for ATV approach was installed
246	2005 Nov 7	5:22	McArthur/Tokarev	EO-12/Quest	First Quest-based EVA since April 2003; camera installed and set up on P1 truss; failed Rotary Joint Motor Controller retrieved; removal and replacement of a remote power controller module on the mobile transporter; jettisoned a Floating Potential Probe
247	2006 Feb 3	5:43	McArthur/Tokarev	EO-12/Pirs	Release of SuitSat-1; retrieval of the Biorisk experiment; photography of a sensor as part of a micrometeoroid experiment; an adapter for a small crane was relocated and on the MT the remaining umbilical was tied off
248	2006 Jun 1	6:31	Vinogradov/Williams J.	EO-13/Pirs	Repair of an Elektron unit vent; Biorisk experiment retrieval; contamination-monitoring device was removed from Zvezda; a malfunction camera was replaced on the MBS
249	2006 Jul 8	7:31	Sellers/Fossum	STS-121/Quest	Test of RMS/OBSS combination as a platform for astronauts to inspect and/or repair a damaged orbiter; in the zenith Interface Umbilical Assembly (IUA) a blade blocker was installed which would protect the undamaged power, data, and video cable (the cable was then rerouted in preparation for the second EVA)

250	2006 Jul 10	6:47	Sellers/Fossum	STS-121/Quest	The full operation of the MT railcar completed; spare pump for station cooling system delivered; at two points during the EVA the SAFER pack on Sellers' suit came loose which required Fossum to secure the packs with safety tethers
251	2006 Jul 12	7:11	Sellers/Fossum	STS-121/Quest	This EVA was focused on further evaluation of tile repair systems; a 20 s infrared image of selected RCC panels on the leading edge of the wing of Discovery was taken; they then tested a tile repair material called NOAX on pre-damaged tiles located inside a test container
252	2006 Aug 3	5:54	Williams J./Reiter	EO-13/Quest	Installed a Floating Potential Measurement Unit on MISSE; a thermal radiator rotary joint controller on S1 truss; and a starboard jumper and spool-positing device on S1; replaced a malfunctioning GPS antenna; replaced a light on the MET handcart; tested an infrared camera for future damage detection in Shuttle RCC tiles; took close-up photography and inspected scratches on the Quest air lock; German EVA
253	2006 Sep 12	6:26	Tanner/Stefanyshyn-Piper	STS-115/Quest	Completed installation of P3/P4 truss onto ISS; configured Solar Alpha Rotary Joint (SARJ)
254	2006 Sep 13	7:11	Burbank/MacLean	STS-115/Quest	Continued installation of P3/P4 truss and the activation of SARJ
255	2006 Sep 9	6:42	Tanner/Stefanyshyn-Piper	STS-115/Quest	Installed radiator onto P3/P4; replaced an S-band radio antenna; installed insulation on another antenna; used IR camera to image Shuttle wing in a test to detect damage
256	2006 Nov 23	5:38	López-Alegría/Tyurin	EO-14/Pirs	Completed "Orbital Golf Shot" sponsored and filmed by Canadian golf company; inspected and photographed a Kurs antenna on Progress; relocation of ATV antenna; BTN neutron experiment installed

(continued)

EVA	Start date of EVA	Time (h:min)	EVA crew members	Mission/Spacecraft	Notes
257	2006 Dec 12	6:36	Curbeam/Fuglesang	STS-116/Quest	Installation of P5 truss; replacement of broken video camera on S1 truss; preparation for relocation of P6 truss to its permanent location; first Swedish EVA
258	2006 Dec 14	5:00	Curbeam/Fuglesang	STS-116/Quest	Reconfiguration of station's electrical wiring; putting P3/P4 channel 2–3 into service; bringing new solar arrays online; relocation of two handcarts; installed thermal cover on station RMS; installed tool bags for future EVA use; second Swedish EVA
259	2006 Dec 16	7:31	Curbeam/Williams S.	STS-116/Quest	Completion of station electrical rewiring; activation of circuits 1 and 4; installation of RMS grapple fixture; placed three bundles of Zvezda debris panels for future installation; assisted in deploying P6 solar array
260	2006 Dec 18	6:38	Curbeam/Fuglesang	STS-116/Quest	Assisted in retracting P6 solar array in preparation for move during STS-120; Curbeam set new EVA record for four EVAs on one Shuttle mission; third Swedish EVA
261	2007 Jan 31	7:55	López-Alegría/Williams S.	SEO-14/Quest	Reconfiguration of one of two coolant loops on Destiny from a temporary to permanent system; connection of a Station-to-Shuttle Power Transfer System (SSPTS) cable; secured starboard radiator of the P6 truss and installed shroud over it; removal of Early Ammonia Servicer on the P6 truss, a get-ahead task for later jettisoning of EAS system
262	2007 Feb 4	7:11	López-Alegría/Williams S.	EO-14/Quest	Second of two cooling loops on Destiny reconfigured to permanent system; completed the work on the EAS at the P6 truss; photo-documented inboard end of P6 starboard solar wing in preparation for retraction during STS-117; continued SSPTS preparation work; removal of sunshade from a data relay device

263	2007 Feb 8	6:40	López-Alegría/Williams S.	EO-14/Quest	Removal of shrouds on P3 truss's Rotary Joint Motor Controllers and Bays 18 and 20; discarded the shrouds from vicinity of station; deployment of Unpressurized Cargo Carrier Assembly Attachment System on upper face of P3; removal of launch locks from P5 truss; connection of four cables of SSPTS to PMA-2 at the forward end of Destiny
264	2007 Feb 22	6:18	López-Alegría/Turin	EO-14/Pirs	Retracted antenna of Progress cargo carrier at aft port of Zvezda; replacement Russian material experiments; photographed Russian SatNav antenna; photographed antenna for the ATV as well as a German robotics experiment
265	2007 May 30	5:25	Yurchikhin/Kotov	EO-15/Pirs	Installation of Zvezda Service Module Debris Protection panels; rerouted GPS antenna cable
266	2007 Jun 6	5:37	Yurchikhin/Kotov	EO-15/Pirs	Ethernet cable installed on Zarya; additional Zvezda debris panels installed; Russian scientific experiment deployed
267	2007 Jun 11	6:15	Reilly/Olivas	STS-117/Quest	Completed installation of S3/S4 truss
268	2007 Jun 13	7:16	Forrester/Swanson	STS-117/Quest	Assisted in retraction of P6 truss; partial failure discovered due to reversed wiring on S3/S4 SARJ
269	2007 Jun 15	7:58	Reilly/Olivas	STS-117/Quest	Repair of Shuttle OMS pod thermal blanket; completion of P6 solar array retraction; installation of hydrogen ventilation valve onto Destiny
270	2007 Jun 17	6:29	Forrester/Swanson	STS-117/Quest	Relocated TV camera from External Stowage Platform on Quest airlock to S3 truss; removal of final SARJ launch restraints; installed computer cable on Unity; opened hydrogen vent valve on Destiny laboratory; tethered two debris panels on Service Module

(continued)

EVA	Start date of EVA	Time (h:min)	EVA crew members	Mission/Spacecraft	Notes
271	2007 Jul 23	7:41	Yurchikhin/Anderson C.	EO-15/Quest	Replacement of components on MT redundant power system; cleaned CBM on nadir port of Unity node; jettisoned ammonia tank and flight support equipment
272	2007 Aug 11	6:17	Mastracchio/Williams	STS-118/Quest	Attached S5 segment of truss; retracted forward radiator on P6 in preparation for moving P6 to the end of the port truss, its final position; third Canadian EVA
273	2007 Aug 13	6:28	Mastracchio/Williams	STS-118/Quest	Installed new CMG onto Z1 truss; stowed older, failed CMG on an External Stowage Platform (ESP-2) for subsequent return to Earth on later Shuttle mission; fourth Canadian EVA
274	2007 Aug 15	5:28	Mastracchio/Anderson C.	STS-118/Quest	Relocation of two CETA carts from left side of MT to right side; relocated antenna base from P6 to P1; installed upgraded communication equipment; slight damage to second layer in Mastracchio did not pose serious damage but he returned to air lock early as a precaution
275	2007 Aug 18	5:02	Williams D./Anderson C.	STS-118/Quest	Retrieved MISSE containers 3 and 4; installed Orbiter Boom Sensor System (OBSS) Boom Stand; installed External Wireless Instrumentation System antenna; secured gimbal locks on Z1; fifth Canadian EVA
276	2007 Oct 26	6:14	Parazynski/Wheelock	STS-120/Quest	Installed Harmony Module in temporary location; retrieved S-band support assembly; preparation for relocation of P6 truss; disconnected P6/Z1 truss segments' fluid lines
277	2007 Oct 28	6:33	Parazynski/Tani	STS-120/Quest	Disconnected the Z1 to P6 umbilical; detached P6 from Z1; configured S1 radiator; inspected S4 SARJ; installed handrails on Harmony Module

278	2007 Oct 30	7:08	Parazynski/Wheelock	STS-120/Quest	Attached P6 to P5; reconfigured S1 following redeployment; inspected port SARJ (found to be in good condition)
279	2007 Nov 3	7:19	Parazynski/Wheelock	STS-120/Quest	Inspected and repaired P6 solar array
280	2007 Nov 9	6:55	Whitson/Malenchenko	EO-16/Quest	Disconnected and stowed SSPTS cables; PMA-2 umbilical stored; temporary stowage of Harmony Node avionics umbilical
281	2007 Nov 20	7:16	Whitson/Tani	EO-16/Quest	Partial completion of external configuration of PMA-2 and Harmony Module with attachment of umbilicals
282	2007 Nov 24	7:04	Whitson/Tani	EO-16/Quest	Completion of umbilical hookup for PMA-2 and Destiny Lab; inspection and photography of starboard SARJ
283	2007 Dec 18	6:56	Whitson/Tani	EO-16/Quest	Inspection of S4 SARJ and Beta Gimbal Assembly (BGA); this EVA marked 100 space walks in support of ISS assembly and maintenance
284	2008 Jan 30	7:10	Whitson/Tani	EO-16/Quest	Replacement of one of the Bearing Motor Roll Ring Modules on a solar array; continued inspection and photography of SARJ
285	2008 Feb 11	7:58	Walheim/Love	STS-122/Quest	Whilst still in the payload bay of the Shuttle a grapple fixture was installed onto the Columbus module and preparatory work on the modules data and electrical connections was carried out; the astronauts also replaced the P1 truss nitrogen (N2) tank
286	2008 Feb 13	6:45	Walheim/Schlegel	STS-122/Quest	Installation of the P1 nitrogen tank assembly and stowage of the older tank assembly in the orbiter payload bay; the astronauts also completed further work on the SSPTS; German EVA
287	2008 Feb 15	7:25	Walheim/Love	STS-122/Quest	Installation of SOLAR telescope and EuTEF facility on to the External Stowage Platform on Columbus; installation of keel pin covers on Columbus; retrieval of failed CMG which was then stowed in orbiter payload bay

(continued)

EVA	Start date of EVA	Time (h:min)	EVA crew members	Mission/Spacecraft	Notes
288	2008 Mar 14	7:01	Linnehan/Reisman	STS-123/Quest	Temporary installation of Japanese Experiment Logistics Module–Pressurized Section on to the nadir of Harmony; commenced assembly of Dextre
289	2008 Mar 16	7:08	Linnehan/Foreman	STS-123/Quest	Continued Dextre assembly with attachment of two arms
290	2008 Mar 17	6:53	Linnehan/Reisman	STS-123/Quest	Competed Dextre assembly; installation of spare station equipment onto an ESP-located Quest air lock; unable to attach MISSE-6 experiment onto Columbus module after latching pins failed to engage
291	2008 Mar 20	6:24	Foreman/Reisman	STS-123/Quest	Replacement of RPC; tested Shuttle thermal tile repair materials and techniques; removed port and nadir CBM launch locks from Harmony Module; cover removed from the left arm of Dextre
292	2008 Mar 22	6:02	Foreman/Reisman	STS-123/Quest	Completed stowage of OBSS onto station; installation of ELM-PS trunnion covers; successful installation of MISSE-6 experiment on the outside of Columbus; five covers removed from starboard SARJ; inspection and photography of SARJ completed
293	2008 Jun 3	6:48	Fossum/Garan	STS-124/ Quest	Transfer of OBSS back to Shuttle; preparation for Kibo Experiment Module–Pressurized Module installation; replaced trundle bearing assembly on SARJ (further inspection performed)
294	2008 Jun 5	7:11	Fossum/Garan	STS-124/Quest	Installation of covers and external TV equipment onto Kibo; get-ahead task for relocation of ELM-PS and depleted nitrogen tank carried out; a failed power supply required removal of TV camera
295	2008 Jun 8	6:33	Fossum/Garan	STS-124/Quest	Removal and replacement of starboard nitrogen tank assembly; completed fitting out of Kibo; reinstalled TV camera after repairs

296	2008 Jul 10	6:18	Volkov S./Kononenko	EO-17/Pirs	Removal pyrotechnic bolt from docked Soyuz TMA
297	2008 Jul 15	6:18	Volkov S./Kononenko	EO-17/Pirs	Installation of Zvezda Service Module; installation of Vsplesk experiment; retrieval of Biorisk experiment; realigned ham radio antenna
298	2008 Sep 27	0:22	Zhai/Liu	Shenzhou 7/ Orbital Module	First Chinese EVA (Zhai); collection of experimental package from the exterior of Shenzhou 7; waved Chinese flag at TV camera; Liu supported activities whilst "standing" in the hatch
299	2008 Nov 18	6:52	Stefanyshyn-Piper/Bowen	STS-126/Quest	Transfer of empty nitrogen tank from ESP-3 to the payload bay of Shuttle; astronauts transferred a new flex hose rotary coupler to ESP-3 for future use; Kibo External Facility berthing mechanism insulation cover removed; commenced cleaning and lubricating starboard SARJ; replacement of 11 trundle bearing assemblies in SARJ
300	2008 Nov 20	6:45	Stefanyshyn-Piper/ Kimbrough	STS-126/Quest	Continued cleaning and lubricating starboard SARJ; relocation of two CETA carts from starboard side of MT to port side; lubrication of end effector A snare bearing of station robotic arm assembly
301	2008 Nov 22	6:75	Stefanyshyn-Piper/Bowen	STS-126/Quest	Completed cleaning and lubrication of all but one of TBAs on the starboard SARJ
302	2008 Nov 24	6:07	Bowen/Kimbrough	STS-126/Quest	Final trundle bearing assembly replacement on starboard SARJ; lubrication of port SARJ; installation of video camera; reinstalled insulation covers on the Kibo External Facility berthing mechanism; installed EVA handrails on Kibo and GPS antenna; photography of training umbilical system cables and radiators; maintenance performed on Kibo robotic arm grounding tab

(continued)

EVA	Start date of EVA	Time (h:min)	EVA crew members	Mission/Spacecraft	Notes
303	2008 Dec 23	5:38	Fincke/Lonchakov	EO-18/Pirs	Installation of electromagnetic energy-measuring device (Langmuir probe) on Pirs; removal of Biorisk experiment; installation of EXPOSE-R experiment package on Zvezda; installation of Impulse experiment
304	2009 Mar 10	4:49	Fincke/Lonchakov	EO-18/Pirs	Installed EXPOSE-R onto a universal science platform located on the Zvezda Module; visual inspection and photography of Russian segment; removal of tape straps from the docking target area of Pirs
305	2009 Mar 19	6:07	Swanson/Arnold	STS-119/Quest	Connected S6 truss to S5 truss; connected S5/S6 umbilicals; released launch restrains and removed keel pins; thermal covers removed and stored; S6 radiator deployed
306	2009 Mar 21	6:30	Swanson/Acaba	STS-119/Quest	Preparation of a worksite for STS-127 activities; unpressurized cargo carrier installed on P3 truss; installation of GPS antenna to Kibo; infrared images of radiator panels on the P1 and S1 trusses taken
307	2009 Mar 23	6:27	Arnold/Acaba	STS-119/Quest	Relocation of a crew equipment cart; lubrication of station arm's grapple snares; attempted to deploy a cargo carrier (failed)
308	2009 May 14	7:20	Grunsfeld/Feustel	STS-125/OV-104	Hubble Service Mission 4; replacement of Wide Field and Planetary Camera 2 with Wide Field Camera 3; replacement of Science Instrument Command and Data Handling Unit; lubrication of three shroud doors; installation of Soft Capture Mechanism for possible grapple by future spacecraft for de-orbit at end of life
309	2009 May 15	7:56	Massimino/Good	STS-125/OV-104	All three gyroscope rate-sensing units replaced; removal of first of two battery unit modules

310	2009 May 16	6:36	Grunsfeld/Feustel	STS-125/OV-104	COSTAR removed and replaced with Cosmic Origins Spectrograph; repaired Advanced Camera for Surveys; performed a number of get-ahead tasks
311	2009 May 17	8:02	Massimino/Good	STS-125/OV-104	Repair of telescope imaging spectrograph
312	2009 May 18	7:02	Grunsfeld/Feustel	STS-125/ OV-104	Final Hubble-related EVA (23rd of series) and final Shuttle air lock EVA; replacement of final battery module and installation of Fine Guidance Sensor No. 3; degraded insulation panels from Bays 8, 5, and 7 removed and replaced by three new outer blanket layers; removed old protective cover around low-gain antenna and reinstalled a new protective cover in its place
313	2009 Jun 5	4:54	Padalka/Barratt	EO-20/Pirs	Preparation of Zvezda Service Module transfer compartment for the arrival of Mini-Research Module 2; installation and photography of docking antenna; photography of Strela-2 crane
314	2009 Jun 10	0:12	Padalka/Barratt	EO-20/Zvezda	IVA in depressurized Zvezda transfer compartment; replacement of a Zvezda hatch with a docking cone; preparation to receive Mini-Research Module 2 (Poisk) which would arrive later in year at zenith port and serve as an additional docking facility for Russian spacecraft
315	2009 Jul 18	5:32	Wolf/Kopra	STS-127/Quest	Japanese Exposure Facility installed; P3 nadir UCCAS deployed; S3 zenith outboard PAS deployment postponed due to time constraints during EVA
316	2009 Jul 20	6:53	Wolf/Marshburn	STS-127/Quest	Transferred ORU from ICC in payload bay of Shuttle to ESP-3; some tasks postponed due to lack of time; 40th anniversary of Apollo 11 lunar EVA
317	2009 Jul 22	5:59	Wolf/Cassidy	STS-127/Quest	Preparation work at Japanese pressurized module; P6 battery replacement (two of six units); high levels of CO_2 in Cassidy's suit

(continued)

EVA	Start date of EVA	Time (h:min)	EVA crew members	Mission/Spacecraft	Notes
318	2009 Jul 24	7:12	Marshburn/Cassidy	STS-127/Quest	Replacement of final four of six P6 batteries
319	2009 Jul 27	4:54	Marshburn/Cassidy	STS-127/Quest	SPDM thermal cover adjustment; reconfiguration of Z1 patch panel; visual inspection equipment installation; postponement of S3 nadir PAS to a later mission
320	2009 Sep 1	6:35	Olivas/Stott	STS-128/Quest	Preparation for replacement of the empty ammonia tank; retrieved MISSE and EuTEF facilities from exterior of Columbus (stowed them in Discovery payload bay for return to Earth)
321	2009 Sep 3	6:39	Olivas/Fuglesang	STS-128/Quest	Exchanged old ammonia tank for a new unit; at approximately 820 kg (1,800 lb) this was the largest mass ever moved by astronauts on an EVA
322	2009 Sep 5	7:01	Olivas/Fuglesang	STS-128/Quest	Preparation work in advance of the arrival of the Tranquility Node on STS-130 in February 2010; in addition to attaching cables for the new node a communications device was replaced; installed a pair of GPS transmitters; fitted a new circuit breaker
323	2009 Nov 19	6:37	Foreman/Satcher	STS-129/Quest	Deployed S3 outboard Payload Attach System; a bracket for the ammonia lines intended for Unity was installed; a spare antenna was installed on the truss; lubrication was applied to the Payload Orbital Replacement Unit Attachment Device grapple mechanism on the Mobile Base System and the hand snares of the RMS on the Kibo
324	2009 Nov 21	6:08	Foreman/Satcher	STS-129/Quest	A variety of installations of small devices: the GATOR (Grappling Adaptor to On-Orbit Railing) bracket to Columbus, an additional ham radio antenna, and truss antenna for the EMU wireless helmet camera; relocated the Floating Potential Measurement Unit; deployed a pair of brackets designed to attach cargo to the station's truss

325	2009 Nov 23	5:42	Satcher/Bresnik	STS-129/Quest	Installation of new High Pressure Gas Tank onto the Quest airlock; installation of MISSE 7A and 7B onto ELC-2; a pair of micrometeoroid shields were strapped to ESP-2; relocated a foot restraint; fitted insulation covers on cameras on the MSS and Canadarm2 end effector; a bolt was released on the Ammonia Tank Assembly; worked on the docking adaptor's heater cables
326	2010 Jan 14	5:44	Surayev/Kotov	EO-22/Pirs	Prepared Poisk for future dockings
327	2010 Feb 12	6:32	Behnken/Patrick	STS-130/Quest	Preparatory work for berthing of Tranquility with Unity Node; spares for SPDM were relocated from Shuttle to station; commenced connection from Tranquility to station
328	2010 Feb 14	5:54	Behnken/Patrick	STS-130/Quest	Plumbing and connections for ammonia supply installed between Destiny, Unity, and Tranquility Modules and covered them in insulation; prepared nadir port for the attachment of Cupola; installed handrails on Tranquility
329	2010 Feb 17	5:48	Behnken/Patrick	STS-130/Quest	Further plumbing connection made between Unity and Tranquility; insulation and launch locks removed from Cupola; handrails installed on Tranquility; get-ahead tasks for future work on Zarya carried out; cables on Unity and S0 truss insulated
330	2010 Apr 9	6:27	Mastracchio/Anderson C.	STS-131/Quest	Replaced new ammonia tank from Shuttle to temporary stowage location; old fluid lines disconnected on S1 truss; Japanese seed experiment retrieved from Kibo; replacement of final gyro on S0 truss; get-ahead tasks performed at Harmony zenith CBM and P1 truss
331	2010 Apr 11	7:26	Mastracchio/Anderson C.	STS-131/Quest	Removal of old ammonia tank from S1 truss (replaced with new tank); electrical connection commenced; old tank temporarily stored

(continued)

EVA	Start date of EVA	Time (h:min)	EVA crew members	Mission/Spacecraft	Notes
332	2010 Apr 13	6:24	Mastracchio/Anderson C.	STS-131/Quest	Connections of new ammonia tank completed; old tank moved to Shuttle payload bay; no longer required micrometeoroid shield from Quest removed for return to Earth; preparation work completed for installing a spacer antenna on Z1 truss; other tasks differed due to delays encountered in securing old tank in payload bay of Shuttle
333	2010 May 17	7:25	Reisman/Bowen	STS-132/Quest	Spare Ku-band antenna installed on Z1 truss; new tool platform installed on Dextre; torque on the bolts holding the replacement batteries to the ICC-VLD cargo carrier broken
334	2010 May 19	7:09	Bowen/Good	STS-132/Quest	Repaired the OBSS on Atlantis; four of six units on P6 batter replaced; removed gimbal locks from Ku-band antenna installed on first EVA of mission
335	2010 May 21	6:46	Good/Reisman	STS-132/Quest	Final two of six P6 battery replacement units completed; installed P4/P5 ammonia jumpers; retrieved spare PDGF from Atlantis and stowed inside Quest airlock; replenishment of EVA tools in toolboxes
336	2010 May 27	6:42	Yurchikhin/Kornienko	EO-24/Pirs	Preparatory work on Rassvet module to utilize Kurs automated rendezvous system; routed and mated cables for command and data handling on both Zvezda and Zarya
337	2010 Aug 7	8:03	Wheelock/Caldwell Dyson	EO-24/Quest	Attempted to replace failed ammonia pump module, but stuck quick-disconnect would not release, delaying completion of task; a "bake-out" was required to ensure ammonia was evaporated prior to reentering the station
338	2010 Aug 11	7:26	Wheelock/Caldwell Dyson	EO-24/Quest	Completion of removal of failed pump module from S1 truss; commenced insulation on replacement pump

339	2010 Aug 16	7:20	Wheelock/Caldwell Dyson	EO-24/Quest	Completed installation of new pump module on S1 truss
340	2010 Nov 15	6:27	Yurchikhin/Skripochka	EO-25/Pirs	Installation of portable multipurpose workstation on Zvezda, struts between Zvezda and Zarya modules, handrail on Pirs and SKK #1-M2 cassette on Poisk module; removed Plasma Pulse Injector Science hardware, EXPOSE-R experiment Kontor science hardware, and TV camera from Rassvet; test experiment on contamination underneath insulation of Russian ISS segment carried out
341	2011 Jan 21	5:23	Kondratyev/Skripochka	EO-26/Pirs	Installation of two experiments on Zvezda nadir side; removed two other experiments; installed TV camera on Rassvet module
342	2011 Feb 16	4:51	Kondratyev/Skripochka	EO-26/Pirs	Two experiments installed on Zvezda module; removal of two material exposure experiment panels from outside of Zvezda module; jettisoned a foot restraint
343	2011 Feb 28	6:34	Bowen/Drew	STS-133/Quest	Connected power extension cable between Unity and Tranquility nodes; moved failed ammonia pump module; installed wedge under camera on S3 truss to allow clearance from newly installed ELC-4; replacement of guide for railcar system; vacuum bottle "filled" for Japanese museum education experiment
344	2011 Mar 02	6:14	Bowen/Drew	STS-133/Quest	Removal of thermal insulation from a platform; attachment bucket on Columbus exchanged; camera assembly installed on Dextre where the robotic electronics platform was also removed; insulation for the light on the cargo cart installed
345	2011 May 20	6:19	Feustel/Chamitoff	STS-134/Quest	Retrieved two MISSE-7 experiments; installed new MISSE-8 experiment on ELC-2; vented nitrogen from ammonia servicer; commenced installation of external wireless communications antenna on Destiny; faulty CO_2 senor in Chamitoff's suit delayed other tasks to later EVAs

(continued)

EVA	Start date of EVA	Time (h:min)	EVA crew members	Mission/Spacecraft	Notes
346	2011 May 22	8:07	Feustel/Fincke	STS-134/Quest	Refilled P6 (P5) radiators with ammonia; completed venting the Early Detection Ammonia System; lubrication of port SARJ as well as parts of Dextre; grapple bars on port radiators installed
347	2011 May 25	6:54	Feustel/Fincke	STS-134/Quest	Grapple fixture installed on Zarya allowing SSRMS to translate to Russian segment for robotic operations; backup power cables to Russian segment installed; completed installation of wireless video system begun during EVA 1
348	2011 May 27	7:24	Fincke/Chamitoff	STS-134/Quest	Stowed OBSS on right side of station truss; retrieved grapple from left side of truss top; replaced the unit on the boom; restraints on one of the Dextre arms were released; replaced the thermal insulation of spare gas tank for the Quest Air Lock; ISS Assembly Complete point reached; final Shuttle crew EVAs
349	2011 Jul 12	6:31	Fossum/Garan	EO-28/Quest	Retrieval of failed ammonia pump module from ESP-2 (secured in orbiter payload bay for return to Earth); removed Robotic Refueling Mission (RRM) payload located in payload bay of Shuttle to a platform on ISS; retrieved material science experiment; released a stuck wire on a power grapple fixture; installed thermal covers on PMA-3; reconfigured some external equipment; final EVA while Shuttle docked with ISS
350	2011 Aug 6	6:23	Volkov S./Samokutyayev	EO-28/Pirs	Deployed ARISSat-1 ham radio satellite; installed packages on the outside of Russian segment; removed unwanted antennas from Poisk; photo-documented exterior of Russian segment; relocation of Strela crane deferred to subsequent EVA
351	2012 Feb 16	6:15	Kononenko/Shkaplerov	EO-30/Pirs	Relocated Strela-1 crane from Pirs to Poisk; installed Vinoslivost Materials Sample Experiment; collected further samples from beneath insulation on Zvezda to search for any living organisms

352	2012 Aug 20	5:51	Padalka/Malenchenko	EO-32/Pirs	Relocated Strela-2 crane from Pirs to Zarya; deployed Sfera-53 passive air density calibration satellite; retrieved various external experiments; installed five micrometeoroid debris shields on exterior of Zvezda
353	2012 Aug 30	8:17	Williams, S./Hoshide	EO-32/Quest	Connected two power cables between U.S. and Russian segments; removed and replaced MBSU 1; difficulty in securing new unit delayed other tasks; third longest EVA in history; Japanese EVA
354	2012 Sep 5	6:28	Williams, S./Hoshide	EO-32/Quest	Completed installation of new MBSU unit; replaced one of cameras on Canadarm2; Sunita Williams surpassed Whitson's EVA record for total spacewalking time for a female
355	2012 Nov 1	6:38	Williams, S./Hoshide	EO-33/Quest	Reconfigured EAQS; demated PVR 2B FQDC; set up TTCR; conducted photo documentary of exterior surfaces and performed SARJ inspections

Appendix D

EVA durations 1965–2012

Name	Country	Total EVAs	Total duration (h:min)
Anatoly Y. Solovyov	U.S.S.R./Russia	16	79:51
Michael E. López-Alegría	U.S.A.	10	67:40
Sergei V. Avdeyev	Russia	13	59:52
John M. Grunsfeld	U.S.A.	8	58:43
Jerry L. Ross	U.S.A.	9	58:18
Joseph R. Tanner	U.S.A.	7	56:09
Sunita L. Williams	U.S.A.	7	50:40
Viktor M. Afanasyev	U.S.S.R./Russia	9	50:05
Stephen L. Smith	U.S.A.	7	49:49
Michael E. Fossum	U.S.A.	7	48:32
Edward M. Fincke	U.S.A.	9	47:46
Stephen G Bowen	U.S.A.	7	47:18
Robert L. Curbeam, Jr.	U.S.A.	7	45:34
Nikolai V. Budarin	Russia	9	44:54
Douglas H. Wheelock	U.S.A.	6	43:30
Yuri I. Onufriyenko	Russia	8	42:43
Richard M. Linnehan	U.S.A.	6	42:43
Andrew J. Feustel	U.S.A.	6	42:18

(*continued*)

Name	Country	Total EVAs	Total duration (h:min)
Scott E. Parazynski	U.S.A.	6	42:04
David A. Wolf	U.S.A.	7	41:57
Talgat A. Musabayev	Russia	8	41:29
Sergei K. Krikalev	U.S.S.R./Russia	8	41:18
Piers J. Sellers	U.S.A.	6	41:10
Garrett E. Reisman	U.S.A.	6	40:31
Peggy A. Whitson	U.S.A.	6	39:49
Daniel M. Tani	U.S.A.	6	39:11
Richard A. Mastracchio	U.S.A.	6	38:30
Clayton C. Anderson	U.S.A.	6	38:28
Vladimir N. Dezhurov	Russia	9	37:23
Rex J. Walheim	U.S.A.	5	36:23
Leroy Chiao	U.S.A.	6	36:17
James H. Newman	U.S.A.	5	35:56
Musa K. Manarov	U.S.S.R./Russia	7	34:34
John D. Olivas	U.S.A.	5	34:28
Heidemarie Stefanyshyn-Piper	U.S.A.	5	34:02
Gennady I. Padalka	Russia	9	33:06
Pavel V. Vinogradov	Russia	8	32:50
Michael J. Foreman	U.S.A.	5	32:19
Anatoly P. Artsebarsky	U.S.S.R.	6	32:09
Arne C. Fuglesang	Sweden	5	31:54
Aleksandr A. Serebrov	U.S.S.R./Russia	10	31:52
Fyodor N. Yurchikhin	Russia	5	31:52
Yuri V. Usachev	Russia	7	30:50
James F. Reilly, II	U.S.A.	5	30:43

Name	Country	Total EVAs	Total duration (h:min)
Yuri I. Malenchenko	Russia	5	30:07
Thomas D. Akers	U.S.A.	4	29:40
Michael J. Massimino	U.S.A.	4	29:32
Michael T. Good	U.S.A.	4	28:53
Leonid D. Kizim	U.S.S.R.	7	28:51
Vladimir A. Solovyov	U.S.S.R.	7	28:51
Ronald J. Garan	U.S.A.	4	27:03
Steven R. Swanson	U.S.A.	4	26:22
F. Story Musgrave	U.S.A.	4	26:19
Mark C. Lee	U.S.A.	4	26:01
Mikhail V. Tyurin	Russia	5	25:31
Patrick G. Forrester	U.S.A.	4	25.30
Jeffrey A. Hoffman	U.S.A.	4	25:02
William S. McArthur, Jr.	U.S.A.	4	24:21
Eugene A. Cernan	U.S.A.	4	24:13
Daniel T. Barry	U.S.A.	4	23:49
Aleksandr Y. Kaleri	Russia	5	23:24
Michael L. Gernhardt	U.S.A.	4	23:16
Harrison H. Schmitt	U.S.A.	4	23:10
Tracy Caldwell Dyson	U.S.A.	3	22:49
James S. Voss	U.S.A.	4	22:45
C. Michael Foale	U.S.A.	4	22:45
Gennady M. Strekalov	U.S.S.R./Russia	6	22:31
Valeri G. Korzun	Russia	4	22:19
Charles M. Duke, Jr.	U.S.A.	4	21:38

(*continued*)

Name	Country	Total EVAs	Total duration (h:min)
Akihiko Hoshide	Japan	3	21:20
Kathryn C. Thornton	U.S.A.	3	21:11
James D. A. Van Hoften	U.S.A.	4	20:45
John W. Young	U.S.A.	3	20:14
David R. Scott	U.S.A.	5	20:14
Stephen K. Robinson	U.S.A.	3	20:05
Soichi Noguchi	Japan	3	20:05
John B. Herrington	U.S.A.	3	19:55
Peter J. K. Wisoff	U.S.A.	3	19:53
Thomas D. Jones	U.S.A.	3	19:49
Aleksandr S. Viktorenko	U.S.S.R./Russia	6	19:42
Winston E. Scott	U.S.A.	3	19:36
Franklin R. L. A Chang-Diaz	U.S.A.	3	19:31
Philippe Perrin	France	3	19:31
Carlos I. Noriega	U.S.A.	3	19:20
James B. Irwin	U.S.A.	4	19:14
Vasily V. V. Tsibliyev	Russia	6	19:10
Jeffrey N. Williams	U.S.A.	3	19:09
Thomas H. Marshburn	U.S.A.	3	18:59
Sergei A. Volkov	Russia	3	18:59
Carl E. Walz	U.S.A.	3	18:55
Oleg D. Kononenko	Russia	3	18.51
Vladimir G. Titov	U.S.S.R./Russia	4	18:47
Gregory J. Harbaugh	U.S.A.	3	18:29
Robert L. Satcher	U.S.A.	3	18:27
Robert L. Behnken	U.S.A.	3	18:14

Name	Country	Total EVAs	Total duration (h:min)
Christopher J. Cassidy	U.S.A.	3	18:05
Daffyd R. Williams	Canada	3	17:47
Richard J. Hieb	U.S.A.	3	17:42
Pierre J. Thuot	U.S.A.	3	17:42
Nicholas J. M. Patrick	U.S.A.	3	17:14
Oleg V. Kotov	Russia	3	16:46
Oleg I. Skripochka	Russia	3	16:41
Gerald P. Carr	U.S.A.	3	15:51
Stanley G. Love	U.S.A.	2	15:23
Edward G. Gibson	U.S.A.	3	15:20
Chris A. Hadfield	Canada	2	14:50
Thomas Reiter	Germany	3	14:16
Lee M. E. Morin	U.S.A.	2	14:07
Gennady M. Manakov	U.S.S.R./Russia	3	13:46
Owen K. Garriott	U.S.A.	3	13:44
Gregory E. Chamitoff	U.S.A.	2	13:43
William R. Pogue	U.S.A.	2	13:31
Joseph M. Acaba	U.S.A.	2	12:57
Robert S. Kimbrough	U.S.A.	2	12:52
Alvin B. Drew	U.S.A.	2	12:45
Takao Doi	Japan	2	12:42
Richard R. Arnold, II	U.S.A.	2	12:34
Bruce McCandless, II	U.S.A.	2	12:12
Robert L. Stewart	U.S.A.	2	12:12
Sherwood C. Spring	U.S.A.	2	12:00

(continued)

Name	Country	Total EVAs	Total duration (h:min)
Daniel W. Bursch	U.S.A.	2	11:46
Joseph P. Allen, IV	U.S.A.	2	11:42
Dale A. Gardner	U.S.A.	2	11:42
William F. Fisher	U.S.A.	2	11:34
Charles Conrad, Jr.	U.S.A.	4	11:33
Valeri I. Tokarev	Russia	2	11:05
Jack R. Lousma	U.S.A.	2	11:01
Jerome Apt	U.S.A.	2	10:49
Aleksandr N. Baladin	U.S.S.R.	2	10:47
Alan L. Bean	U.S.A.	3	10:30
Yuri V. Lonchakov	Russia	2	10:27
Yuri V. Romanenko	U.S.S.R.	4	10:16
Linda M. Godwin	U.S.A.	2	10:14
Dmitri Y. Kondratyev	Russia	2	10:14
Aleksandr A. Volkov	U.S.S.R.	2	10:09
Aleksandr F. Poleshchuk	Russia	2	9:58
Salizhan S. Sharipov	Russia	2	9:58
Kenneth D. Bowersox	U.S.A.	2	9:46
Don R. Pettit	U.S.A.	2	9:46
Alan B. Shepard, Jr.	U.S.A.	2	9:22
Edgar D. Mitchell	U.S.A.	2	9:22
George D. Nelson	U.S.A.	2	9:13
Susan J. Helms	U.S.A.	1	8:56
Aleksandr I. Laveikin	U.S.S.R.	3	8:48
Vladimir A. Dzhanibekov	U.S.S.R.	2	8:35
Claude Nicollier	Switzerland	1	8:10

Name	Country	Total EVAs	Total duration (h:min)
Buzz Aldrin	U.S.A.	4	8:09
Daniel Burbank	U.S.A.	1	7:11
Steven MacLean	Canada	1	7:11
Vladimir A. Lyakhov	U.S.S.R.	3	7:08
Carl J. Meade	U.S.A.	1	6:51
Hans W. Schlegel	German	1	6:45
Mikhail B. Kornienko	Russia	1	6:42
Nicole P. Stott	U.S.A.	1	6:35
Aleksandr M. Samokutyayev	Russian	1	6:23
Andrew S. W. Thomas	U.S.A.	1	6:21
Paul W. Richards	U.S.A.	1	6:21
Jean-Pierre Haigneré	France	1	6:19
Anton N. Shkaplerov	Russia	1	6:15
Edward Tsang Lu	U.S.A.	1	6:14
Michael R. U. Clifford	U.S.A.	1	6:02
Jean-Loup J. M. Chrétien	France	1	5:57
Tamara E. Jernigan	U.S.A.	1	5:55
G. David Low	U.S.A.	1	5:50
Aleksandr P. Alexandrov	U.S.S.R.	2	5:45
Maxim V. Surayev	Russia	1	5:44
Randolf J. Bresnik	U.S.A.	1	5:42
Timothy L. Kopra	U.S.A.	1	5:32
Sergei V. Treshev	Russia	1	5:21
Michael R. Barratt	U.S.A.	2	5:06
Frank L. Culbertson, Jr.	U.S.A.	1	5:05

(continued)

Name	Country	Total EVAs	Total duration (h:min)
Svetlana Y. Savitskaya	U.S.S.R.	1	5:00
Viktor P. Savinykh	U.S.S.R.	1	5:00
Jerry M. Linenger	U.S.A.	1	4:57
John L. Phillips	U.S.A.	1	4:57
Sergei V. Zaletin	Russia	1	4:52
Bernard A. Harris, Jr.	U.S.A.	1	4:37
Donald H. Peterson	U.S.A.	1	4:17
Yuri P. Gidzenko	Russia	2	3:35
Kathryn D. Sullivan	U.S.A.	1	3:27
David C. Leestma	U.S.A.	1	3:27
Joseph P. Kerwin	U.S.A.	1	3:25
S. David Griggs	U.S.A.	1	3:00
Richard F. Gordon, Jr.	U.S.A.	2	2:41
Anatoly N. Berezovoi	U.S.S.R.	1	2:33
Valentin V. Lebedev	U.S.S.R.	1	2:33
Neil A. Armstrong	U.S.A.	1	2:31
Paul J. Weitz	U.S.A.	2	2:21
Vladimir V. Kovalyonok	U.S.S.R.	1	2:05
Aleksandr S. Ivanchenkov	U.S.S.R.	1	2:05
Michael Collins	U.S.A.	2	1:29
Georgi M. Grechko	U.S.S.R.	1	1:28
Thomas K. Mattingly, II	U.S.A.	1	1:24
Valeri V. Ryumin	U.S.S.R.	1	1:23
Russell L. Schweickart	U.S.A.	1	1:07
Ronald E. Evans, Jr.	U.S.A.	1	1:06
Alfred M. Worden, Jr.	U.S.A.	1	0:39

Name	Country	Total EVAs	Total duration (h:min)
Yevgeny V. Khrunov	U.S.S.R.	1	0:37
Alexei S. Yeliseyev	U.S.S.R.	1	0:37
Zhai Zhigang	China	1	0:22
Liu Boming	China	1	0:22
Edward H. White, II	U.S.A.	1	0:21
Alexei A. Leonov	U.S.S.R.	1	0:12

Appendix E

A selected timeline 1961–2012

1961 Apr May Aug	Yuri Gagarin becomes the first person to fly in space and completes one orbit Alan Shepard becomes the first American in space on a suborbital flight Gherman Titov is launched on the first 24 h mission (of 17 orbits)	
1962 Feb Jul Aug	John Glenn becomes the first American to orbit the Earth (with 3 orbits) First X-15 flight to exceed 50 miles (Robert White) Andrian Nikolayev sets new endurance record (3 da 22 h)	
1963 Jun Aug	Valeri Bykovsky sets new endurance record (4 da 23 h) Valentina Tereshkova becomes first woman in space (2 da 22 h) Highest X-15 flight (66.75 miles) (Joseph Walker)	
1964 Oct	First multiperson space crew (three cosmonauts on Voskhod 1 First civilians in space	
1965 Mar Mar Jun Aug Dec	Alexei Leonov becomes first person to walk in space First U.S. multiperson crew (two astronauts on Gemini 3) Ed White becomes first American to walk in space Gemini 5 sets new endurance record (7 da 22 h) Cooper becomes first person to orbit Earth a second time Gemini 7 sets new endurance record (13 da 18 h) First space rendezvous (Gemini 6 with Gemini 7)	
1966 Mar Sep	First space docking (Gemini 8 with Agena unmanned target vehicle) Gemini 11 attains highest altitude of Earth orbital manned flight (850 miles)	
1967 Jan Apr	Three Apollo 1 astronauts killed in pad fire Soyuz 1 pilot Vladimir Komarov killed during landing phase	

(continued)

383

1967 Oct Nov	(*cont.*) X-15 fastest flight (4,520 mph or Mach 6.7 by Pete Knight) X-15 pilot Michael Adams is killed in crash of #3 aircraft after reaching 50.4 miles
1968 Aug Oct Dec	Thirteenth and final X-15 astro-flight First three-man Apollo flight (Apollo 7) Schirra becomes first person to make three orbital space flights Apollo 8 becomes first lunar orbital mission
1969 Jan Mar May Jul Oct Nov	Soyuz 5/4 first manned docking and crew transfer (by EVA) Manned test of LM in Earth orbit (Apollo 9) Manned test of LM in lunar orbit (Apollo 10) First manned lunar landing (Apollo 11) First three-spacecraft (Soyuz 6, 7, 8) operations Second manned lunar landing (Apollo 12)
1970 Apr Jun	Apollo 13 aborted lunar landing mission Lovell becomes first to fly in space four times Soyuz 9 cosmonauts set new endurance record (17 da 16 h)
1971 Feb Apr Jun Jul	Third manned lunar landing (Apollo 14) Launch of world's first space station (Salyut, which de-orbits October 1971) First space station (Salyut) crew Killed during entry phase (Soyuz 11) Fourth manned lunar landing (Apollo 15)
1972 Apr Dec	Fifth manned lunar landing (Apollo 16) Sixth and final (Apollo) manned lunar landing (Apollo 17)
1973 Apr May Jul Nov	Salyut 2 (Almaz) fails in orbit (de-orbits in 26 days) Launch of unmanned Skylab (re-enters Jul 1979) First Skylab crew sets new endurance record of 28 days Second Skylab crew increases endurance record (59 days 11 h) Third and final Skylab crew increases endurance record (84 da 1 h)
1974 Jun Jul Dec	Launch of Salyut 3 (Almaz), which de-orbits January 1975 First successful Soviet space station mission (Soyuz 14) Launch of Salyut 4 (de-orbits February 1977)
1975 Apr Jul	Soyuz 18-1 crew survive launch abort Soyuz 19 and Apollo dock in space (first international mission)

1977 Sep Dec	Salyut 6 launched (de-orbits July 1982) First Salyut 6 resident crew set new endurance record (96 da 10 h)
1978 Jan Mar Jun	First Soyuz exchange mission (Soyuz 27 for Soyuz 26) First Soviet Interkosmos mission (Czechoslovakian) First non-Soviet, non-American person in space (Remek) Second Salyut 6 crew sets new endurance record (139 da 14 h)
1979 Feb	Third Salyut 6 resident crew increases endurance record (175 da)
1980 Apr Jun	Fourth Salyut 6 resident crew increases endurance record (187 da 20 h) First manned flight of Soyuz T variant
1981 Apr Nov	First Shuttle launch (Columbia STS-1) on 20th anniversary of Gagarin's flight John Young becomes first to make five space flights. First return to space by manned spacecraft (Columbia STS-2)
1982 Apr May Nov	Salyut 7 launched (de-orbits February 1991) First Salyut 7 resident crew sets new endurance record (211 da 9 h) First "operational" Shuttle mission (STS-5) is also the first four-person launch
1983 Apr Jun Sep Nov	First flight of Challenger Sally Ride becomes first U.S. woman in space during STS-7 First five-person launch Soyuz T10-1 launchpad abort First Spacelab mission (STS-9) First six-person launch John Young flies record sixth mission
1984 Feb Feb Jul Aug Oct	First use of MMU (STS-41-B) on untethered space walks Third Salyut 7 resident crew sets new endurance record (236 da 22 h) Svetlana Savitskaya becomes the first woman to walk in space (Soyuz T12/Salyut 7) First flight of Discovery (STS-41-D) First seven-person launch (STS-41-G) Kathy Sullivan becomes first American woman to walk in space
1985 Jan Jul Oct Oct	First classified DOD Shuttle mission (STS-51C) First Shuttle abort-to-orbit profile (STS-51F) First flight of Atlantis (STS-51J) First eight-person launch (STS-61A)

(continued)

1986 Jan Feb Mar	Challenger and its crew of seven lost 73 seconds after launch (STS-51L) Mir core module launched unmanned First resident crew to Mir (Soyuz T-15)
1987 Feb Dec	Second Mir resident crew sets new endurance record (326 da 11 h) First manned Soyuz TM variant First flight of over a year as third Mir resident crew sets endurance record (365 da 22 h)
1988 Sep	Shuttle return-to-flight mission (STS-26)
1990 Apr	Hubble Space Telescope deployment (STS-31)
1992 May	First flight of Endeavour (STS-49)
1993 Dec	First Hubble Service Mission (STS-61)
1994 Jan Feb	Valeri Polyakov sets new endurance record (437 da 17 h) for one mission (lands March 1995) First Russian cosmonaut to fly on Shuttle (Krikalev during STS-60)
1995 Feb Mar Jul Nov	First Shuttle–Mir rendezvous (STS-63/Mir) Eileen Collins becomes first female Shuttle pilot First American launched on Soyuz (Thagard on Soyuz TM-21) First Shuttle docking with Mir (STS-71, Thagard down) Second Shuttle–Mir docking (STS-74)
1996 Mar Sep Nov	Third Shuttle–Mir docking (STS-76, Lucid up) Fourth Shuttle–Mir docking (STS-79, Lucid down, Blaha up) Longest Shuttle mission (17 da 15 h, STS-80) Musgrave becomes only astronaut to fly all five orbiters
1997 Jan Feb May Jun Sep	Fifth Shuttle–Mir docking (STS-81, Blaha down, Linenger up) Second Hubble Service Mission (STS-82) Sixth Shuttle–Mir docking (STS-81, Linenger down, Foale up) Collision between unmanned Progress vessel and Mir space station damages Spektr module Seventh Shuttle–Mir docking (STS-86, Foale down, Wolf up)

1998	
Jan	Eighth Shuttle–Mir docking (STS-89, Wolf down, Thomas up)
Jun	Ninth and final Shuttle–Mir docking (STS-91, Thomas down)
Oct	John Glenn returns to space aged 77 (36 years after his first space flight)
Nov	First ISS element launched (Zarya FGB)
Dec	First ISS Shuttle mission (STS-88)
1999	
Jul	Eileen Collins becomes first female U.S. mission commander (STS-93)
Aug	Mir vacated for first time in 10 years
Dec	Third Hubble Service Mission (STS-103)
2000	
Apr	Last (28th) Mir resident crew (72 da)
Oct	First ISS resident crew launched
2001	
Mar	Mir space station de-orbits after 15 years' service
Apr	Dennis Tito becomes first space flight participant or "tourist"
2002	
Mar	Fourth Hubble Service Mission (STS-109)
Apr	Jerry Ross becomes first person to fly seven missions in space
Oct	First manned flight of Soyuz TMA
2003	
Feb	Columbia and crew of seven lost during entry phase of mission STS-107
Apr	ISS assumes two-person caretaker crews
Oct	First Chinese manned space flight (Shenzhou 5)
	Yang Liwei becomes first Chinese national in space
2004	
Sep	SpaceShipOne flies to 337,500 ft (102.87 km)
Oct	SpaceShipOne flies to 367,442 ft (111.99 km) and wins $10 million X-Prize
2005	
Jul	First post-Columbia (STS-107) Shuttle return-to-flight mission (STS-114)
Oct	First Chinese two-man space flight (Shenzhou 6)
2006	
Jul	Second post Columbia (STS-107) Shuttle return-to-flight mission (STS-121)
Aug	ISS returns to three-person capability
Sep	Resumption of ISS construction (STS-115)
2007	
Aug	Barbara Morgan becomes the first educator astronaut (teacher) in space (STS-118)
Oct	Peggy Whitson becomes first female space station commander (TMA-11/ISS-16)

(*continued*)

2008	
Apr	Sergei Volkov becomes first son of a cosmonaut (Aleksandr Volkov) to fly in space
Sep	Zhai Zhigang becomes first Chinese to perform EVA (Shenzhou 7)
Oct	Richard Garriott becomes first son of an astronaut (Owen Garriott) to fly in space
2009	
May	Final Hubble Service Mission (SM4)
	Final Shuttle-based EVAs (STS-125)
May	Six-person ISS resident crew operations commence (ISS-20)
Nov	Final ISS crew member to return by Shuttle (Stott on STS-129)
2010	
Oct	Maiden launch of Soyuz TMA-M spacecraft
2011	
Feb	Final flight (39th) of OV-103 Discovery (STS-133)
Apr	Soyuz TMA-21 (call sign Gagarin) celebrates 50 years of human space flight
May	Final (25th) flight of OV-105 Endeavour
	Final EVAs by Shuttle crew members (STS-134)
	ISS assembly complete
Jul	Final Shuttle mission (STS-135)
	Final flight (33rd) of OV-104 Atlantis
2012	
Jun	First Chinese space station crew (Shenzhou 9, which lasted 13 days)
	First Chinese female citizen in space (Liu Yang)

Bibliography

The authors have referred to their own extensive archives in the compilation of this book. In addition, the following publications and resources were of great help in assembling the data.

Space agencies

The press kits, new releases, and mission information from NASA, ESA, CSA, RKK-Energiya, JAXA (NASDA), CNES, and Novosti have been invaluable resources for many years.

Since the 1990s additional information has been gained from various websites around the world; the principal ones for each main space agency are

- American Space Agency (NASA) *http://www.nasa.gov/*
- Russian Space Agency (RSA) *http://www.federalspace.ru/?lang=en*
- European Space Agency (ESA) *http://www.esa.int/esaCP/index.html*
- Canadian Space Agency (CSA) *http://www.asc-csa.gc.ca/eng/default.asp*
- Japanese Space Agency (JAXA) *http://www.jaxa.jp/index_e.html*
- Chinese Manned Spaceflight Engineering *http://en.cmse.gov.cn/*

Other useful websites

- S.P. Korolev Rocket and Space Corporation Energiya *http://www.energia.ru/english/*
- CollectSpace.com *http://www.collectspace.com/*
- NASA Spaceflight.com *http://www.nasaspaceflight.com/*
- NASA Watch *http://nasawatch.com/*
- CBS News *http://www.cbsnews.com/network/news/space/home/index.html*
- Dragon in Space (China) *http://www.dragoninspace.com*

- Xinhua News (China) *http://www.chinaview.cn*
- CCTV (China) *http://english.cctc.cn/01/index/shtml*
- Encyclopedia Astronautica *http://www.astronautix.com/*
- Space Facts (German website written in English) *http://www.spacefacts.de/*

Magazines

- *Flight International* from 1961
- *Aviation Week and Space Technology* from 1961
- BIS *Spaceflight* from 1961
- *Soviet Weekly/Soviet News* 1961–1990
- *Orbiter*, Astro Info Service, 1984–1992
- *Zenit*, Astro Info Service, 1985–1991
- *ESA Bulletin* from 1975

British Interplanetary Society books

- *History of Mir 1986–2000*, edited by Rex Hall 2000
- *Mir: The Final Year Supplement*, edited by Rex Hall 2001
- *The International Space Station: From Imagination to Reality*, Vol. 1, edited by Rex Hall 2002
- *The International Space Station: From Imagination to Reality*, Vol. 2, edited by Rex Hall 2005

NASA reports

- *NASA Astronautics and Aeronautics*, various volumes, 1961–2009
- *Mir Hardware Heritage*, David S. F. Portree, NASA RP-1357, March 1995.
- *Walking to Olympus: An EVA Chronology*, David S. F. Portree and Robert C. Trevino, NASA Monograph in Aerospace history #7, October 1997

NASA histories

1966   *This New Ocean: A History of Project Mercury*, NASA SP-4201

1977   *On the Shoulders of Titans: A History of Project Gemini*, NASA SP-4203

1978   *The Partnership: A history of the Apollo-Soyuz Test Project*, NASA SP-4209

1979   *Chariots for Apollo: A History of Manned Lunar Spacecraft*, NASA SP-4205

1983   *Living and Working in Space: A History of Skylab*, NASA SP-4208

1989   *Where No Man Has Gone Before: A History of Apollo Lunar Exploration Missions*, NASA SP-4214

2000   *Challenge to Apollo: The Soviet Union and the Space Race 1945–1974*, Asif Siddiqi, NASA SP-2000-4408

Other books

1980 *Handbook of Soviet Manned Space Flight*, Nicholas L. Johnson, Vol. 48, AAS Science and Technology Series
1981 *The History of Manned Space Flight*, David Baker
1987 *Heroes in Space: From Gagarin to Challenger*, Peter Bond
1988 *Space Shuttle Log: The First 25 Flights*, Gene Gurney and Jeff Forte
1988 *Soviet Manned Space Program*, Phillip Clark
1989 *The Illustrated Encyclopedia of Space Technology*, Ken Gatland
1990 *Almanac of Soviet Manned Space Flight*, Dennis Newkirk
1992 *At the Edge of Space: The X-15 Flight Program*, Milton O. Thompson
1999 *Who's Who in Space: The ISS Edition*, Michael Cassutt
2001 *Space Shuttle: The History of the National Space Transportation System*, Dennis Jenkins
2007 *Into that Silent Sea*, Francis French, Outward Odyssey Series, Nebraska Press
2007 *In the Shadow of the Moon*, Francis French, Outward Odyssey Series, Nebraska Press
2008 *Homesteading Space*, David Hitt, Outward Odyssey Series, Nebraska Press
2010 *Footprints in the Dust*, Colin Burgess, Outward Odyssey Series, Nebraska Press

Springer-Praxis Space Science Series
(with extensive references and bibliographies for further reading):

1999 *Exploring the Moon: The Apollo Expeditions*, David M. Harland
2000 *Disasters and Accidents in Manned Spaceflight*, David J. Shayler
2000 *The Challenges of Human Space Exploration*, Marsha Freeman
2001 *Russia in Space: The Failed Frontier*, Brian Harvey
2001 *The Rocket Men: Vostok and Voskhod, the First Soviet Manned Spaceflights*, Rex Hall and David J. Shayler
2001 *Skylab: America's Space Station*, David J. Shayler
2001 *Gemini: Steps to the Moon*, David J. Shayler
2001 *Project Mercury: NASA's First Manned Space Programme*, John Catchpole
2002 *The Continuing Story of the International Space Station*, Peter Bond
2002 *Creating the International Space Station*, David M. Harland and John E. Catchpole
2002 *Apollo: Lost and Forgotten Missions*, David J. Shayler
2003 *Soyuz: A Universal Spacecraft*, Rex Hall and David J. Shayler
2004 *China's Space Program: From Concept to Manned Spaceflight*, Brian Harvey
2004 *Walking in Space*, David J. Shayler
2004 *The Story of the Space Shuttle*, David M. Harland
2005 *The Story of Space Station Mir*, David M. Harland
2005 *Women in Space: Following Valentina*, David J. Shayler and Ian Moule

2005   *Space Shuttle Columbia: Her Missions and Crews*, Ben Evans

2005   *Russia's Cosmonauts: Inside the Yuri Gagarin Training Center*, Rex Hall, David J. Shayler, and Bert Vis

2006   *Apollo: The Definitive Source Book*, Richard W. Orloff and David M. Harland

2006   *NASA Scientist Astronauts*, Colin Burgess and David J. Shayler

2006   *The First Men on the Moon: The Story of Apollo 11*, David M. Harland

2007   *Energiya-Buran: The Soviet Space Shuttle*, Bart Hendrickx and Bert Vis

2007   *Soviet and Russian Lunar Exploration*, Brian Harvey

2007   *Praxis Manned Spaceflight Log 1961–2006*, Tim Furniss and David J. Shayler with Michael D. Shayler

2007   *On the Moon: The Apollo Journals*, Grant Heiken and Eric Jones

2007   *The Story of the Manned Space Stations*, Philip Baker

2008   *Salyut—The First Space Station: Triumph and Tragedy*, Grujica S. Ivanovich

2009   *The First Soviet Cosmonaut Team*, Colin Burgess and Rex Hall

2009   *Escaping the Bonds of Earth: The Fifties and the Sixties*, Ben Evans

2009   *Paving the Way for Apollo*, David M. Harland

2011   *How Apollo Flew to the Moon*, W. David Woods

2011   *Apollo 12: On the Ocean of Storms*, David Harland

2011   *Selecting the Mercury Seven: The Search for America's First Astronauts*, Colin Burgess

2012   *At Home in Space: The Late Seventies into the Eighties*, Ben Evans

2012   *Tragedy and Triumph in Orbit: The Eighties and Early Nineties*, Ben Evans

Printed by Printforce, the Netherlands